THE WORLD AS WE KNOW IT

The World as We Know It

FROM NATURAL PHILOSOPHY TO MODERN SCIENCE

PETER DEAR

PRINCETON UNIVERSITY PRESS
PRINCETON & OXFORD

Copyright © 2025 by Princeton University Press

Princeton University Press is committed to the protection of copyright and the intellectual property our authors entrust to us. Copyright promotes the progress and integrity of knowledge created by humans. By engaging with an authorized copy of this work, you are supporting creators and the global exchange of ideas. As this work is protected by copyright, any reproduction or distribution of it in any form for any purpose requires permission; permission requests should be sent to permissions@press.princeton.edu. Ingestion of any PUP IP for any AI purposes is strictly prohibited.

Published by Princeton University Press
41 William Street, Princeton, New Jersey 08540
99 Banbury Road, Oxford OX2 6JX

press.princeton.edu

GPSR Authorized Representative: Easy Access System Europe - Mustamäe tee 50, 10621 Tallinn, Estonia, gpsr.requests@easproject.com

All Rights Reserved

Library of Congress Control Number: 2024953019

ISBN 9780691235844
ISBN (e-book) 9780691235851

British Library Cataloging-in-Publication Data is available

Editorial: Eric Crahan, Rebecca Binnie
Production Editorial: Elizabeth Byrd
Jacket: Chris Ferrante
Production: Danielle Amatucci
Publicity: Alyssa Sanford (US), Carmen Jimenez (UK)

Jacket Credits: Courtesy of Albert Einstein Archives / Hebrew University of Jerusalem, British Library / Bridgeman Images, Cambridge University Library / Darwin Online, Library of Congress, Science History Institute, and Smithsonian Libraries and Archives.

Printed in the United States of America

10 9 8 7 6 5 4 3 2 1

For Theo and Celia, who should feel no obligation to read it.

CONTENTS

Introduction: Natural Philosophy and the Sciences ... 1

1. Divine Order: Isaac Newton and Physico-Theology ... 9
2. Celestial Order and Universal Gravity ... 26
3. Mixed Mathematics and Probability ... 45
4. Inventories of Electricity ... 59
5. Organization: Living Things ... 75
6. Cleaning up Chemistry: The Classification of Matter ... 88
7. Laplace, Revolutionary Order, and the Invention of Mathematical Physics ... 110

Entr'acte: Institutions and Pedagogy ... 123

8. Classification and Extinction: Cuvier and Natural History in the Early Nineteenth Century ... 132
9. Darwin's Taxonomy: Geology and the Organization of Life ... 147
10. Evolution and Scientific Naturalism ... 162
11. Thermodynamics and Modern Physics ... 181
12. Chance and Determinism: New Models of Science ... 196

13	Electromagnetism, Action at a Distance, and Aether	213
14	The Chemical Use of Atoms	238
15	Laboratories of the Heavens: Physics in the Observatory	256
16	New Modes of Natural Philosophy	285
	Conclusion. The World We Have Gained . . . and Lost	305

Acknowledgments 313

Notes 315

Bibliography 345

Index 361

THE WORLD AS WE KNOW IT

INTRODUCTION

Natural Philosophy and the Sciences

Science as a Product of History

A central means of understanding a human enterprise is to grasp it historically: How did it get this way? Modern science owes its principal intellectual and social traditions to cultural practices developed during the past three or four centuries in European or "Western" contexts. Global expansion, emulation, and modification, occurring especially in the twentieth century, subsequently yielded the worldwide technical endeavors familiar to us in the present. Many ingredients in the development of modern science have come from non-European sources, of course; much of the enterprise itself, however, has owed its dynamic to European projects, especially those of capitalist and imperial expansion. Not all large historical movements creating the present are purely benign, including those producing so celebrated an enterprise as science. Understanding comes from recognizing that the history of science is not about awarding accolades and prizes, but about self-awareness. For many years now there has been a good deal of discussion among historians on the question of why science became established as an ongoing enterprise in early modern Europe (or "the West") while similar practices in other cultural regions of the world did not become established and institutionalized in the same way.[1] But the question seems much less pressing as soon as one ceases to assume that there is a natural kind called "science" that floats free of its historical specificities. Perhaps it is no more than a word people use that contingently links together a variety of activities and knowledge claims under a common designation. Perhaps science is not a "thing" waiting to be discovered by many distinct societies.

This book is about why present-day chemists trace their enterprise back to Lavoisier; why physicists trace theirs, via Einstein and Maxwell, to Newton; why biologists trace theirs to Darwin; and why all of them work in institutional settings that resemble ones that appeared originally in nineteenth-century Germany. These historical connections help us understand the character of what scientists do. But we should not be misled by these connections: we cannot assume that everyone linked in these ways has always been engaged in the same project. Science itself has changed and been reconfigured over the centuries in ways that have been enormously consequential. Before the nineteenth century, one might even argue that our kind of science, often called Western, did not exist.[2]

Science today bears strong marks of its roots in the nineteenth century. Many—perhaps most—of the characteristic features of science as a cultural institution had appeared by 1900. Much has happened since then, of course, but the enterprise of science and the role of the scientist in 1900 remain fully recognizable today. This striking continuity appears most clearly in the terminology we use: we speak of *science* and *scientists* unproblematically; everyone thinks they know what those words mean without ever having to think much about them. Above all, science is and was a *profession* associated with university training, degrees, and self-regulating professional societies. The historical continuity of such markers over the last century or two is striking. But the nineteenth-century creation of science relates also to its intellectual content: scientists promulgate knowledge claims about the world for which their institutional credibility serves to secure assent from the rest of us. And we often describe those scientific knowledge claims as "discoveries," unless we (whether rival scientists or interested nonscientists) wish to deny or contradict them. But the authority of science usually makes serious opposition to claimed discoveries very difficult to maintain.[3]

Science and Natural Philosophy

The modern sense of the word *science* in English, as well as the English word *scientist*, only came into general use in the second half of the nineteenth century; the word *scientist* itself was first coined in the 1830s.[4] Before that time, so-called men of science (to use a common nineteenth-century term) were often described as pursuing "natural philosophy."[5] There are still chairs of natural philosophy at the Universities of Edinburgh and Glasgow (among others) in Scotland, at the University of Cambridge in England, and elsewhere, often founded in the eighteenth century and with their subject areas corresponding nowadays to physics. The term *science* then typically referred to a specific formal discipline (often mathematical in structure), and never to a universalizing science covering all subjects, as now.[6] "The science of . . ." was the typical form of expression for the word; one might refer collectively to "the sciences."

Natural philosophy had a long pedigree.[7] The term in Latin, still the dominant European language of learning at the start of the eighteenth century, was usually *philosophia naturalis*, or its almost-synonym *physica*. Natural philosophy, with its home in the universities, had been particularly associated since the Middle Ages with the texts of the ancient Greek philosopher Aristotle; its name served to distinguish it from moral philosophy, or ethics. By the eighteenth century, this academic, Aristotelian picture had been modified and challenged in many ways, but the central conception of natural philosophy remained: it was an intellectual discipline that sought to understand and explain the world and its behaviors rather than to employ those behaviors for practical ends. But operational dimensions were also present and often celebrated as part of natural philosophy's virtues. Most notably, the projects and ideas of the great English statesman of the early seventeenth century, Francis Bacon, attempted to reform natural philosophy from its foundations by using operational efficacy as a criterion of truthfulness, not merely a possible advantageous spin-off. Bacon's stance was subsequently adopted by the Royal Society of London and by natural philosophers (especially those independent of universities) throughout Europe, and it represented a considerable shift in enterprises dedicated to knowledge of nature. Natural philosophy remained as a way of understanding the world, but it was increasingly coupled to instrumental uses as, so to speak, the other side of the coin. The characteristic shape of something we recognize as modern science dates from this period, the seventeenth and eighteenth centuries, while never abandoning the intellectual goals of natural philosophy as part of its essential character. Henceforward, a mutually justifying hybrid of natural philosophy and instrumental technique (the latter often borrowed from the classical mathematical disciplines) moved to the fore in European knowledge practices and began to be called *Science*.[8]

The Scientific Revolution

In 1700, at the start of the period discussed in this book, Europe had just undergone what is sometimes known as the Scientific Revolution, a label applied to aspects of the intellectual and cultural changes of the sixteenth and seventeenth centuries relating to knowledge of nature. Conventionally, the Scientific Revolution is taken to run from Copernicus's ideas about the motion of the earth around the sun through to Isaac Newton's work on universal gravitation, but its importance amounts to more than just changes in accepted views about the structure of the universe.[9]

Those changes had, to be sure, resulted in many novelties: an indefinitely large universe with the stars as other suns, novel mechanical concepts that included inertia and the inverse square law of gravitation, and many others. But underlying these conceptual innovations were even more basic changes in attitudes that made the new world picture acceptable. By the late seventeenth century, two of the most basic changes in outlook concerned the questions of what counted as a proper explanation of a natural phenomenon and what counted as the proper way to investigate phenomena.

By 1700, proper explanations were taken by many (though not all) prominent natural philosophers—such as Isaac Newton, Robert Boyle, and G. W. Leibniz—to be *mechanical*. But disagreement remained about what that word really meant.[10] The universe was seen as some kind of gigantic machine, and that image determined how it would be understood. If the universe were a mechanical device, different kinds of explanations of phenomena might be given than if it were seen as a living thing, some kind of organic being. Mechanical explanations relied on seeing the world as made up of dead, inert matter that followed determinate mathematical laws. Newton's theory of universal gravitation was criticized by natural philosophers in France and elsewhere on the grounds that Newton's theory failed to explain gravitational attraction in terms of matter pushing on matter (the acme of mechanical force transmission); in a larger sense it, too, was seen by its adherents as mechanical. A Newtonian action-at-a-distance view of gravity simply noted an empirical regularity without committing to some particular means through which the regularity was performed. But in combination with Newton's laws of motion and collision, the account given in Newton's great Latin treatise of 1687, the *Principia* (in English, *The Mathematical Principles of Natural Philosophy*), remained "mechanical" for many insofar as it provided a determinate, predictive set of mathematical rules that supposedly described the motion of all bodies in the universe and the forces that produced that motion. From this perspective, *mechanical* meant an explanation in terms of forces and resultant motions, regardless of the source of the forces. Either way, philosophers routinely expressed their mechanical perspective through a metaphor of the world as a clock and God as the clockmaker.[11]

The clockwork metaphor effectively determined what counted as an adequate explanation. But natural philosophers needed to persuade each other that their particular explanation of some phenomenon was not simply possible because it was consistent with mechanistic principles; they also wanted to show that their explanations were *true*. By 1700, a special form of observation had become a technique of choice among natural philosophers: experimentation.[12] A well-designed experiment created a situation that would result in a different behavior depending on the underlying mechanism that produced it. Given prior assumptions about the universe being a mechanical system, it was reckoned that experiments could narrow down the possibilities so as to approach the truth about a phenomenon even if there was no guarantee that all possible alternative explanations had been taken into account.[13]

Natural philosophy in 1700, then, saw nature as a kind of machine, made up of particles of dead matter interacting with one other. The basic procedure for learning about how that machine worked was by inference from observation and especially experiment. More broadly, natural philosophy concerned itself with understanding the world, and the world was God's creation. Indeed, it has been plausibly argued that this divine component serves to define natural philosophy as a field, distinguishing it from other enterprises like "science."[14] While something of an exaggeration, this view warns against regarding natural philosophy as simply an obsolete terminological alternative to "science."

This book traces natural philosophy as it mutated into science in its recognizable modern form in the nineteenth century. But that mutation occurred only because of the natural philosophers whose work drove it. Who were they? What sorts of communities did they form? And what approaches to knowledge did these communities promote? There is much left out of the following account, and many historians would no doubt claim, with justice, that vitally important matters are ignored or given short shrift (I am particularly conscious of my frequent resort to Western European scientific work). But selection must be made; I have chosen topics, people, and settings that seem to me to represent a coherent and meaningful development of a worldview that frames our present sense of what we think we believe about our home, and that represent the most active scientific communities of their time. The goal is not to ascribe credit to particular figures but to discuss some of the most consequential ideas and approaches found in the natural philosophy of modern science. A variety of similar stories could equally well be told—in particular, the intersecting history of medicine, which might easily generate a parallel and distinct narrative of similar scope—but I am fond of this one.[15]

Natural Philosophers

Natural philosophers in the eighteenth century were not usually university professors; the practice of natural philosophy did not typically take place in universities. Research was not a recognized part of a university's function, and if occasionally someone who was a university teacher took an interest in the pursuit of natural philosophy, it was generally a purely private interest.[16] Research of any kind did not become an integral part of a university's function until the nineteenth century.

Rather, the social focus for the sciences in 1700 was the scientific society, whether a private voluntary affair or an academy under state patronage. Such groups had started to appear in the seventeenth century, with the principal examples, which served as models for others, being the Royal Society of London and the Royal Academy of Sciences in Paris, both founded in the 1660s. Their names reveal a kind of official status deriving from royal patronage. Nonetheless, the Royal Society received no financial benefit from the Crown, and was in effect a self-governing gentleman's club. The Paris Academy, on the other hand, included salaried positions and other material benefits (such as an astronomical observatory) paid for by the state. A similar kind of academy was started in Berlin by the Prussian monarchy in 1700, and Tsar Peter the Great founded his own academy of sciences in St. Petersburg in 1724.[17]

Such societies (and there were many) were the foci of natural philosophical activity outside the pedagogically oriented universities and colleges. Publications by natural philosophers would typically appear in a journal like the *Philosophical Transactions of the Royal Society* or in the *Memoirs* of the Royal Academy of Sciences in Paris, or as books bearing the approval of those groups. Scientific life in large part centered on these regional societies. It should not be thought, though, that they reflected new social arrangements suited to some novel and forward-looking enterprise. These societies or academies tended to be modeled on preexisting social institutions, such as law courts, universities, parliaments, or societies of accredited medical practitioners.

Society members were typically from reasonably well-to-do backgrounds—not usually from the nobility, nor from the working class (besides the occasional instrument maker). They were educated people, perhaps with independent incomes; in England, very often people who might be called members of the gentry, often including clergymen. So the societies comprised "gentlemen." Indeed, very few women were allowed to play this philosophical game, just as there were no female lawyers or members of parliament. When some women engaged in natural philosophy in the seventeenth and eighteenth centuries, they were kept on the sidelines of the organized activity of the scientific academies.[18]

At the beginning of the eighteenth century, then, the new natural philosophy was accorded a status that made it a public enterprise along the same lines as law or medicine or established religion. And the biggest name associated with it soon came to be that of the president of the Royal Society of London, Sir Isaac Newton. After a period of debate and contestation, Newton's picture of the universe, its content and structure, and its relationship to God the Creator, became the primary reference for natural philosophers throughout Europe. It might seem obvious to us now that Newton's world picture should have acquired such dominance—after all, we might say, he more or less got it right, in a way that gave rise to enormously successful research trajectories leading to the physical sciences of the present. And we would not be altogether wrong to do so. But what we must try to achieve, if we are to take historical knowledge seriously, is to see those trajectories as the products of decisions and judgments made by particular kinds of people in specific circumstances, and to understand their scientific enterprises as attempts to answer questions about the world that are not always the same as ours. The natural world that people in the past conjured up was created from the cultural materials available to them; it was not something they simply found and reported. The world as we know it was made by people in history, and we have inherited it from them to make our future.

1

Divine Order

ISAAC NEWTON
AND PHYSICO-THEOLOGY

FROM THE point of view not just of the history of science, but of European intellectual history generally, the eighteenth century has sometimes been seen as the century of Newton. Certainly Isaac Newton's shadow hangs over many aspects of thought in that period. But what does this amount to in practice? Natural philosophers frequently invoked Newton's name and commonly used the term *Newtonian*. But identifying a central, coherent meaning for "Newtonianism" can be a bit of a challenge since it encompassed so many diverse ideas and approaches.[1]

The very fact that natural philosophers frequently invoked Newton's name, however, is itself significant. Although identifying somebody as a Newtonian (generally from self-identification) reveals little in detail about what they actually believed or did, knowing about some of the different things the label could mean is a good way of getting a general sense of the period. The fact is that Newton's name carried enormous prestige in the eighteenth century; natural philosophers liked to claim that their own ideas were, perhaps even despite appearances, really Newtonian. As a consequence, many versions and interpretations of Newton, often very different, were developed.

Isaac Newton in England

Newton's prestige related especially to perceptions of his achievement in his most important treatise, the *Mathematical Principles of Natural Philosophy*, or *Principia*, of 1687. This volume laid down the elements of Newton's mechanics, the inverse square law of gravitation, and his entire world system, made up of an infinite space with lumps of gravitating matter dotted about in it. Newton came to be seen in some quarters as almost superhuman on the basis of that work. A famous epigram by the early eighteenth-century English poet Alexander Pope gives a window on how Newton was seen in educated England:

> NATURE and Nature's laws lay hid in Night:
> God said, *Let* NEWTON *be!* and all was Light.[2]

Pope's excess reflects some of the public adulation that Newton received in his lifetime. When he died in 1727, Newton was given an elaborate state funeral in Westminster Abbey as a mark of his heroic stature.

The achievement that inspired such adoration was this: the seventeenth century had produced Johannes Kepler's laws of planetary motion—the specification of the mathematical form of planetary orbits as ellipses with the sun at one focus, and associated descriptive laws governing speeds. There was in addition Galileo's law for falling bodies, whereby the speed of fall is proportional to time elapsed. Newton had managed to derive both sets of laws from a unified set of axioms. The monument installed in the Abbey in 1731 alludes to this aspect of Newton's work by displaying a simple measuring scale (steelyard) representing the solar system, the sun counterbalancing the six planets to signal the mechanical relationship between them, mediated by gravity.[3]

Newton's were considerable mathematical achievements, but in the eyes of some natural philosophers in Continental Europe, there was still a lot to be desired. The main criticism that Newton received when the *Principia* first appeared was that the law of universal gravitation, although it seemed to fit the phenomena, failed to provide any kind of mechanical *explanation* of gravity. Newton appeared to portray gravity as action at a distance, whereby one body is responsible for exerting a

FIGURE 1.1. Isaac Newton's tomb in Westminster Abbey. The detail commemorates various aspects of Newton's work. Copyright: Dean and Chapter of Westminster, London.

FIGURE 1.1. (*continued*).

force upon another body across the distance separating them with no intervening action to mediate the phenomenon. Such a view was not immediately acceptable to Continental natural philosophers (nor was it unproblematic for Newton himself, as private letters and later writings indicate, although Newton was noncommittal in the *Principia* itself).[4] Critics thought that action could be communicated between bodies only through material interactions of some kind, such as pressure or collision.

Leading members of the French Academy of Sciences, such as Christiaan Huygens and Nicolas Malebranche, who were perfectly mathematically competent to understand Newton's work, accepted Newton's mathematical demonstrations in the *Principia*, but regarded them as providing phenomena to be explained rather than being satisfactory explanations in themselves.[5] Newton had established the existence of an inverse square law attractive force between masses, but had not, they complained, explained its causes. The latter would have identified Newton's achievement as true natural philosophy, conformable to the Aristotelian definition. Newton's early critics thought that he had failed to complete the job by giving a causal explanation of gravity. For them, the true desideratum would have been an account of gravitational force in terms of matter pushing against matter. There was nothing superhuman about coming up with a new, more comprehensive mathematical description.

Philosophers like Huygens adhered to a particular kind of mechanistic natural philosophy, the contact-action version, that was dominant (especially in France) by the end of the seventeenth century. It had been invented by René Descartes back in the 1630s and '40s, and accordingly bears the name *Cartesianism*. Descartes's was a world not made up of particles exerting force across empty space, like Newton's, but of space entirely filled with matter. Planetary motion in our solar system, for example, resulted from the behavior of a great whirlpool or vortex of "subtle matter" sweeping the planets around the central sun like apples floating in a barrel.[6]

Newton gave later Continental Cartesian natural philosophers a lot of work to do: they had to explain universal gravitation, with its inverse square law, in terms of vortices. Although they ultimately failed, the work on fluid mechanics in the early eighteenth century of people like the mathematically inclined Bernoulli clan from Switzerland and Leonhard Euler in Germany was prompted by this program of trying to get a Cartesian vortex universe to work in quantitative terms.

The Cartesian contact-action ideal of mechanical explanation remained dominant in Continental natural philosophy until the 1730s and '40s, after which it was replaced by the Newtonian physics of forces and particles, including action at a distance. In England, on the other hand, Newton's ideas had been widely accepted shortly after publication. The reasons are perhaps less than obvious: one cannot attribute acceptance to some self-evident correctness, since to many competent judges there was none. But insistence on contact-action explanations was less entrenched in England than it was on the Continent. The fellows of the Royal Society were more ready to accept a system of the universe framed in terms of incorporeal forces, *if* that system and those forces were claimed to have been derived from observed phenomena and experimental results. In a sense, the ideal of empiricism was more important for them than was the ideal of a mechanical explanation based on contact action.[7] A more immediate reason

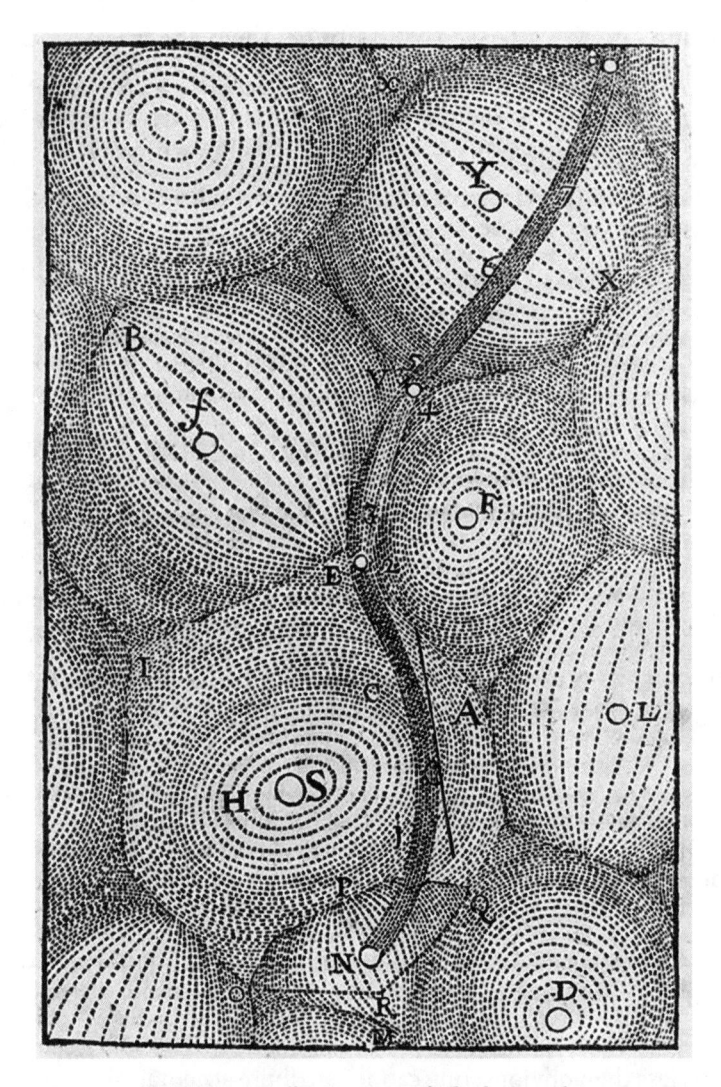

FIGURE 1.2. Cartesian stellar vortices, of which our solar system is one. From Descartes's *Principles of Philosophy* (1644).

for the success of Newton's ideas in England, however, was their value in the late seventeenth and early eighteenth centuries in the service of orthodox religious and political doctrines.

Newton's *Principia* appeared in 1687, and in the 1690s his natural philosophy was already being cited from Anglican pulpits—the established state church—as evidence of God and His rule over Creation. In 1691, the year of his death, another leading fellow of the Royal Society, Robert Boyle, established in his will the Boyle Lectureship. This created an annual series of lectures "for proving," according to Boyle's will, "the Christian Religion, against notorious Infidels."[8]

The first Boyle lecturer was the scholar (and soon-to-be master of Trinity College, Cambridge) Richard Bentley. Newton himself coached Bentley in aspects of the *Principia* relevant to his Boyle lectures. In a letter to Bentley in 1692, Newton observed:

> When I wrote my treatise about our Systeme I had an eye upon such Principles as might work with considering men for the beleife of a Deity & nothing can rejoyce me more then to find it usefull for that purpose But if I have done the publick any service this way 'tis due to nothing but industry & a patient thought.[9]

A number of the subsequent Boyle lecturers in the 1690s and the first decade of the eighteenth century also used the Newtonian picture of the world as evidence for the existence and providence of God. The published lectures soon became one of the chief ways that educated people in England in the early eighteenth century learned about Newton's system of the world as something they should favor. That authors of these influential lectures adopted Newton's ideas so eagerly therefore needs explanation. Why was Newtonianism, Newton's picture of the world, so successful among such people?[10]

The early Boyle lecturers who used Newton's work usually lacked a deep understanding of it. Unlike mathematically competent readers (and critics) elsewhere in Europe, who largely accepted Newton's mathematical results but criticized his natural-philosophical interpretations, these English proponents of Newton took the mathematics for granted, using it as a pretext for their acceptance of his philosophical inferences regarding his "system of the world." However, they were not simply adopting this new picture of nature on trust and then using it as support for their theological positions. Acceptance of Newton's technical results about the inverse square law, dynamics, and orbital motions by no means necessarily entailed acceptance of particular interpretations of gravity as action at a distance, and these technical results certainly failed to demonstrate his ideas about the relationship of God to the natural world. But an ideologically interested writer or lecturer could try to represent them in that way.

To a large extent, the rapid success of Newtonianism in England was a product rather than a cause of this quickly established liaison between the *Principia* and religious orthodoxy. Anglican clergy picked up Newton's ideas deliberately rather than reacting to a general acceptance of them: Newton's ideas were brand new when the churchmen adopted them, still awaiting general approval.[11]

Above all, the place of God in Newton's universe suited Anglican orthodoxy. Among the principal targets of the Boyle lectures were political radicals who threatened the framework of parliamentary and church authority. That framework had been set up in 1688 after Parliament deposed King James II for being a Catholic and invited William of Orange from the Netherlands to become king on condition that he recognize the ultimate power of Parliament—the form of a so-called constitutional monarchy. But this settlement of what had been long-running political battles in England was not universally accepted: political radicals still yearned to extend political power even more widely, unhappy with it simply being shifted from the king to Parliament, because Parliament represented only the wealthy, usually land-owning, classes.

The crucial political assumption that the radicals denied was that society should be ordered as a hierarchy, with rulers at the top and subjects at the bottom. The Boyle lecturers and other orthodox churchmen were supporters of the new political settlement, not least because it included the hierarchy of the Anglican Church as part of the apparatus of the state. Radical challenges involved arguing against the proposition that society should be ordered in such a way that people *knew their places*—not just because this proposition was held to be politically wise, but because it was supposedly part of God's design.

The Boyle lecturer for 1704–05 was Samuel Clarke, a close associate of Newton's who again reminds us that political uses of Newton's work were not independent of the master himself. Clarke put the argument like this:

> All inanimate and all irrational beings, by the necessity of their nature, constantly obey the laws of their creation; and tend regularly to the ends for which they were appointed. How monstrous then is it, that reasonable creatures, merely because they are not necessitated should abuse that glorious privilege of liberty, by which they are exalted in dignity above the rest of God's creation, to make themselves the alone unreasonable and disorderly part of the universe![12]

Clarke's point is that humans and their society are part of Creation, and society should therefore be governed by the same basic principles of order as nature itself. The consequence is an argument for the political status quo.

Newton had explicitly placed God as the ruler of the universe: God was "Lord God *Pantokrator*," with dominion over all things.[13] One of the consequences for Newton was that gravitation could not be an essential property of matter. Instead, he wanted to portray it as a property imposed on matter in some way from the outside. In the 1690s, Newton privately proposed that perhaps gravitational behavior was brought about by the direct agency of God: God simply wills that matter should be mutually attracted according to the inverse square law. At other times he postulated possible intermediate physical mechanisms to account for gravity, but he always stuck fast to the idea that gravity should not be regarded as a property deriving from matter itself; its activity (as he saw it) was rooted, whether directly or indirectly, in God. Otherwise there would be sources of activity truly distinct from God, which would invite idolatry.[14]

This was not abstract theology. One of the most prominent radicals in England in the late seventeenth and early eighteenth centuries, John Toland, put forward exactly the opposite view. The fact that Toland bothered to talk about the natural world at all, given that he was centrally concerned with political questions, again shows how important people at this time found the analogy between natural and social orders. In a book published in 1704, Toland claimed that Newton's work showed that gravity *was* an essential property of matter, definitive of its very nature, not imposed by some external agency.[15]

So for the Newtonian guardians of orthodoxy, God ruled the universe by imposing His will upon it; gravity was a law of nature willed by God on inert, passive matter. Similarly, in human society, laws are imposed by those at the top on the rest of the population, which has no say in the matter. For a radical like John Toland, by contrast, gravitational behavior should not be seen as being extrinsic to matter, conferred by the mysterious workings of divine will; rather, it is intrinsic to the nature of matter itself. Similarly in society, Toland believed, laws are not properly to be imposed on people, but should reside in and derive from the people themselves, essential properties of human beings in society.

Newtonianism's rapid acceptance in England among people who were often incapable of understanding the mathematical and technical aspects of Newton's work was in significant part due to its usefulness in promoting orthodox religion and supporting the established political order (despite Toland's attempts to subvert it). Newton fully approved of all this, because for him, natural philosophy was always about God, from whom no part or aspect of the universe was independent. But as the case of Toland shows, not everyone saw things in the same way. These were dangerous waters, however, and Newtonians (including Newton himself and his successor as Lucasian Professor of Mathematics at Cambridge, William Whiston) ran considerable risks in maintaining theological positions supposedly derived from natural philosophy rather than established ecclesiastical authorities.

Newton on the Continent

The French Enlightenment thinker Voltaire noted and made light of systematic differences between Newtonianism and the natural philosophy of Cartesianism, identifying them as associated with their countries of origin. In 1734 he published an account of his sojourn in England in the late 1720s (during which he attended Newton's funeral in Westminster Abbey). Voltaire was a leading figure in turning philosophical France, or at least Paris, from being predominantly Cartesian to being predominantly Newtonian.[16] He wrote the first popularization of Newtonian philosophy to appear in France, the *Elements of Newton's Philosophy*, in 1738. It summarizes, for the nonmathematician, Newton's system of the world, the doctrine of gravitational attraction, and Newton's experimental work on light and colors. But Voltaire's enthusiasm for Newton also rested on the religious implications that he saw in Newton's philosophy. Instead of using Newton's universe to show the omnipresence of God, Voltaire used its mathematical determinacy and law-like structure to support his own position of deism. Deism, a doctrine also favored by John Toland, is the belief that, although God initially created the universe, He then left it alone to run according to the immutable laws of nature (also divinely created and ordained), much like a clock that never runs down. The position rejected ideas of divine providence, the idea that prayers have any efficacy, and in fact most of the central tenets of Christianity. By emphasizing Newtonianism's picture of an entirely deterministic universe governed by laws, Voltaire saw himself as undermining the French Catholic Church, which he saw as a bastion of ignorance, superstition, and social privilege.[17]

Another aspect of Newton's appeal in eighteenth-century France was its value for mathematicians. Mathematicians in the Paris Academy of Sciences vigorously defended Newton against the Cartesians in the 1730s, although the translator of Newton's *Principia* into French in this period, the prominent philosopher and metaphysician Émilie du Châtelet, was, being a woman, never admitted to the Academy—despite her reputation and having been trained by the luminaries Pierre-Louis Maupertuis and Alexis Clairaut.[18] Some mathematicians who were in the Academy, like Jean Lerond D'Alembert, were attracted to the mathematical possibilities of Newton's worldview. They worked on a completely theoretical, mathematical mechanics, based on Newton's laws of motion and gravitation, that was known as *rational mechanics*, a term applied by Newton himself. By accepting Newton's view of a world composed of forces and particles, they gave themselves countless interesting new problems to solve. Their skills therefore claimed for them a central role as experts in understanding the nature of a universe framed in those Newtonian terms.

But the wider appeal, and the overall convincing quality, of Newton's physical universe lay to a great extent in the aspects that Voltaire particularly promoted. D'Alembert, the mathematician, was attracted by the political uses of Newton. These two aspects—the interest for mathematicians and the broader social implications—brought about a practical victory for Newtonianism over Cartesianism in the French Academy of Sciences during the 1730s and '40s.[19]

Although Voltaire is the most notable example of Newton's ideological appeal in France, he was not a member of the Academy. But he was an important member of a liberal intellectual reform movement that included Academy members and was at the forefront of promoting Newtonian philosophy and related ideas in France throughout the century. The members of this group are known collectively as the *philosophes*, and they can be broadly characterized as pamphleteers and professional writers. Like Voltaire, who was one of the earliest, they saw themselves as the vanguard of liberal reform and enlightened rationalism in eighteenth-century France.

The most important product of this movement was the publication, between 1751 and 1772, of the seventeen-volume *Encyclopédie ou Dictionnaire raisonné des sciences, des arts et des métiers*.[20] It covered subjects from mathematics, law, and religion to natural philosophy and trades of all kinds, and one of its editors was D'Alembert, the Newtonian mathematician. The articles were written by scores of contributors, including Voltaire, and they share a similar general orientation. That orientation had put the *Encyclopédie*'s other editor, Denis Diderot, in prison for a while, and the *Encyclopédie* itself ceased appearing for a time because of trouble with the authorities. The work is liberal, opposed to the institutions of the absolutist French state, and extremely rough on the Catholic Church. It is Newtonian in its conception of the structure and operations of the universe, as well as Lockean: The English philosopher John Locke had represented his approach to understanding as indebted to Newton's compatible view that all natural knowledge comes from the senses, and the two names, Newton and Locke, often went hand in hand in the eighteenth century as explicators of this supposedly antidogmatic empiricism.[21]

It remained the case, however, that in the eighteenth century the term *Newtonian* could mean many different things, and people used Newton's name to lend respectability to a wide variety of ideas that they wanted to have taken seriously. Hence the very different political uses of Newton's work provided by the examples discussed earlier. One of the chief resources for would-be Newtonians was not the *Principia* but Newton's other major publication, the 1704 treatise called *Opticks*. Written originally in English instead of Latin, it was also full of experiments instead of arduous mathematics. That gave it a great deal of appeal to amateur enthusiasts in natural philosophy, especially the bulk of the Royal Society membership in England. Even better, from their point of view, were the speculations that Newton presented at the end of the work, which he called the "Queries."

These queries, which were expanded in successive editions of the *Opticks*, are about such questions as the ultimate structure of matter and the nature of various physical effects in matter, including chemical phenomena, the refraction of light, and electrical effects. Matter was made up of minute particles, or corpuscles. In some of the queries Newton suggests explanations in terms of short-range attractive and repulsive forces between corpuscles, similar to gravity. In others he plays with the idea of subtle aethers, fine fluid substances themselves made up of mutually repelling corpuscles, that might pervade the pores of ordinary matter to produce various effects, especially those of electricity. These discussions are largely qualitative speculation, so Newton called them *queries* to distinguish them from what he took to be demonstratively provable.

In fact, the very inconsistency and vagueness of the suggestions in the queries meant that people conducting nonmathematical, empirical investigations of nature could use Newton's authority for their own explanations of experimental effects, such as electrical and chemical phenomena, in terms of corpuscles, aethers, and short-range forces. The widespread acceptance of Newtonian themes in the Netherlands (then called the United Provinces) in the early eighteenth century took the form of an acceptance of Newton's theory of matter (together with the empiricism that was often held to underpin it) and the pursuit of the mathematical project of the *Principia*. Chemists, physicians, and philosophers such as Herman Boerhaave (see chapter 6) and Bernard Nieuwentijt established diverse "Newtonian" approaches to natural philosophy on a par with the English themselves.[22]

In many ways, the mathematical Newton of the *Principia* and the more experimental and speculative Newton of the *Opticks* and "Queries" correspond to two distinct styles of Newtonian natural philosophy in the eighteenth century.[23] Certainly, just as mathematical mechanics in the eighteenth century owed much of its technical content to other mathematicians, such as Leibniz and the Bernoullis, rather than to Newton, so experimental work on electricity and chemistry also drew on traditions quite distinct from Newton's.[24] But the eighteenth century is appropriately called the "century of Newton" not just because of his role in shaping intellectual life in that period but also because of the prestige of his name, which could be used to lend respectability to others' work, whether experimental or mathematical.

Natural Theology

The early association of Newton's philosophy with particular theological doctrines, seen in the Boyle lectures, ensured that British Newtonianism was for a long time integrated with a conviction of the designful nature of the world. Newton spoke explicitly of these matters in both the *Principia* and the *Opticks*, arguing for the arbitrary, and hence carefully chosen, features of the solar system that could not be explained by the physical laws governing it—laws revealed by Newton's own work:

> While Comets move in very excentrick Orbs in all manner of Positions, blind Fate could never make all the Planets move one and the same way in Orbs concentrick, some inconsiderable Irregularities excepted which may have risen from the mutual Actions of Comets and Planets upon one another, and which will be apt to increase, till this System wants a Reformation. Such a wonderful Uniformity in the Planetary System must be allowed the Effect of Choice.[25]

Only the choices of an intelligent and designful Creator could account for these features of the solar system, according to Newton; the arrangement of the solar system proved that God had designed it. There was no physical reason in Newton's mechanics why planetary orbits should be more or less circular instead of wildly elliptical, or why all the planets should orbit the sun in the same direction and in nearly the same plane, instead of their orbits being tilted all over the place. The orderliness of the universe bore witness to divine action in its establishment; the blind laws of nature by themselves were inadequate to produce it. In the 1713 second edition of the *Principia*, Newton summarized, "This most excellent system of the sun, planets, and comets could not have arisen without the design and dominion of an intelligent and powerful being."[26]

This style of argumentation was fundamentally apologetic, aimed at establishing theological points concerning God through the medium of natural philosophy. As such, it resembled some of the Boyle Lectures, and indeed the best example is the lecture series for 1711–12, given by the Reverend William Derham, a fellow of the Royal Society. These lectures were soon published in book form as Derham's *Physico-Theology: Or, A Demonstration of the Being and Attributes of God, from His Works of Creation*, and was followed in 1714 by *Astro-Theology: Or, A Demonstration of the Being and Attributes of God, From a Survey of the Heavens*.[27] Both books were popular successes, going through numerous editions over several decades. They used what came to be known as the "argument from design," whereby features of the created world are held to display purposefulness in their structures or operations, much like the designful aspects of the solar system adduced by Newton. If such features could be shown to be highly unlikely to occur by chance, the inference to an intelligent and beneficent Creator was straightforward, or so Derham believed. This style of argument he called "Physico-Theological," following Robert Boyle; the field is also known as *natural theology*, and had a fine career in Britain well into the nineteenth century.[28] The first of the two volumes Derham devoted to the "terraqueous globe" of the earth, commemorating the Aristotelian elements of earth and water (saying nothing of the air), while the second was devoted to the heavens.[29] Most of the terrestrial part addressed animals (including Man), which provided a wealth of apparently functional contrivances to bear witness to providential design. Who but an infinitely wise and beneficent Creator could have done this?

After describing the eyes in various animals and the contrivances that serve them, Derham remarks:

> Thus have I survey'd this first Sense of Animals [i.e., sight], I may say, in a cursory, not accurate, strict manner, considering the Prodigious Workmanship thereof; but so as abundantly to demonstrate it to be the contrivance, the Work of no less a Being than the Infinite, Wise, and Indulgent Creator. For none less could compose so admirable an Organ, so adapt all its Parts, so adjust it to all Occasions, so nicely provide for every Use, and for every Emergency: In a Word, none less than GOD, could I say, thus contrive, order, and provide an Organ, as magnificent and curious, as the Sense is useful.[30]

In other places Derham discussed quadrupeds, making special note of the elephant's trunk: "A Member so admirably contrived, so curiously wrought, and with so great Agility and Readiness, applied by that unweildly [sic] Creature to all its several Occasions, that I take it to be a manifest Instance of the Creator's Workmanship."[31] Derham's perspective owes much to anatomical practice, rooted in the work of the ancient anatomist and physician Galen, which itself reflects the philosophy of Aristotle, who asserted purpose and design in all things, but especially animals.[32] Derham stressed the attribution of this designfulness to God.

These examples of the argument from design for the existence and providence of God replicate the strategy of the naturalist John Ray in works such as *The Wisdom of God Manifested in the Works of the Creation*.[33] The implied premise is always that such apparently fitting arrangements in nature could not have arisen by chance but need to be explained by intelligent contrivance attributable to God. One might imagine, in the case instanced by Newton regarding the orientation of planetary orbits, that some rough estimate of probability could have been made so as to show the overwhelming likelihood that the state of affairs was due not to chance but to divine arrangement. Probabilistic arguments for God's design were already being proposed by others, however, as Derham knew (see chapter 3).

Newtonian natural philosophy was a conservative, or traditional, endeavor. Natural philosophy continued to be assessed in relation to natural theology, whether or not theological questions were explicitly addressed. Nature was God's Creation, and that understanding was so fundamental a part of natural philosophy that it underpinned the assumptions that generated the premises of scientific arguments. Without God, philosophical questions would have had no motivation.

This is what it meant to be a natural philosopher for many in the early eighteenth century, and not only in England. Regardless of theological nuance, Nature must be understood as God's Creation. And yet one could not claim to be able to know God's mind. How, then, to argue for evidence of God's design or intentions? The Swedish naturalist Carl Linnaeus hinted at the difficulty when he wrote of himself, "God has suffered him to peep into His secret cabinet."[34] God need not always be so generous, but natural philosophy could not proceed without Him.

2

Celestial Order and Universal Gravity

ONE OF the most ambitious areas of investigation that eighteenth-century natural philosophers took on was the enormous one of determining the overall structure of the universe—what we now call *cosmology*. In fact, the word in its modern sense was invented in the eighteenth century and owes its central dynamic to Newtonian convictions. The questions that eighteenth-century cosmology addressed depended on new understandings of the heavens that had grown up during the course of the previous century, especially regarding assumptions about how the universe ought to work given the new Newtonian worldview. There had been precise ideas long before the eighteenth century about the structure of the universe, but after Newton there was a completely new framework that changed the nature of the issues.

Newton on Stellar Stability

Among three fundamental assumptions shared by most natural philosophers by 1700 was that the universe is indefinitely large—not necessarily infinite in extent, but in that it made no sense to think of the universe as having set limits. Everyone, following Descartes as well as Newton, was prepared to admit that three-dimensional space could very well go on forever in all directions.

Additionally, nearly everyone accepted the idea that stars are other suns. This became a standard view during the seventeenth century following Descartes, its most forthright advocate. It is a little surprising that this view became so widely established in as short a time as it did; neither Kepler nor Galileo had subscribed to it, for example, and it had little obvious observational support. It had been promoted in the fifteenth century by Cardinal Nicholas of Cusa, to little fanfare—Nicholas was not even an astronomer. Its subsequent rapid adoption in the seventeenth century must owe a great deal to a broader acceptance of Cartesian norms of physical explanation, which made Descartes's accounts of vortical planetary motion around stellar centers seem part and parcel of wider celestial and cosmological ideas. Descartes proposed that the motion of the planets around the sun (taking for granted the century-old Copernican system of celestial motion) should be understood in terms of a whirlpool of subtle matter that swept them around a center. This center was the focus of outward centrifugal pressure in Descartes's mechanical system—a pressure experienced as the blazing light of the sun. In this immense universe there were countless such systems, each with its glowing central sun, seen by us as distant stars. Descartes intended this picture to be explanatory rather than simply descriptive.[1]

The third assumption, the basis for cosmological theorizing, was due to Isaac Newton: the idea of universal gravitation. Newton had really been sticking his neck out when he claimed that the inverse square law of gravitational force that apparently linked terrestrial gravity and the celestial motions of bodies in the solar system actually applied to all (ordinary) matter throughout the universe, including that of stars.[2] All Newton could do on the basis of direct observation and experimentation was show that one could quantitatively identify the force acting on heavy bodies at the earth's surface with the force attracting planets to the sun and moons to planets, assuming the correctness of his analytical assumptions. So the most that could be claimed with direct evidence was that gravitation seemed to apply to celestial bodies in the solar system whose motion could be observed, from which forces could therefore be inferred. But Newton went ahead and said that it applied to *all* bodies in the universe, even those displaying no measurable motion.

No motion had ever been observed among the stars, as judged by available star catalogues, ancient and modern, dating back to Ptolemy's catalogue from the second century A.D. Newton nonetheless rose above such considerations in search of higher truths. His methodological principles ("Rules," as he finally called them) gave him license to extend properties found in all observable bodies to all bodies whatsoever. Stars did not display motion relative to one another ("proper motions") from which to infer their mutual gravitational attractions, but bodies in the solar system clearly did, which allowed their gravitational properties to be made manifest. And so, nature being "conformable to herself," as Newton famously said in Query 31, gravitation should also be expected among the stars.

Newton therefore held that gravitational properties are universal, possessed by all material bodies—a position that should be accepted at least until evidence appeared to the contrary. This had certain consequences for further investigation that provided a handle for understanding aspects of the world that would otherwise be inaccessible. Assuming the existence of gravity among the stars created physical problems about the structure of the heavens, the solutions to which could be treated as actual knowledge even though they depended on apparently insecure foundations. Building on conjectures about the world would, if progress seemed to result, serve to confirm those beliefs even in the absence of formal confirmation. After all, science is in large part about acting in the world.[3]

The central physical problem that resulted from Newton's arguments, which Newton himself confronted, was this: If the stars are gravitational bodies, meaning that they all attract one another, then why don't they collapse together? Bearing in mind the fact that the stars *had never been seen to change positions* as far back as records went, the problem as Newton saw it was to explain how the stars could maintain their fixed positions spread out through space.

Newton came up with a possible solution to the problem in the 1690s, in manuscript drafts for a projected new edition of the *Principia*. He imagined a universe populated by a uniform distribution of stars stretching out indefinitely in all directions. The gravitational attraction on any particular star from all the other stars surrounding it would be

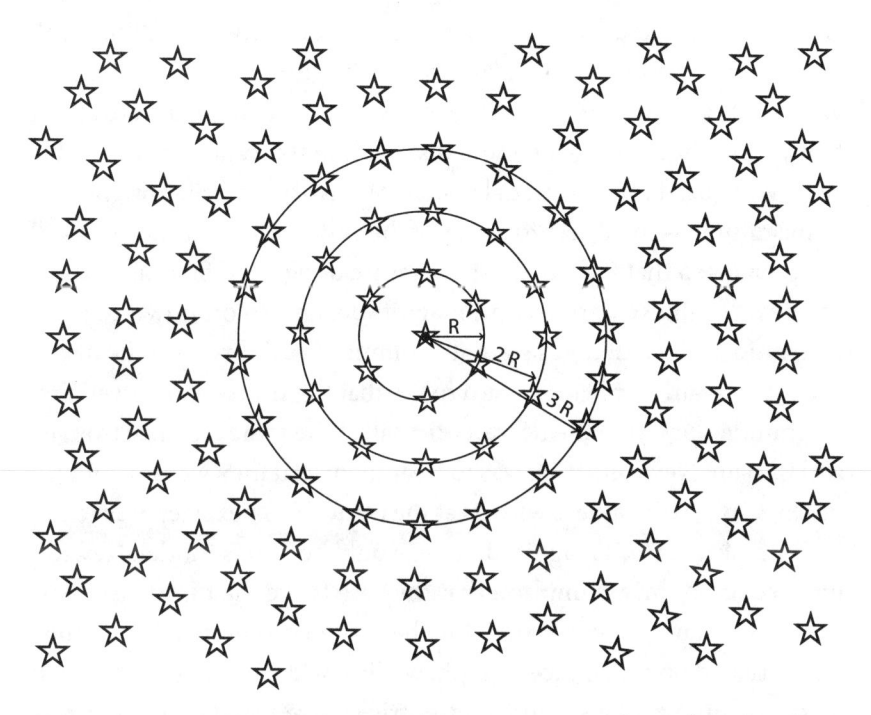

FIGURE 2.1. The spheres drawn around the sun represent the uniform distribution of stars throughout space and enable easy calculation of the numbers of stars of different brightnesses as seen from the vicinity of the sun.

practically equal in all directions, and therefore cancel out: there would be no net gravitational force acting to draw each star from its position.[4]

Yet Newton recognized a signal objection to this idea, focused on the simple observational point that the stars do not appear to be uniformly distributed across the sky. Even if the irregular groupings of stars as we see them, such as the constellations, were dismissed as anomalies or minor fluctuations in an overall average uniformity, it turned out to be difficult to argue that evidence of stellar distances, whether from each other or from ourselves, confirmed the doctrine of uniform distribution of stars throughout space.

In his draft texts on the question, Newton tried to argue that the number of stars of different brightnesses fit the assumption of uniform spatial distribution. He based his inference on the working assumption that, on average and without too much variation between extremes, all stars are of the same *absolute* or intrinsic brightness, so differences in *apparent* brightness are solely due to distance: brighter stars are closer to us, while fainter stars are more distant. The star catalogues available to Newton used the traditional system of classifying stellar brightness, or magnitude—the direct forebear of the system that astronomers still use today—so that the brightest stars were designated first magnitude, the next-brightest stars second magnitude, and so on down to sixth magnitude, which denotes stars at the limit of visibility with the naked eye.[5] It was fairly straightforward to say that the thirteen observed first magnitude stars were the nearest ones, all at the same distance from us and the sun. Newton's trick was to treat them as if they were all located on the surface of a sphere with us at the center. His next step was to ask how far away second magnitude stars would be in this model. Newton proceeded on the assumption that they are twice the distance of first magnitude stars and could therefore be treated geometrically as if they were situated on the surface of a sphere with twice the radius of the first. Since the surface of a second, concentric sphere with twice the radius of the first is four times as large (the square of the radius), one would expect, if the stars are indeed evenly distributed, to find four times the number of second magnitude stars. Third magnitude stars will be three times the distance away, implying nine times more of them than of first magnitude stars, and so on.

Newton's draft of this argument listed the predicted number of stars for each magnitude as calculated from this simple model. He then left a blank space on the page to put in the observed numbers from catalogue data alongside the predicted numbers—he evidently lacked that information while writing. His idea was to show that the predicted and observed numbers were very similar, which would serve to support the validity of the assumptions from which the predicted numbers and the model generating them were derived, central among them being the assumption of uniform distribution. Strikingly, following the incomplete table, Newton noted how closely the predicted and actual numbers align, even though the latter were then unknown to him. Newton was not one to contemplate failure.

When he subsequently compiled the observational figures, however, Newton found that the fit was very poor. In particular, the fainter (and hypothetically more distant) stars were far too plentiful. Accordingly, he attempted to massage his data in search of a better fit. Ignoring stars of the fifth and sixth magnitudes made the disparities look less egregious, for example. More promisingly, it occurred to him that there was really no compelling reason (he surely knew this all along) to say that third magnitude stars are exactly three times further away than first. He found that if you shift them a bit further out you can get closer to the observed figures; the further out (and hence larger) the surface of the sphere accommodating uniformly distributed third magnitude stars, the more there would be. In the final draft of this material, Newton suggested, along similar lines, that sixth magnitude stars are eight or nine times further away than first (nowadays the figure would by convention be ten times), with stars of the other magnitudes arranged appropriately in between.

Unsurprisingly, this material never appeared in print. As soon as Newton allowed himself to play around with the presumed distances of stars of particular magnitudes to get the results to fit better his favored assumption of uniform distribution, he threw away any claim to be testing that assumption. When a second edition of the *Principia* finally appeared in 1713, Newton simply backed off from the problem. He wrote that the effect of stars collapsing is too slow to be observable because of the enormous distances between the stars that God has wisely instituted.[6]

Newtonians on Stellar Stability

Consequently, eighteenth-century Newtonians inherited a ready-made problem of accounting for the large-scale structure of the universe in terms of stellar distribution and the effects of gravity. The question was: How would the universe have to be arranged so as to be consistent with universal gravitation while at the same time not violating appearances too flagrantly?

A number of English Newtonians in the early part of the century addressed the problem through speculative cosmologies invented with it in mind. Some used versions of Newton's argument that stars were distributed evenly. In the Royal Society's *Philosophical Transactions* in 1720–21, Edmund Halley (of comet fame) published a paper arguing that an infinite uniform distribution of stars would be stable, because everything would settle into equilibrium: with no privileged center there could be no coalescence.[7]

An argument for a regular, organized—if not necessarily uniform—distribution of stars had already been put forward by William Derham, the Newtonian clergyman and Boyle lecturer. Derham addressed the question in his 1714 work *Astro-Theology*, a follow-up to the previous year's very popular *Physico-Theology*, itself a version of his Boyle Lectures (1711–12). Derham argued that while the stars seem to be dotted irregularly across the sky, this is probably an illusion due to our vantage point on earth, much as the paths of the planets do not appear to us to be simple and ordered even though, when properly understood, they turn out to be very well organized. Nothing less could be expected from the wise Creator who built harmony into the world.[8]

In 1721 another early Newtonian author, John Keill, also praising the heavens as the manifest handiwork of God, explained the different brightnesses of stars as a function of their distance.

> Hence arise the distribution of *Stars*, according to their Order and Dignity, into *Classes*; the first Class containing those which are nearest to us, are called *Stars* of the first Magnitude; those that are next to them, are *Stars* of the second Magnitude: The third Class comprehends them of the third Magnitude, and so forth, 'till we come to the *Stars* of the sixth Magnitude, which comprehends the smallest *Stars* that can be discerned with the bare Eye.[9]

Unlike Newton or Halley, however, Keill forebore to relate this picture to one designed to vindicate a claim of uniform distribution, whether or not as a solution to the inconvenient consequences of stellar gravitation. But without stellar gravitation, there was little obvious point to the assumption that stellar magnitude was a function of distance.

In any case, not all early English Newtonians argued for uniform stellar distribution. Nonetheless, they all worried about the problem of squaring the appearance of stars with a belief in gravity among them; this last was nonnegotiable, given its centrality in Newtonian philosophy.[10] And God necessarily remained an important part of their arguments.

William Whiston, Newton's successor as Lucasian Professor of Mathematics at Cambridge (until he was ejected for religious heterodoxy), addressed many of the foregoing issues directly in *Astronomical Principles of Religion* in 1717.[11] On the stars he observed, "This entire grand System of Things is subject to this Power of Gravity; and ... that Power of Gravity has its Effects as well among the Fixed Stars, with their several Systems, as in our Planetary and Cometary Worlds, about the Sun."[12] Given that there is no center for the sun and stars to orbit and thereby maintain a dynamic stability like that of our solar system,

> it follows, that the several [solar] Systems, with their several Fixed Stars or Suns, do naturally and constantly, unless a Miraculous Power interposes to hinder it, approach nearer and nearer to the common Center of all their Gravity; and that in a sufficient Number of Years, they will actually meet in the same common Center, to the utter Destruction of the whole Universe.[13]

Gravity itself bears witness to God's action in the material world. "Since this Power [sc. gravity] has been demonstrated to be Immechanical, and beyond the Abilities of all Mechanical Agents; 'tis certain that the Author of this Power is an Immaterial or Spiritual Being, present in, and penetrating the whole Universe."[14]

The apparent immobility of nonetheless mutually gravitating stars was therefore the central cosmological problem for English Newtonians in the first decades of the eighteenth century. In addition, attempts to make sense of stellar gravitation usually assumed that gravitational attraction is exerted between stars that are at rest with respect to each other. Although Whiston recognized an alternative, dynamical possibility that might explain why stars do not fall into each other, he rejected it without discussion: no doubt stars appear to be stationary with respect to one another, and if they are collapsing upon one another, they do so very slowly over immense distances, as Newton had said in the General Scholium.

But, ironically, great distances and long periods of time also allowed the possibility of unobserved dynamic orbital stability. Planets did not fall into the sun, or into each other, because they were moving, and Newtonian mechanics accounted for how, in effect, inertia could balance gravity to yield orbital motion about a common center. Newtonians were slow to suggest that something similar might be going on with the stars. There was no reason why stars should not move in a Newtonian universe, but the prepossession that the stars were, as the usual expression had it, *fixed* (as distinct from the moving planetary bodies) seems to have held sway for a good while despite Halley's suggestion that discrepancies between Ptolemy's star chart from classical antiquity and modern measurements of stellar positions could be due to proper motions by the stars in question.[15] But as other astronomers gradually began to echo Halley's claims, the view took hold that stars do indeed display motion among themselves.

Dynamic Stability and the Developing Heavens

Despite these departures, it was not until the 1750s that there emerged a properly worked-out cosmology involving the idea that there might be orbital motion of stars about a gravitational center or centers that would maintain overall stability. Its author, the German philosopher Immanuel Kant, was no astronomer. He based his ideas on his general knowledge of the appearances of the heavens as well as the speculations of the French mathematician Pierre-Louis Maupertuis and a slightly garbled version of a book by an Englishman called Thomas Wright.[16]

Kant's initial idea was to explain the appearance of the Milky Way. Seen by the naked eye as a faint milky band stretching around the night sky, it had been understood as the optical consequence of densely packed faint stars since Galileo's telescopic discoveries of the early seventeenth century. Kant suggested that the sun lies within a vast system of stars that extends out around it like a disk—think of a frisbee. A viewer looking out in the plane of that disk would see an expanse of stars stretching away indefinitely, crowded together into the distance, whereas looking in other directions would reveal empty, starless space beyond the mass of stars. Kant borrowed this idea from a review published in a German periodical in 1751 of Wright's book *An Original Theory or New Hypothesis of the Universe* (1750). Wright's conception was actually somewhat different from the impression that Kant drew from the periodical, since Wright, with theological purposes in mind, sketched a picture of stars arranged in a shell of modest thickness around a center, with the sun as part of the shell. The Milky Way effect resulted, Wright suggested, from looking out tangentially to the curve of the shell rather than looking through a disk of stars. The latter, however, was what Kant took away from the account that he read.[17]

Kant took the disk idea and combined it with Maupertuis's discussion of small, faint cloudy areas in the sky. These objects, known as nebulae (*nebula* being a Latin term for cloud), often showed an apparently elliptical form. Maupertuis drew on published reports in the Royal Society's *Philosophical Transactions* by William Derham. Derham's catalogue of nebulae, published in 1733, noted that some appeared round or oval. Maupertuis took this to imply the existence of faint ellipsoids in the heavens—bloated stars that bulged at the equator owing to axial rotation.

Kant suggested that Maupertuis's ellipsoidal stars might really be very distant collections of stars too far away to be resolved into their individual stellar components. These systems, rather than being ellipsoid, would be approximately flat and disk-shaped, hence appearing elliptical if seen at an angle with respect to the observer on earth. Someone located inside such a system should therefore see a "Milky Way" effect as suggested by Wright. Kant could then suggest that we ourselves live inside one of these disk-shaped stellar systems dotted throughout the heavens. He analogized Maupertuis's elliptical figures to other Milky Ways.[18]

The physical, as opposed to observational, underpinning of Kant's theory was simply gravitational. These stellar disks owed their shape to rotation about their centers and avoided gravitational collapse, by general analogy to the solar system, by the counteracting centrifugal tendencies produced by the rotation. Kant supposed that these Milky Ways, these "galaxies," might themselves be gravitationally linked into

FIGURE 2.2. A modern photograph of the Andromeda nebula (M31). Kant suggested that such apparently elliptical shapes were due to the angle of presentation of a disk of stars. Photograph used with permission from Willem Jan Drijfhout, AstroWorld Creations (September 20, 2020).

FIGURE 2.3. A modern photograph of the Pleiades: A gravitationally linked star cluster? Photograph by Antonio Ferretti and Attilio Bruzzone (February 13, 2024).

even higher-order systems. He even speculated about how this kind of universe could have come into existence from an original uniform distribution of shapeless matter, and how it may continue to develop in the future. He also suggested that gravitational condensation had produced the solar system, a proposal subsequently promoted by the French mathematician Laplace and known as the nebular hypothesis.[19]

Kant proposed a dynamic and continually evolving universe, structured by gravity, in contrast to previous eighteenth-century theories of static gravitational universes that tried to explain away an *absence* of expected gravitational effects. However, not many copies of Kant's book, published anonymously in 1755, entered circulation, because the bookseller went bankrupt. Consequently it remained little-known for several decades. Its subsequent notoriety followed the appearance of similar ideas put forward by the British astronomer William Herschel, which prompted German attempts to claim retrospective priority for Kant.[20]

Despite the dynamic options offered by a Newtonian universe, attempts to explain away the apparent immobility of the stars continued at midcentury. One of the more radical speculated that gravity, though universal, might not strictly follow the inverse square law. Instead, the force might decrease with distance so as eventually to drop to nothing and then even become a gradually increasing force of repulsion. The difference between this and strict inverse-square attractive force was, in this view, simply too slight to be measured among bodies in the relatively small confines of the solar system. This idea, especially promoted by the Croatian Jesuit priest Roger Boscovich, attempted to replace particles as the basic constituents of matter with distance forces associated with mathematical points—no space-filling solids remained, just the forces by which they are inferred.[21] Boscovich explicitly related his ideas to the question of the immobility of the stars.

> It is usually objected to the universal gravitation of Newton, that in accordance with it the fixed stars should by their mutual attraction approach one another, & in time all cohere into one mass. Others reply to this, that the universe is indefinitely extended, & therefore that any one fixed star is equally drawn in all directions.[22]

Rejecting this view because of its unacceptable implication of an infinite universe, Boscovich also provisionally rejected the otherwise acceptable view that the apparent immobility of the stars is due to the great distances between them and the long periods of time that would be required to observe their motion. Such a view, he said, would imply the gradual decay and destruction of the "universe of corporeal nature."[23]

> That this is not the case cannot be absolutely proved; & yet a Theory which opens up a possible way to avoid this universal ruin, in the way that my Theory does, would seem to be more in agreement with the idea of Divine Providence. For it may be that . . . the last arc of my curve, which represents [solar] gravity . . . will depart [when far beyond the limits of the solar system] very considerably from the hyperbola having its ordinates the reciprocals of the squares of the distances, & once more will cut the axis [i.e., become a repulsion].[24]

Once again, the appropriate characteristics of a providential God are used to establish the acceptability of a natural-philosophical proposition. Newton and God were seen almost as scientific collaborators.

John Michell's Newtonian Astrophysics

Although a Newtonian picture of the universe was perfectly orthodox throughout Europe by the middle of the eighteenth century, the assumption that gravity existed among the stars still had no observational evidence in its favor. Even proper motions among the stars, which had been sometimes argued for on evidentiary grounds by Halley and others, were not such as to reveal forces between bodies in the stellar heavens. The first positive attempt to show the effects of stellar gravitation was made in 1767 by an English clergyman, John Michell, in the Royal Society's *Philosophical Transactions*, as part of his general project for finding ways to determine the absolute brightnesses (magnitudes) of the stars and thence to infer their distances by comparison with their apparent brightnesses as seen from the earth. The central new idea of the paper in 1767 concerned double stars and star groupings generally.[25]

In effect, Michell proceeded from the default assumption that stars are dotted throughout the heavens at random so that overall they might be taken to be roughly evenly spread across the sky. With no further grounds for judging the matter, it was to be expected that there might sometimes be stars that appear close together in the sky but are in reality at greatly differing distances from us; they just happen to be in almost the same line of sight. No pair of apparently close-together stars had ever been observed to change their relative positions as a sign that they were orbiting around each other. There was therefore no observational evidence of double stars that were actually gravitational systems—true binaries.[26]

Nonetheless, if you assumed that the cosmos was Newtonian, as everyone did, the possibility of true binary stars as opposed to mere optical doubles was something to bear in mind. Michell reckoned that he could give evidence, on the basis of observation, that they really did exist, despite the absence of detected motion.

> The argument, I intend to make use of, in order to prove this, is of that kind, which infers either design, or some general law, from a general analogy, and the greatness of the odds against things having been in the present situation, if it was not owing to some such cause.[27]

So Michell set about computing probabilities. If he could show that there were more close double stars than would be predicted by random distribution, then he would have direct evidence that some doubles were genuine and presumably gravitationally linked binaries, not just chance line-of-sight doubles. Thus he argued that the odds of the components of the double star Beta Capricorni being as close as they are by chance were eighty to one against. Expanding his argument beyond doubles, Michell also came up with the figure of five hundred thousand to one against the probability that the apparent closeness to one another of the six main stars of the Pleiades cluster is owing simply to chance.

On this basis, Michell argued that stars are probably always associated in groups or clusters, including the sun and its brighter (hence likely closer) stars. He concluded that the stars are really collected together in clusters in some places, where they form a kind of systems, whilst in others there are either few or none of them, to whatever cause this may be owing, whether to their mutual gravitation, or to some other law or appointment of the Creator.[28]

Michell's extension is notable: Newtonian gravity is but a subset of distance forces generally, all with their pedigrees in Newton's speculations in Query 31 of the *Opticks*.[29]

Michell's investigations in Newtonian astrophysics focused on determining absolute stellar magnitudes as a way of estimating stellar distances and involved ideas about the nature of Newtonian-style light particles that would be subject to gravitational slowing as they fled the star that ejected them. In a 1784 paper Michell tried to detect refractive differences between double stars that would also be due to the differing masses of the source stars producing differing speeds of their respective light particles.[30] That this ambitious project yielded few positive results cannot diminish its character as a mature research program in Newtonian physical science.

William Herschel and the Natural History of the Heavens

By far the most important eighteenth-century work in stellar astronomy and cosmology was done by William Herschel, much of it with the active collaboration of his sister Caroline.[31] Herschel tends to be remembered as a great observer and telescope maker, which he was—he discovered the planet Uranus in 1781, for example. But he was at least as keen a theorist, and his observational work was done not simply to map the heavens for its own sake but to develop an observational component to his physical ideas, which were characteristically Newtonian in the same way as Michell's.

Herschel was German by birth, from Hanover, like the eighteenth-century Georgian kings of England. In 1757, a few decades after another Hanoverian, the composer George Frideric Handel, Herschel went to England as a young man in search of better things. He was trained in music and made his living as an organist. Settled in Bath, in the west of England, in the 1770s, and already interested in natural philosophy, he also began to read books on astronomy. Herschel taught himself to make telescopes—large reflectors that focused light using parabolic mirrors instead of lenses—and became so good that by the 1780s his telescopes were second to none for resolving power (discernment of fine detail) and light-collecting power (ability to show faint objects).

The kind of astronomy that interested Herschel differed from what most observational astronomers concerned themselves with. Most astronomers at that time used telescopes to make extremely precise positional measurements of planets, comets, or stars, locating them very precisely in the sky at very precise times. Such work generally used modest instruments. Large telescopes that could detect very faint objects were usually unnecessary; well-made instruments that could yield high precision in measurements (usually meaning modestly sized refractors) were the norm for this so-called positional astronomy.

But Herschel was interested in examining stars, star clusters, and nebulae so as to learn about the large-scale structure of the heavens and the kinds of formations found in it. For such work he needed telescopes that could render visible very faint objects and provide high resolution to enable the visual separation of individual, closely packed stars in stellar clusters. Throughout most of his career, Herschel primarily used reflecting telescopes equipped with mirrors twelve and eighteen inches in diameter, although he also built larger, more cumbersome instruments. Herschel's observational astronomy was directed toward answering physical questions about the operation of the heavens rather than toward producing detailed descriptions of celestial appearances and motion. He wanted to study the heavens like a philosopher, asking *why* and *how*, not just *what*.

Herschel's major work concerned the structure and evolution of stellar systems. In investigating nebulae during the 1780s, he found that with his large telescopes he was able to resolve into stars many nebulae that other people had been able to see only as milky patches of light. He began to develop the idea that *all* nebulae, even the ones he had not succeeded in resolving, were really star clusters, and their apparent cloudiness was due to the optical inadequacy of the telescope. That led to the idea, much like Kant's, that the Milky Way was the optical effect of our local star system, and that large unresolved nebulae were other major star systems beyond our own.

This rejection of genuine nebulosity in favor of the optical effect of huge distant star systems developed alongside Herschel's ideas on the development and evolution of star systems based on his observations of globular clusters. He argued that these apparent agglomerations of stars must be gravitationally related, using the same kinds of probabilistic arguments as Michell's well-known work.

By the early 1790s, however, Herschel's observations of so-called planetary nebulae began to persuade him that maybe there was such a thing as genuine nebulosity after all. Planetary nebulae take their name from the fact that through low-power telescopes they look a little like planets because they appear as small disks rather than as stellar points. They typically present as cloudy disks, but Herschel found one with a star clearly present at its center. It seemed implausible to Herschel that the cloudiness was due simply to hitherto unresolved clouds of much tinier stars. This looked too much like a star with a vaporous shell around it.

Herschel consequently proposed that stars originally form from cloudy material—as with planetary nebulae—before the gravitational formation of star systems, which ultimately produce globular clusters. But such a view barred the argument that hitherto unresolved nebulae are always large star systems beyond our Milky Way system. Nonetheless, his work continued to revolve around ways of gauging stellar distances and cosmic dimensions, attempting to follow the evolution of the heavens. Herschel regarded different kinds of nebulae and star groupings as moments in stellar development that represented distinct species of stellar formation. He described systematizing these stages of development as a kind of "natural history" of the heavens. He interpreted these snapshots as sequences that allowed a reconstruction of

FIGURE 2.4. A modern photograph of a globular cluster. Photograph taken by NASA Hubble (2017). Available under Creative Commons Attribution 2.0 Generic License.

the temporal history of heavenly structures—all powered by gravitational forces, and perhaps unknown repulsive forces as well to combat gravitational collapse.[32] Variety in the heavens displayed a systematic order rooted in time.

At the end of the eighteenth century, then, Herschel had established a detailed, observationally based, dynamic picture of an evolving Newtonian universe. It was a remarkable extension of the basic Newtonian physical assumptions, and it set the stage for later, nineteenth-century developments. This extension had followed from certain decisions (including some of Newton's) to take as provisional truth assumptions that were more interested conjecture than empirically based belief. For Newton and many of his early eighteenth-century acolytes, the credibility of these assumptions rested on theological convictions that had deeper foundations than natural philosophy itself.[33] God's constancy and activity made possible a true natural philosophy. Herschel's universe, a century after Newton, gave detailed form to observational results perceptible in their classificatory variety as the seeds of many possible Newtonian universes.

3

Mixed Mathematics and Probability

MICHELL AND Herschel followed the lead of other mathematicians in using probabilistic arguments. The use of mathematics in worldly and natural-philosophical concerns had long been part of mathematics, even part of what mathematics was understood to be. Modern physics descends directly from those practices.

Mixed and Pure Mathematics

In the eighteenth century, mathematics was not conceptualized in terms of a categorical distinction between pure and applied; that familiar modern dichotomy was born in the nineteenth century. Before the nineteenth century, the word *mathematics* encompassed a much broader field than it did later, and the main distinction between types of mathematics was pure versus mixed.[1] "Pure" mathematics comprised geometry and arithmetic, the latter including algebra and the infinitesimal calculus, known jointly as analysis. Pure mathematics dealt with the relations and properties of quantity in general—general magnitude, whether pure number or pure extension, independent of any particular things to be counted or measured.

"Mixed" mathematics, by contrast, included any field that focused on the quantitative characteristics of things. It included geometrical optics, mechanics, and astronomy (also called geometric astronomy, as opposed to qualitative stargazing), but it also included music, meaning acoustics and harmonics, by this time talked about in terms of frequencies of soundwaves. Even practical endeavors like navigation, shipbuilding, and fortification counted as mixed mathematics, because they used mathematically formulated principles. In the case of fortification, the principles were associated with those of classical architecture, which also used geometrical principles and procedures. Artillery counted, too; ballistics (as pioneered by Galileo: ways of calculating the ranges of guns at different elevations, taking into account matters such as the effects of air resistance). The eighteenth century also witnessed the rise of a new branch of mixed mathematics, the calculus of probabilities, a mathematical science applying broadly to nature, or insurance statistics, or, especially, gambling games.[2]

Each of these fields could properly be called mathematics, not only the so-called pure subjects; they were all regarded as doing basically the same kinds of things. William Herschel first became seriously interested in astronomy in the 1770s after he read a book on the optics of telescopes by an author he had already read, and liked, on music. That the same author, Robert Smith, had written books on both music and astronomical optics is unsurprising, because both subjects counted as mixed mathematical sciences.[3]

Mixed mathematics was a way of understanding the relationship of mathematics to natural philosophy that had different implications from the later category of applied mathematics. The idea of applied mathematics since the nineteenth century implies that a finished piece of pure mathematics can be plugged into a problem to do with real things. The implication is that the piece of pure mathematics has been developed independently of any relation to things in the world (depending on one's views on the nature of pure magnitude). But the idea of mixed mathematical sciences in the eighteenth century—an idea that goes back to Aristotle—took a different path. It held that the mathematics of real things is not distinct from the things themselves. Theoretical mechanics, for example, the rational mechanics of Newton and most eighteenth-century mathematicians after him, was the mathematical treatment of forces and masses and motions; its conceptual development, with the invention of appropriate mathematical techniques and ideas associated with infinitesimal calculus, often took place with specific reference to *mechanical* problems. It was not a matter of pure mathematics being worked on independently and then being applied to mechanics whenever it seemed relevant.

The term *mixed mathematics* indicated the involvement of quantities of something and therefore had to include considerations to do with the nature of the actual objects themselves (such as mass, which could be measured but was distinct from continuous and discrete magnitude per se). The extension of the term is most strikingly represented by a characteristically eighteenth-century branch of mixed mathematics mentioned in the previous chapter: probability, or what came to be called most often "the calculus of probabilities." It illustrates strikingly the idea of mixed mathematics and how people in the eighteenth century thought that mathematics could help to create real knowledge about the world—to discover new things. The belief that mathematics somehow expresses the fundamental structure of the universe owes much to that perspective on mixed mathematics, where mathematics was tailored to a world suited to fit it.

The Calculus of Probabilities: Mathematics as Apologetics

Mathematical probability had started out in a serious way in the 1650s, with work by the French mathematicians Blaise Pascal and Pierre de Fermat and the Dutchman Christiaan Huygens. They were each interested in finding ways of calculating appropriate stakes for simple gambling games, focusing on coin tossing. By the early eighteenth century several mathematicians were writing papers on aspects of probability. The first large-scale treatise on the subject was *Ars conjectandi*, the art of conjecturing, by the Swiss mathematician Jakob Bernoulli, published posthumously in 1713.[4] Bernoulli chiefly addressed coin-tossing games, dice games, and so forth, which were the usual topics. But his ultimate aim was to point the way toward the application of probabilities to human affairs in general: legal judgments, economic and political decisions, and any matters that eluded certainty but where there was an interest in increasing as much as possible the chances of being right. That was what the calculus of probabilities was all about: finding ways of controlling human judgment in practical affairs.

In general, probability focused on determining the best course of action to take in given circumstances, such as how much to wager in a dice game or how to decide a legal case on the basis of numbers of witnesses. In practice, of course, real human affairs were a lot messier than dice games and a lot harder to turn into mixed mathematics. But that conception of what probability was all about—making judgments—was what made it mixed mathematics: it was about practical things in the world rather than abstract quantity.

Typically, in analyses of coin-tossing games, a mathematician assumed at the outset a fundamental underlying probability—a one-in-two chance for heads and the same for tails, as with a so-called ideal coin. In developing the calculus of probabilities, such a coin was posited as a premise of the argument, rather than giving consideration to an actual coin, and the mathematician then determined what would follow from that assumption. Statistical probability, by contrast, was something rather different, and was called in the eighteenth century *inverse probabilities*. It represented a different sort of probability from that applying to coin-tossing games because instead of initially assuming an underlying probability governing each individual event, it began as though completely ignorant of any kind of underlying probability and tried to infer one from data, as from many trials of such events.

When John Michell used mathematical probabilistic arguments to establish his claims about gravity among the stars, therefore, such reasoning was not unprecedented. Newton's much earlier argument about the improbability of the solar system having the order it displays by chance had treated a similar point as self-evident, but Newton had made no attempt to calculate relevant probabilities. Others, however, were prepared to take on the job. The trend in England for probabilistic arguments to show God's presence was prompted specifically by a paper published in the Royal Society's *Philosophical Transactions* for 1710 by

```
n=0                    1
n=1                  1   1
n=2                1   2   1
n=3              1   3   3   1
n=4            1   4   6   4   1
n=5          1   5  10  10   5   1
n=6        1   6  15  20  15   6   1
n=7      1   7  21  35  35  21   7   1
```

The numbers in the triangle correspond to the coefficients of the terms in binomial expansions for each line; e.g., for the line n=5 above, expanding $(M + F)^5$ we have

$$M^5 + 5M^4F + 10M^3F^2 + 10M^2F^3 + 5MF^4 + F^5$$

This tells us, for example, that the probability of getting, on any single collective coin-toss, exactly 3 Ms and 2 Fs is 10/32, where 32 is the total number of possibilities.

FIGURE 3.1. Pascal's triangle: each number is the sum of the two nearest numbers above it.

John Arbuthnot, physician to Queen Anne, satirist (a contemporary of Jonathan Swift), and amateur mathematician. His paper is titled "An Argument for Divine Providence, Taken from the Constant Regularity Observed in the Births of Both Sexes."[5] Arbuthnot sets up the argument mathematically by considering a coin-tossing game. Such an approach had become routine in probabilistic argument since it had been developed by Pascal and Huygens.[6] Consider a set of a certain number of coins (in effect, two-sided dice), each of which has its sides marked respectively M and F. Now consider the likely outcomes of tossing all the coins and adding up the resulting numbers of displayed Ms and Fs.

Arbuthnot says that the relative probabilities of different outcomes are given by the binomial coefficients of the expansion $(M+F)^n$, where n is the number of coins. That yields the result that the probability of getting, from tossing n coins, exactly k Ms and $(n-k)$ Fs is given by the coefficient of the term $M^k F^{n-k}$, divided by the total number of different possibilities, 2^n. Arbuthnot then goes on to say that, as the number of coins, n, increases, so the likelihood of any given outcome (coefficient/2^n) gets smaller. And so he says, returning at last to the original topic, "Consequently (supposing M to denote Male and F Female), in the vast Number of Mortals, there would be but a small part of all the possible Chances, for its happening at any assignable time, that an equal Number of Males and Females should be born."[7] In other words, the chance of getting, in a given year, exactly 50 percent males and 50 percent females is minuscule.

Arbuthnot now starts trying to turn this to account by arguing that the same general conclusions apply to any proportion merely *approximating* 50:50. Since the chances of getting 50:50 or any given proportion close to it are themselves very small, it is extremely unlikely, he asserts, that there will never be an outcome very far from 50:50. He now addresses the available data, consisting of the Bills of Mortality for London, which gave the annual death rates over many decades and which ought to give similar sex proportions as might be expected from birth figures, as well as the London christening records for the previous eighty-two years. If the sex ratios were governed by chance, says Arbuthnot, one would expect on the basis of his probability analysis to find individual years showing large deviations from a 50:50 proportion.

However, the empirical evidence showed no such large deviations; the proportions were always very close to equal, from which Arbuthnot concludes that the consistent regularity, year in and year out, of the 50:50 sex ratio cannot be due to mere chance. God must be deliberately maintaining the ratio Himself, which is why the paper is an "Argument for Divine Providence."

Arbuthnot goes on to point out that, in fact, the actual ratio gives a slight but consistent preponderance of males (about eighteen in thirty-five), again so improbably uniform that it cannot, he says, be due to mere chance. It holds for every one of the previous eighty-two years, and the odds against that happening by chance, assuming fifty-fifty odds on that outcome in any single year, are 1 in 2^{82}. God must be maintaining such improbable consistency, and He does it, Arbuthnot has no doubt, to compensate for the rather more risky lives that men lead compared to women.[8]

The most important aspect of Arbuthnot's paper is its central ambition: to use mathematics in a theological argument. Mathematics in this period was routinely regarded as the pinnacle of certain knowledge. The term *mathematical certainty* described the highest and most unquestionable knowledge possible. Thus, by showing how this newly developing branch of mathematics, the calculus of probabilities, could be applied to arguments for the existence and providence of God, Arbuthnot indicated a way of bolstering theological arguments with the most prestigious kind of demonstration available. And so, regardless of the evident flaws in his argument, Arbuthnot's paper had an enormous amount of influence especially with English writers in simply giving them the idea of using probability calculations for such purposes.[9]

William Derham's *Physico-Theology* was first published in 1713, just a couple of years after Arbuthnot's *Philosophical Transactions* paper, and the subtitle of Derham's book reminds us of its central concern: *A Demonstration of the Being and Attributes of God from His Works of Creation*. Derham picks up on Arbuthnot's argument as part of his "demonstration," including the wisdom of having a slight preponderance of males over females: "This Surplusage of Males is very useful for the Supplies of War, the Seas and other expences of the Men above the Women. That this is the Work of the Divine Providence and not a Matter of Chance, is well made out by the very Laws of Chance, by a person able to do it, the ingenious and learned Dr. *Arbuthnot*."[10] Derham evidently did not regard himself as a person "able to do it," but he had the next best thing: there was nothing like the authority of a mathematician.

In fact, had Derham but known it, Arbuthnot's argument had already been undermined by Bernoulli's s *Ars conjectandi*. Bernoulli's book presented for the first time a basic principle relating to the derivation of statistical probabilities: the "law of large numbers," as the nineteenth-century French mathematician Siméon Denis Poisson was to dub it. It applied to situations that can be modeled on that of having an urn containing a very large number of balls of two different colors that are otherwise physically identical. A ball is removed from the urn and its color noted; it is then returned to the urn and mixed back in. The process is then repeated many times over. Bernoulli asserted that, as the number of drawings increases, the proportion of drawn balls of one color to the number of drawn balls of the other color will get closer and closer to the actual proportion of the two colors in the urn itself, and consequently to the probability of the next ball to be drawn being of one color or the other.

In other words, this is a kind of limit concept: it justifies on a formal basis the inference of probabilities from statistical data. In the urn example, if, of one hundred balls drawn from the urn, approximately twenty-five have been white and seventy-five have been black, one may justifiably infer that the probability of the next ball being black is three in four.[11]

We can be confident that Bernoulli neither invented nor proved this kind of inference; he just formalized and mathematized it by stating it as an explicit *principle*. This is a common feature of formal mathematical and scientific inference in this period, when so-called scientific explanations were supposed to be deduced from fundamental principles that were not themselves demonstrable. Such principles were the simplest assertions that could be made in their subject domain, and were therefore not derivable from still simpler statements (which by definition did not exist). Yet stating such foundational principles gave them a status that reflected on inferences drawn from them; the reliability of mathematical or scientific conclusions relied on the underlying foundations being (in practice) unquestioned. Such seems to have been the case for the law of large numbers.

But the fact that, by the mathematical standards that were being created for probability at the time, Arbuthnot's reasoning had been demonstrably erroneous failed to lessen the enthusiasm of more mathematically competent English writers than Derham for his approach. Abraham de Moivre was a Protestant (Huguenot) French mathematician and fellow of the Royal Society living in England who in 1718 published a book called *The Doctrine of Chances*, with a dedication to Isaac Newton. His dedication advertises his sympathy for the physico-theological project by speaking of learning from Newton's philosophy "how to collect, by a just Calculation, the Evidences of exquisite Wisdom and Design, which appear in the *Phenomena* of Nature throughout the Universe."[12] In the third edition of the book, published posthumously in 1756, De Moivre defended Arbuthnot's approach to the birth records against the criticism of another of the Bernoullis, Nicholas, who had said that there was nothing contrary to chance in the steady birth-rate ratio to which Arbuthnot had drawn attention. The apparent fact that eighteen males were born out of every thirty-five total, rather than exactly one out of every two, should not astonish us, Nicholas Bernoulli said, because exactly the same proportions could be obtained from rolling a thirty-five-sided die with eighteen white and seventeen black sides (otherwise identical) a large number of times, in accordance with Jakob Bernoulli's precept. Consequently, according to Nicholas, the birth data provided nothing that required chance to be augmented by God's providential interference.[13]

De Moivre's response said that Arbuthnot had been right in spirit if not in detail, since to get those results one would deliberately have to make such a thirty-five sided die, and the very form of the die itself would thus provide evidence of intelligent design. Therefore, whatever analogous structural causes produce the birth ratio show the same thing; the birth ratio still counts as evidence of God's purposeful design, even if not in quite the way originally suggested by Arbuthnot.[14] De Moivre wrote, "As if we were shewn a number of Dice, each with 18 white and 17 black faces, which is Mr. *Bernoulli's* supposition, we should not doubt but that those Dice had been made by some Artist; and that their form was not owing to *Chance*, but was adapted to the particular purpose he had in View."[15] These are "Laws according to which Events happen," De Moivre held, and "it is no less evident from Observation, that those Laws serve to wise, useful and beneficent purposes; to preserve the stedfast Order of the Universe, to propagate the several Species of Beings, and furnish to the sentient Kind such degrees of happiness as are suited to their State."[16]

All these English philosophical concerns with finding demonstrable evidence for an intelligent and benevolent God, both Creator and superintendent of the universe, followed Newton's general theological orientation to natural philosophy. Recall that Newton was President of the Royal Society at the time of Arbuthnot's publication in the *Philosophical Transactions*. But they did it through the medium of mixed mathematics, which revealed God's immanent role.

Mixed Mathematics and the St. Petersburg Paradox

Because probability was a branch of mixed mathematics, it carried with it no conception of anything like pure mathematical probability. The validity of its procedures was related to, and in some sense assessed by, the things or situations with which its procedures were concerned. Again, this is how it differed from later applied mathematics, seen as plugging independent pieces of pure mathematics into concrete problems.

Perhaps the clearest and most interesting example from eighteenth-century probability theory of this interactive use of mathematics in creating knowledge about the world is the so-called St. Petersburg paradox. It shows how apparently purely mathematical results had to be assessed on the basis of nonmathematical experience.

The St. Petersburg paradox took its name from the Russian Academy of Sciences in St. Petersburg, one of the eighteenth-century royal scientific academies that were modeled on the Paris Academy of Sciences. It was founded in 1724 by Tsar Peter the Great, and like the Parisian prototype it put out its own annual "memoirs" of research papers by its members. Peter stocked his academy with people from all over Europe, the best he could find and attract to Russia with large salaries, one of whom was another of the mathematical Bernoulli family, Daniel. It was his paper in the *Memoirs* for 1730–31 (not actually published until 1738) that gave the St. Petersburg paradox its name, although the paradox itself was first suggested by Nicholas Bernoulli in 1713.[17] It counts as a paradox not because it leads to a logical contradiction but because apparently sound assumptions lead to apparently absurd results.

The problem is to do with a simple gambling game. Two players, *A* and *B*, are playing a coin-tossing game. If the coin comes up heads on the first toss, *B* gives *A* one ducat (or whatever unit of currency you like). If it doesn't come up heads until the second toss, *B* gives *A* two ducats; if it takes three tosses, *B* gives *A* four ducats, and so on: the amount of money that *B* has to pay to *A* *doubles* for each extra toss of the coin until it comes up heads.

In general, then, if no head appears until the nth toss, *B* pays *A* $2n - 1$ ducats. Now the question is: How much should *A* pay to *B* for the privilege of playing this game? The standard criterion in such problems was that the conditions ought to yield a fair game, where chances of gain and loss are equal on both sides. This means that *A* ought to be paying *B*, for the privilege of playing the game, an amount of money equal to *A*'s expectation of monetary return from that game.

The standard way of calculating this expectation, ever since mathematical probability had first been developed in the 1650s for dealing with gambling games, was to set it equal to the probability of each possible outcome multiplied by the value of that particular outcome. That was regarded as the obviously correct way of formalizing these kinds of games, because it makes sense for the simplest cases. If you call heads or tails on a single coin toss and you'll get two dollars if you're right, it's a fair game if you pay one dollar to play, because the probability of you being right (assuming a fair coin) is one in two—half.

The paradox in the St. Petersburg game lay in the fact that if A's expectation is computed on this standard basis, the conclusion is that A must pay B an *infinite* amount of money to play. This is because the expectation would be calculated like this:

$$E = 1/2(1) + 1/4(2) + 1/8(4) + \ldots + 1/2^n(2^{n-1}) + \ldots$$

Since there is a real, although vanishingly small, possibility of an endless sequence of tails, the sum of this series is infinite. All the elements in it are equal to each other—one-half—because the decreasing probability of longer and longer strings of tails is exactly balanced by the increasing payoff for such strings. So A's total expectation going into the game (remember, a single iteration) is an infinite amount. And that should constitute a fair game.

It was clear to everyone that something was wrong here. The idea that anyone in their right mind would pay an infinite amount of money, or even more than just a few dollars or ducats, was clearly ridiculous. And yet this value was computed on the basis of one of the most fundamental principles of eighteenth-century mathematical probability.

How did eighteenth-century mathematicians end up in such a difficulty? The calculus of probabilities was seen as being about controlling human judgment in situations of uncertainty. Accordingly, the fundamental assumption was that the results of proofs in mathematical probability ought to agree with the judgments of what the eighteenth century called the "reasonable man"—that is, someone with good judgment. The St. Petersburg problem was therefore a paradox because the very science that was supposed to capture and formalize the essence of good judgment contradicted, to a spectacular degree, what any reasonable man would do in such an instance.

Daniel Bernoulli's way out, suggested in his St. Petersburg Academy paper in 1738, caused a lot more trouble than the original publication of the paradox twenty-five years before. Mathematicians had been quietly managing to ignore it, which was a practical solution, but Daniel Bernoulli brought attention back to it and raised questions about the established way of doing probability by virtue of the originality of his solution. It relied on redefining the concept of expectation. He proposed distinguishing between two different kinds of expectation. One, called *mathematical expectation*, was defined by the usual mathematical definition of probability times value. A different kind, which he called *moral expectation*, applied to actual reasonable human behavior. Moral expectation depended on the circumstances of the individual player of the game—circumstances that would necessarily color the judgments of a reasonable man—because those judgments are the real subject of this form of probability; they are what it aims to model.

Which is not to say that Bernoulli said that such matters were indeterminate or that every case is different. He wanted to be able to calculate moral expectation just as much as ordinary mathematical expectation. So he made it relative to an individual's current wealth: the richer a player is, the more they can reasonably afford to risk. Bernoulli gave an example of a poor man holding a lottery ticket with a one-in-two chance of winning twenty thousand ducats. If that man is prudent, he will be willing to sell the ticket for nine thousand ducats even though his mathematical expectation is ten thousand. A very rich man, on the other hand, would be foolish not to buy it at that price.

The idea of moral expectation involved measuring value not in a simple monetary sense but in terms of what Daniel Bernoulli called *utility*—how much difference a certain amount of money will make to a particular individual. The more money you already have, the less losing a set amount will matter to you, and the greater the amount you will have to win in order for it to make you significantly happier. Bernoulli expressed this formally as

$$dy \text{ [increment of utility]} = b(dx/x) \text{ where } x \text{ is current wealth.}$$

And that took care of the paradox quite handily, given an appropriate, empirically determined value for the constant b.[18]

The main critic of this solution was the French mathematician, Newtonian, and *philosophe* Jean D'Alembert, one of the editors of the antiestablishment *Encyclopédie* in eighteenth-century France. He said that Daniel's was an ad hoc solution that failed to get to the heart of the difficulty. But just as with Daniel Bernoulli's approach, D'Alembert's solution concerned itself with practical situations and assessments of "reasonable" behavior. D'Alembert's idea was that if, for example, a reasonable man tossed a coin a hundred times and it really did come down tails every time—like one of the higher terms in the series—then any reasonable man would infer that the coin is biased. Consequently, mathematical probability would not be applicable, because its use presupposed that the coin used was equiprobably fair.[19] D'Alembert illustrated the point with an analogous example, again stressing the role of ignorance or knowledge regarding outcomes in real cases:

> We suppose that letters thrown on a board make the word *Constantinopolitanensibus*, and that an ignorant person is asked if those characters were thrown at random [*au hasard*] or not; he would reply that there is every appearance that they were thrown at random. But if one posed the same question to someone who knew the existence of Constantinople, and who knew Latin, he would reply on the contrary that there is every chance, and even that it is certain, that this arrangement is not the result of chance. The first person doesn't know, while the second person does know, that the arrangement of these letters is such that there was almost certainly the operation of an intelligent cause.[20]

He immediately applied the lessons of this analysis to the St. Petersburg problem:

> It's the same with the game with which we are concerned. The experience and knowledge that we have of the laws of nature teach us that the same outcome never occurs many times in a row; and it's by virtue of this acquired knowledge that we place in doubt the repetition of heads or tails a large consecutive number of times.[21]

D'Alembert's approach to the paradox, stressing empirical knowledge of actual cases, makes sense only in the context of eighteenth-century mixed mathematics. It was a mathematics tailored to the real things that it discussed—in this case, the behavior of the reasonable man. The essence of mixed mathematics was that it was about *things in the world*, not just the application of abstract mathematics to theoretical situations. The practice of mixed mathematics always involved discussing objects that had quantitative properties.

Mathematized experimental physics developed in the later eighteenth and nineteenth centuries in large part through this tradition of mixed mathematics, with the mathematics being embedded in the experimental behaviors themselves. The kinds of inventories of likelihood that we have already seen with Michell's (and Herschel's) arguments about star groupings, or Arbuthnot's survey of birth data, established the sensibilities that would render probabilistic inference an acceptable part of philosophical procedures and would undergird statistical physics in the nineteenth century. Our indeterminate quantum world is the offspring of those ideas.

4

Inventories of Electricity

A GOOD case can be made that the specialty we know as "physics" did not emerge full-fledged before the 1840s: characteristic institutional, pedagogical, and theoretical components of modern physics were previously incompletely formed. Even the word itself, as a synonym for natural philosophy in the ancient Greek sense, had referred to the study of all aspects of nature, both living and nonliving, and bore no disciplinary specificity of the kind emergent in the nineteenth century. The birth of modern physics occurred in the context of the appearance, in the opening decades of the nineteenth century, of a new set of experimentally produced phenomena. These had sufficient similarities or relationships to one another, and to already established topics, that it became possible to constitute them all as a new and distinct scientific discipline that relied on a common body of teachable techniques.[1]

The Amber Effect

One of the fields of study that made up the nineteenth-century discipline of physics had only recently come to prominence. In 1700 electricity had possessed very little conceptual precision, but a century later it had a complex profile of concepts and behaviors that placed it close to the center of a broad new area of physical phenomena soon to be elaborated as "physics." The invention of that profile followed explicitly Newtonian themes and concepts.

In 1700 electricity was understood in much the same way as in 1600, when the Englishman William Gilbert invented the word, comparing it to magnetism. The word *electricity* comes from the Greek *elektron*, which means amber, the hardened tree resin. Lumps of amber had been known since antiquity to attract light pieces of straw, paper, and so forth if vigorously rubbed. A rather desultory interest in this amber effect throughout the seventeenth century produced few additions to this list of "electrical" phenomena; it was, for example, noticed that bits of paper attracted to the amber bounced off again as soon as they touched it, but there was no agreement about whether this was significant—perhaps it was a repulsion following contact, but perhaps it was purely mechanical. Electricity was basically understood as a restricted *attraction* phenomenon.[2]

In the 1660s Robert Boyle, a member of the Royal Society, made a study of a wide variety of materials other than amber that also exhibited the attraction effect, but he had little to say about the nature and causes of the effect itself. On the infrequent occasions when the natural philosophers in the Royal Society and their foreign counterparts considered the matter, they tended to explain electrical effects in terms of the action of an invisible effluvium or cloud of particles surrounding the electrified body, which mechanically pushed or impelled light objects toward it in some unspecified (but always "mechanical") way.

In consequence, by the end of the century electricity was still just a curiosity. More importantly, it remained a phenomenon of attraction, sometimes perhaps with a corresponding repulsion, but nothing more. A list of its characteristic properties was therefore quite dull, and philosophers continued to pay little attention to it.[3] It was a small assortment of properties, not a specific natural phenomenon to be investigated.

Matters began to change in 1706 when Francis Hauksbee, who was employed as the Royal Society's experimental demonstrator, brought to one of the society's meetings a tube of glass thirty inches long and some pieces of leaf brass (beaten brass, analogous to gold leaf). The effects Hauksbee produced with this simple apparatus were much better than those usually achieved with electrified amber, wax, or small pieces of glass and pieces of paper. The glass tube could be electrified more vigorously, and the leaf brass produced a much more evident effect. In particular, the repulsion effect after the pieces touched the glass seemed unequivocal.[4]

Hauksbee's work had two important consequences. One was that by improving techniques and renewing interest in electrical effects, Hauksbee had provided a promising and entertaining field of study for English amateur experimentalists, encouraging new electrical research. The other was that Hauksbee had aroused the interest of the Royal Society's president, Isaac Newton.

In the second edition of *Opticks*, published in 1706, in some speculations on matter in the Queries, Newton wrote about aethers—expansive clouds of mutually repelling particles that interacted with ordinary matter through various short-range forces. He suggests that this kind of aether might be responsible for electrical effects.[5] In the second edition of the *Principia*, in 1713, he made some brief remarks about the possibility of a kind of subtle elastic fluid pervading the pores of ordinary matter that he supposed to be responsible for electrical effects.[6]

Newton's speculations were stimulated in large part by Hauksbee's work. And Newton's prestige in the eighteenth century was so high that anything in which he showed an interest immediately acquired philosophical legitimacy. Thanks to Newton, Hauksbee's work served to popularize the study of electricity and to encourage a particular line of interpretation. On the other hand, as regards what Hauksbee's work achieved in understanding electrical phenomena, very little had been accomplished. Hauksbee had amplified some electrical effects, including the vigorous electrification of spinning evacuated glass globes; he also drew attention to the appearance of light inside the spinning globes. But these were still parlor tricks with no real theoretical component. Crucially, there was little extension of the concept or meaning of electricity beyond the simple attraction-repulsion effect.

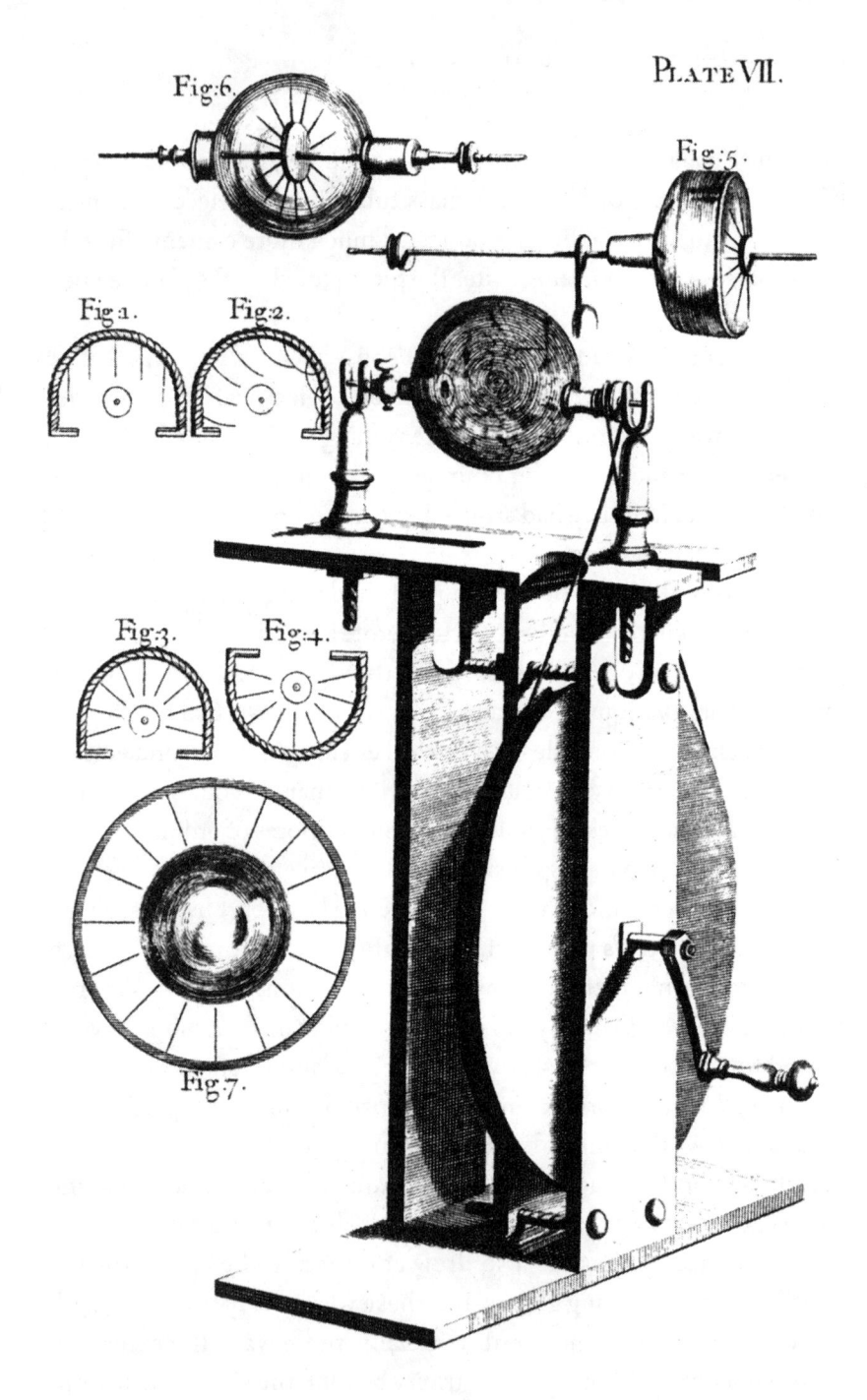

FIGURE 4.1. Francis Hauksbee's electrical apparatus, designed to maximize frictional electrification (1719). *Source*: Science History Images/Alamy Stock Photo.

New Meanings for Electricity

Not until 1729 was there a real extension beyond the central attraction-repulsion effects in basic notions of what counted as electrical phenomena. That came with the work of Stephen Gray, whose investigations were published in the Royal Society's *Philosophical Transactions*. Gray added to the existing list of electrical properties the phenomenon of what he called the *communication* of electricity.[7]

Gray, like Hauksbee, used a long glass tube: "Its Length is three Feet five inches, and near one Inch two Tenths in Diameter. . . . To each End I fitted a Cork, to keep the Dust out when the Tube was not in use."[8] The first novel effect he noticed was, he claimed, a matter of pure accident. Playing around with the tube, he found that when it had become electrified by rubbing, a feather was more strongly attracted to the cork stuck in the end of the tube than to the glass tube itself. He decided that this phenomenon was due to a communication from glass to cork of what he called the "attractive Vertue."[9]

Gray then set about seeing how far this virtue could be communicated along lines of packthread (stout string). He found that he could manage to send it several hundred feet with the string connecting to an electrified glass tube He obtained the attraction effect in such cases by fixing something at the far end of the thread that would draw bits of leaf brass—among other things, an ivory ball, a kettle, and (he curiously explained) a forty-seven-pound charity boy.[10] In the course of these investigations he made further discoveries. When setting up his long lines of packthread, Gray had to support them in some way. At first he tried suspending them from fine silk threads, on the supposition that their thinness would make it harder for the electric virtue to escape along them before reaching the end of the line, and in fact this worked.[11] On the other hand, he found that very fine brass wire used as a support for the packthread resulted in no communication to the far end, leading him to decide that it was the material of the supporting threads, not their thickness, that mattered. In due course he found other substances that did as well as silk, including resin and glass itself—all things that were known to be electrifiable.

Gray only ever explained his observations (conventionally described with a fair amount of circumstantial detail) in qualitative terms, talking about the transmission of electrical effluvia without explaining how these effluvia produced attraction effects. Neither did he draw any formal distinction between what we would call conductors and nonconductors. Nonetheless, he had moved the experimental meaning of electricity beyond simple attraction-repulsion effects.

Gray's published investigations soon caught the attention of a French member of the Royal Academy of Sciences in Paris, Charles Dufay. In 1733 and '34 Dufay published reports in the academy's *Memoirs* of experiments done to confirm Gray's work.[12] In the course of his investigations, Dufay found, among other things, that moistening the line of packthread improved the communication, and he managed to achieve a maximum distance of transmission of 1,256 (Paris) feet.

Dufay also came up with new ideas and developed exemplary phenomena that illustrated those ideas. He conceived of a unified, formal, analytical account of the behavior of gold leaf, which he used instead of leaf brass, when it is first attracted and then, after contact, repelled by an electrified glass tube.

He described the behaviors thus: The unelectrified gold leaf is initially attracted to the electrified tube. When it touches the tube it becomes electrified by communication, as one would expect from Gray's experiments. Only at that point, after the gold leaf becomes electrified, is it repelled by the tube. The difference between the attraction and the repulsion effects depended, therefore, on whether the gold leaf was unelectrified or electrified. The cycle was completed when the electrified gold leaf touched some other object in the room to which it lost its recently acquired electrification. It was then in a state to be again attracted to the glass.

Notice that Dufay's account (which is exactly that: an accounting) is a description of various behaviors.[13] It gives no explanation or justification for what happens at each stage. But the account does more than just describe unexplained behaviors: Dufay's universalized descriptions of behaviors *systematize* the phenomena in such a way as to facilitate predictions of untried procedures, given his classification of their various components. Dufay's account relies on the newly developed idea of the communication of electric virtue, as introduced by Gray, which for Gray was an electrical property. For Dufay, communication was a phenomenon inferred from the behaviors that he described; it was not itself one of those behaviors. Dufay told his readers why certain behaviors occur, but the *why* remains at the level of phenomena and their regularities. Underlying causes are not addressed. Even the concept of electrification is defined phenomenologically, in relation to circumstances of attraction and repulsion.

Dufay's generalizations are really rules governing electrical behaviors. The integration of those rules forms part of a gradual refining of concepts and an exploitation of their explanatory potential. Dufay's account, in fact, *consists* of rules, the priority of which amounts to natural-philosophical explanation.

But Dufay soon found anomalies in his rules. Since the gold leaf was repelled from the glass tube after touching it and becoming electrified, the rule seemed to be that electrified bodies repel one another. But in the course of various experiments, Dufay found—contrary, he said, to what he expected—that when he brought an electrified rod of resin or wax, rather than glass, near a piece of gold leaf that had already been electrified by contact with a glass tube, instead of being repelled, the leaf was strongly attracted to the rod. That clearly went against the general rule of electrified bodies repelling each other.

Dufay resolved this problem by multiplying the entities in his experiments: electrifiable bodies would now come in two kinds. He tried a wide range of materials that were known to be electrifiable (recall that Boyle had inventoried such materials) and identified one set of substances, including rock crystal, that behaved like glass when they were electrified by rubbing, and another set of materials, including gums and resins and wax, that behaved in an apparently opposite way as judged by their attraction and repulsion effects. His accounting, or explanation involved saying that there were two different kinds of electricity, corresponding to the two types of material: vitreous, exemplified by glass, and resinous, exemplified by amber or wax. The new idea was that like electrifications repel, but unlike electrifications attract. A body electrified by communication, like the gold leaf touching the rod, took on whichever kind of electricity characterized the communicating body.

An additional conceptual element was put in place in 1739, back in England. Hauksbee's successor as experimental demonstrator to the Royal Society, Jean Desaguliers (the son of French Huguenot refugees), was the first to distinguish clearly between what we would call *conductors* and *insulators*. Desaguliers's distinction was between materials that could be electrified by rubbing, which he called *electrics* (including both glass and wax), and materials that could not be electrified by rubbing but could receive electric virtue by communication. He called those *non-electrics*, and they included, most notably, metals.[14]

By 1740 electrical research had acquired a set of interrelated conceptual elements and rules that included the idea of the communication of electricity, the existence of two kinds of electricity, rules for attraction and repulsion, and the distinction between electrics and non-electrics. Physical explanations for all this were considerably less clear. Dufay sometimes talked about Cartesian-style vortices of subtle matter whirling around electrified bodies, while the English talked vaguely about effluvia or clouds of particles and their production from ordinary matter by friction. There was very little integration between those ideas and the analytical concepts and rules that had been developed for the practical management of the experimental phenomena themselves.

Magnifying Effects

In the first half of the 1740s, in Germany, thanks to a Leipzig instructor named G. M. Bose, an electrical machine came into use that was capable of producing much larger effects than those obtained through rubbing by hand a glass vessel. The machine used a glass globe fixed on a horizontal axle to allow high-speed rotation by turning a wheel connected to it by a belt. The globe became electrified by the operator placing a hand on it or by a leather cushion contacting the globe. Hauksbee had sometimes used a similar apparatus, but the new machine went a stage further. Hanging over the machine from nonconducting threads from the ceiling was a gun barrel, and from one end of the barrel hung a metal

FIGURE 4.2. Bose's new kind of electrical machine. From Abbé Nollet, *Leçons de Physique* (1767).

chain that brushed against the spinning glass globe. The barrel, called the "prime conductor," drew off the electricity from the globe, and sparks and electrical effects generally could be taken from it as the machine was running.[15]

In 1746 Pieter van Musschenbroek, a professor at the University of Leiden in Holland, invented something that soon came to be called the "Leyden jar." Musschenbroek, working from a Dutch Newtonian perspective, wanted to store electrical effluvia. He tried to do so by accumulating them in a glass jar filled with water, which was known by that time to transmit sparks. He put the jar onto an insulating support, because he thought that the effluvia might otherwise be lost through the walls of the jar, whereas he wanted to trap them. He ran a wire from the prime conductor of his machine into the water, ran the machine, stopped it again, and put his finger to the prime conductor so as to draw off in the form of a spark the electricity that ought to be contained in the jar, which was still directly connected to the prime conductor. Nothing happened.

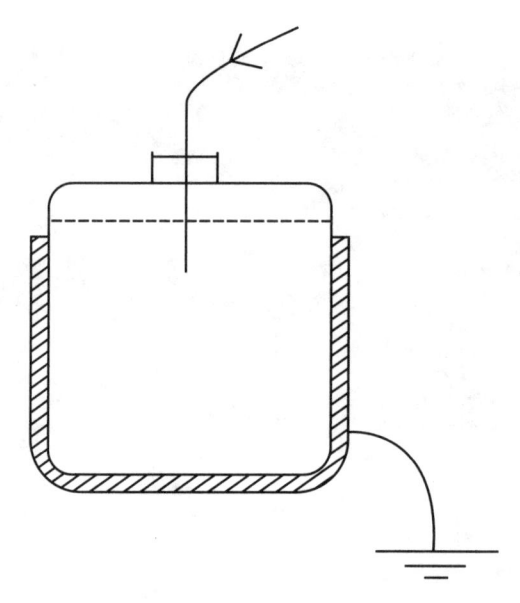

FIGURE 4.3. The Leyden jar.

Not long after, a friend of Musschenbroek's, left unsupervised with all the apparatus, tried the same thing, except that he held the jar in his hand, failing to realize that it was supposed to be put on an insulating support. When he put his hand near the prime conductor, he got a considerable shock. It now seemed that a Leyden jar could be electrified only when the outside of the jar was *not* insulated from its surroundings. It then lost its electrification if the wire was connected to the outside surface of the jar.[16] The shaded area in figure 4.3 represents what was at first the hand of the experimenter, until the Englishman William Watson found that sheathing the outside with metal to increase the area of contact with the glass was more effective.[17]

The large shocks and sparks that could be obtained from the combination of machine and Leyden jar made electricity an even more popular entertainment than it had been heretofore, when doing things like electrifying cutlery to surprise dinner guests had become the rage in some quarters. A party piece to entertain the French king, Louis XV, involved discharging a Leyden jar through a line of 140 courtiers all holding hands. On another occasion 200 Carthusian monks were used;

the abbé Nollet, the natural philosopher who presented it, said: "It is singular to see the multitude of different gestures, and to hear the instantaneous exclamations of those surprised by the shock."[18]

The new apparatus cried out for explanation. William Watson quickly came up with some ideas in 1746, which he described to the Royal Society. The association of sparks and electrical machines had already made it common practice in England, as well as the Netherlands and parts of Germany, to describe electricity as electrical fire, a subtle fluid substance sometimes identified with ordinary fire regarded as a special kind of element. The idea tied in well with Newton's ideas about aethers in the Queries; William Watson thought of electricity as a Newtonian elastic aether composed of mutually repelling particles.[19]

But Watson's contributions to electrical knowledge in the 1740s took a more profound turn than just tinkering with the Leyden jar or engaging in speculations on matter borrowed from Newton. He proposed the crucial idea that the electrical machine, rather than producing electricity from the substance of the glass globe, was actually pumping electrical "fire" up from the ground. He found that by setting up the machine and its operator on an insulating platform, sparking and attraction effects were prevented almost entirely. His interpretation was that the machine had been prevented by the insulating barrier from pumping electricity up from the ground. The earth was thus in effect a kind of electrical reservoir or sink. This was a striking idea, although it appeared to say nothing about how a Leyden jar worked.[20]

When a powerful theory of the Leyden jar appeared, it immediately provided a large measure of integration to the miscellany of conceptual elements and model behaviors that had come to constitute electrical research by the 1740s. The person who achieved this new theory was Benjamin Franklin.

Drawing Electricity Together: Franklin

The center of Franklin's natural-philosophical world was the Royal Society in London. It was bad enough for an English natural philosopher to be stuck out in the provinces away from London and news of the latest ideas, but Franklin, the Philadelphian, had the Atlantic Ocean between him and his correspondents, and news of the electrical machines, and the Leyden jar. But even though information took a while to reach him, he did make sure that he got it. In particular, he was quite familiar with Watson's work.

Franklin pursued his interest in what he too called "electrical fire" with some like-minded enthusiasts in Philadelphia. He concentrated initially not on the usual, classic phenomena of attraction and repulsion, but on the production of sparks. In 1747 Franklin sent his correspondent in London, Peter Collinson, a report on his first experiments made with a good glass tube that Collinson had sent him for the purpose from England, where they could be obtained from makers of scientific instruments. Franklin described in the letter the most significant of these experiments, diagrammed in figure 4.4. Experimenters A and B are standing on insulating cakes of wax; C is in contact with the ground. A rubs the glass tube; as he does so, B draws off sparks from the electrified tube. When A stops, C can draw off a spark from either A or B. If instead A and B touch, a bigger spark will pass between them than would go from either of them to C. If A and B had been in contact while the tube was being rubbed, no sparking at all would occur.[21]

Franklin explained these observations by saying first that in rubbing the tube, A supplies to it electrical fire from his own body, thereby incurring a deficit in his own natural quantity of electrical fire—he thereby becomes "electrized minus." This was a remarkable new idea, implying that all bodies have a natural quantity of electrical fire in them at all times and that electrical effects are about changes in this quantity. Next, B draws the fire from the tube and thereby becomes "electrized plus," having augmented his natural quantity of electrical fire. Meanwhile, C is simply standing on the ground, connected to the earth, so whenever he draws sparks the electrical fire passes through him to or from the earth itself. This scheme explains all the observations, including the larger spark between A and B than between either and C; the disparity of fire is larger between A, with less than his normal quantity, and B, with more, than between either of them and C, who has his proper (intermediate) amount.[22]

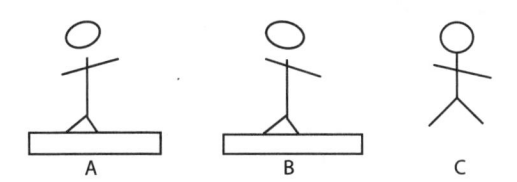

FIGURE 4.4. Franklin's electrical balancing of the books.

Simple as this might seem, Franklin had produced a novel way of looking at electrical fire. Like Watson, Franklin saw electrification as a *transfer* of electricity rather than its creation, and, again drawing on Watson, he regarded the earth as a kind of electrical sink. The difference is that he conceptualized it in terms of the conservation of the quantity of electrical fire in the system.[23] In effect, Franklin, the Philadelphia businessman, balanced the books, with the earth as his bank.

One further point of interest: Franklin described the initial stage of electrification as a supplying of electrical fire from person A to the glass tube; the rest of the analysis follows the movement of that fire. But there was no reason to represent the first stage as the transfer of fire from A to the tube; he could just as legitimately have spoken of the fire going from the tube to A—recall that A is isolated from the earth in either case. When B is described as drawing sparks from the tube, one could then alternatively speak of sparks being drawn by the tube from B, which would account just as well for all of Franklin's observations. The fact that Franklin chose not to see things in this inverse way is the ultimate reason why we nowadays speak of electrons (an alien concept for the eighteenth century) as having a negative rather than positive charge.

Between 1747 and 1749, Franklin applied these ideas to the mysterious behavior of the Leyden jar, the bottle that couldn't be filled with electricity if it was insulated. In order to do so, Franklin needed to take one more conceptual step, which he was able to do because, like Watson, he thought of electrical fire as a Newtonian aether, made up of mutually repelling particles. Figure 4.5 illustrates how he decided it works: Electrical fire enters the jar via the wire from the prime conductor. The electricity does not reside in the water—Franklin found that the jar would remain electrified even if the water were changed out—but instead sits on the inside

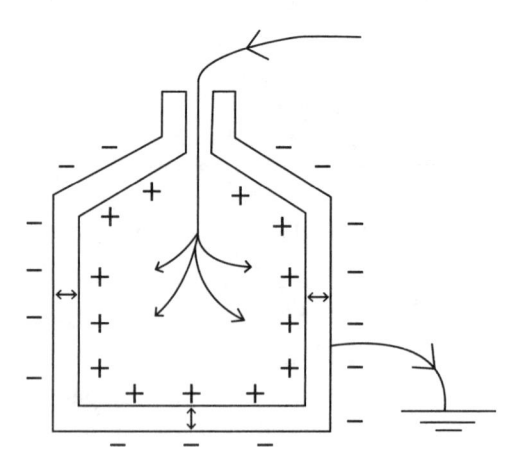

FIGURE 4.5. Franklin's understanding of the Leyden jar in terms of plus and minus charges.

surface of the glass. That surface now has an excess of electrical fire and is said to be charged plus, or positive.

Now, since the particles of the fire are understood to constitute a Newtonian aether, they are taken to be mutually repelling. Consequently, the electrical fire on the inside surface of the jar repels particles of fire on the outside surface; the natural quantity of fire on the outside of the glass will tend to be pushed away. If the outer surface is connected to the earth, that fire can flow away, leaving a deficit of fire—a negative charge—on the outside. That explains why the jar needs to be earthed in order to charge. Finally, the jar is discharged by connecting the inside and outside of the jar so that the positive and negative charges equilibrate.[24]

There were two new ideas underlying Franklin's Leyden jar account. First, in a sense, Franklin invented electrical induction with his idea of the driving off of electrical fire (sometimes "fluid") from the outside of the jar by the electrical fire collecting on the inside, thereby inducing a charge on the outer surface. (A similar idea had appeared in some of Gray's work, to little notice). Additionally, Franklin's idea of positive and negative electrification, based on the notion that all bodies have an intrinsic amount of electricity that can be augmented or depleted, reinterpreted Dufay's two kinds of electricity, vitreous and resinous. Most important of all, Franklin's central idea concerns balancing *quantities* of electrical fire. This is essentially a bookkeeping approach to electricity, entirely understandable from a professional printer working in the increasingly commercial, mercantile society of Philadelphia in the mid-eighteenth century.

Franklin's subsequent work on electricity attempted to clarify and augment the concepts brought together by the Leyden jar. In an attempt at comprehensiveness, he wanted to account for the interaction of atmospheres of electrical fire supposedly surrounding charged bodies so as to produce attraction and repulsion effects; this did not work out so well.[25] It was the Leyden jar conceptualization that integrated disparate concepts into a coordinated system of ideas and acted as a kind of material paradigm for electricity; the role of effluvial clouds of mutually repelling particles (the Newtonian picture) in the production of macroscale attractions and repulsions between bodies was never adequately explained. But the theoretical model had already done its work in rendering the Leyden jar intelligible.

This was the beginning of the end of qualitative experimental work on electricity. Franklin's analysis of the Leyden jar enabled him to make and conceptualize a condenser consisting of two sheets of glass in place of the inside and outside surfaces of a jar.[26] Joseph Priestley in England (in his *History of Electricity* of 1767) and then the French engineer Charles-Augustin Coulomb (in the 1780s) established to the satisfaction of most that the attraction of opposite electrifications followed a good Newtonian inverse-square law of attraction; empirical confirmation was almost beside the point.[27] Electrical phenomena were now, toward the end of the century, being treated according to the model of Newtonian mathematical (rational) mechanics, with electrical attraction and repulsion being viewed as action at a distance.

Certain fundamental questions of natural philosophy remained unresolved. Well into the nineteenth century there remained ambiguity about whether electrical effects should be attributed to one electrical fluid (Franklin's preference) or to two such subtle material agents, composed of mutually repelling subtle particles. The two-fluid theory descended from Dufay's vitreous and resinous electricities, which Franklin had translated into plus or minus electrifications that depended on whether the charged body had more or less than its "normal" quantity of electrical fire. But distinguishing between one and two fluids led to no clear differences in experimental outcomes.[28] Furthermore, the original phenomenon that had first identified electricity, attraction (with or without repulsion), gave Franklin's one-fluid approach some difficulty: he tried to explain attraction/repulsion in terms of electrical fire surrounding positively electrified bodies in the form of particle-clouds, where the clouds surrounding the two bodies would interact with one another. This could account for the mutual repulsion of two positively charged bodies, where the two clouds would push away from each other. But negatively charged bodies, taken by Franklin to *lack* such clouds, also repelled, even though there was nothing evident that could bring such interference about.[29]

Despite such gaps in his ideas on electricity and on the physical character of its experimental phenomena, Franklin's coordination of what had previously been miscellaneous qualitative concepts enabled a shift from experimental to mathematical electricity. But he did it with the help of Newtonian philosophical ideas about aethers.

5

Organization

LIVING THINGS

NATURAL PHILOSOPHY encompassed the entire natural world; the rules governing living beings were in principle the same as those governing the nonliving. In Aristotle's view, it may even have been the case that the principles of living beings set the standard for all else.[1] But in the post-Aristotelian world of the eighteenth century, explanatory principles were in flux. The qualitative variety in nature escaped attempts to trap it under simple categories of matter and its modes or the behaviors investigated by rational mechanics. The point is particularly evident in the case of living things. There, the approach was predominantly one of classification, which was the most characteristic part of eighteenth-century life sciences as a whole. Far more activity in creating knowledge about living things went into description and classification than into areas like the experimental physiology pursued in the following century. In fact, in this period, classification was a central activity in most areas of the study of nature, not just ones to do with living things. Wherever there was apparent diversity in nature, natural philosophers wanted to classify it, reducing it to some kind of order.[2] (Recall William Herschel and the "natural history of the heavens," applicable to various kinds of nebulae.)

Taxonomy and Life in the Eighteenth Century

Among plants and animals, diversity was the rule; the imperative was not to reduce diversity to underlying simplicity, but to organize diversity to elaborate order. Various schemes of natural-historical classification had been developed since the sixteenth century. European exploration in the New World and elsewhere around the globe since the late fifteenth century had led to ever-increasing numbers of new species, especially of plants, but also animals, being brought back to Europe. These new species left naturalists with the job of making some order out of them—cataloguing them, in effect. In the eighteenth century, naturalists such as the Englishman Joseph Banks began to be included in naval expeditions with the express purpose of collecting new specimens.[3]

The impetus for classification in natural history in this period amounted in part to a cataloguing problem: naturalists needed a simple, straightforward procedure for filing newfound species into an organized system. Because they were dealing with such a large and ever-expanding number of plants and animals—to talk of "species" here would risk begging the question, since species themselves represent classificatory divisions—taxonomists from the sixteenth century onward needed to develop categories with a fine level of discrimination. It would be inconvenient to group too many species into the same genus, for example. Imagine going to a library in search of a particular book, but on consulting the catalogue finding that the call number for that title, and many others, was simply "book"; this might be unhelpful, and finer distinctions would be necessary. Conversely, if every individual was distinguished by its own set of characteristics, there would be no basis for comparing or contrasting it with anything else; everything would be unique.

The foundations of any kind of classification in natural history came from ordinary language—general categories like *grasses* or *cats*. But the quantity of new material in the early modern period meant that ordinary observation and ordinary language would not be precise enough. The general attempt to set up schemes to accommodate the rapidly increasing number of plant and animal species becoming known in Europe produced in the seventeenth century more and more refined systems of classification. These schemes, starting with the sixteenth-century work of an Italian named Andreas Cesalpino, generally used as their basic criteria for plants the parts related to reproduction: flowers and fruit.[4] The philosophical justification for such a focus, like so much else, derived from Aristotle: reproduction was central to the transmission of form (equivalent to species) from one generation to the next. Combined with the practical convenience of evaluating and enumerating sexual characters, this approach possessed a prima facie legitimacy that found wide acceptance. In zoology, too, taxonomists found reproductive characteristics useful. The distinction between four-legged reptiles and other kinds of quadrupeds, for instance, could be justified by the fact that reptiles lay eggs. The approach was much more useful for plants, however, because of the classifiable diversities in flowers and fruit.

This was the approach used by the most important taxonomist of the eighteenth century, the Swede Carl Linnaeus. Linnaeus's systems became standard by the second half of the eighteenth century and went on to form the basis for modern taxonomic systems. Linnaeus (b. 1707) spent part of his early career in Holland, where he began to publish works on classification, although he spent the later part of his career back in Sweden, at Uppsala. His major book was the *Systema naturae*, the system of nature. The first edition appeared in 1735 and the tenth and definitive edition (later taken up as the origin of all subsequent systematics) in 1758. Linnaeus's systems became more widely adopted than anyone else's, more for practical than theoretical reasons. What Linnaeus had developed was a system that made it easier to decide, when confronted by yet another new plant from the tropics, how to categorize it in relation to other plants. In other words, it made pigeonholing new specimens much simpler.

Linnaeus's system provided a foundation continuous with taxonomic practice today, even with great changes in the meaning of classification in subsequent centuries. Modern taxonomy involves the broad structure of Linnaeus's taxonomic categories and the way that species are named. To a great extent, Linnaeus borrowed his categories from contemporary practices, but it was a version of his particular deployment of them that became generally used. The approach is hierarchical and resembles a system of nested boxes. The specimen to be classified is first placed in the most capacious category for its properties. That category in turn is divided according to certain criteria into subcategories, and the specimen again is located in its appropriate subcategory. Take a cat, for example, to be classified appropriately in the class *mammalia*. That class contains various subdivisions called orders, one of which, *carnivora* (for meat eaters), is deemed appropriate for cats. There are now further subdivisions into families, families subdivided into genera (plural of *genus*), and genera subdivided into species.[5] Then, on the basis of identifying properties, our cat might take its place as lion, tiger, wildcat, domestic cat, or perhaps some new species differing from all of those.

Vertebrates

Class:	Order:	Family:	Genus:	Species:
Mammalia	Carnivora	Felidae	Felis	–Felis leo
				–Lynx
				–Tigris
				–Catus

Determination of class depended on the broadest structural features; Linnaeus chose suckling as his characteristic for mammals.[6] For plants the crucial features for class were the number and positions of stamens in the flower, while the order depended on the number and positions of the pistils; for genus, Linnaeus relied on other, more particular features. He admitted quite freely that his system was artificial. What he meant was that the criteria used in classifying—the particular characteristics that determined where a particular plant should go—were justified, ultimately, by convenience rather than by theoretical considerations. The same went for his system of zoological classification (although he was primarily a botanist), where he used convenient, countable characters such as toes, teeth, and mammaries—things that could be examined easily and that showed some degree of regularity in number and form, with not so much diversity that there was an insufficient basis for comparison, but not so much uniformity that there was an insufficient basis for distinction. Enumerable things were especially useful.

Although Linnaeus called his system, with its choice of criteria for classification, "artificial," he did so only because he, and virtually all other taxonomists, really wanted a so-called *natural* system. What that meant was a classificatory system that not only provided a convenient means of cataloguing things, but also expressed true affinities between kinds of things in nature. Putting two species of plant in the same genus on the basis of their similar flowers was convenient, but it left open the question of whether it was the right way to group plants that were closely allied in nature. Linnaeus wasn't so sure that it was, and he worked at developing a "natural" system of botanical classification, although he never succeeded to his own satisfaction. Almost everyone else wanted a natural system as well, but used Linnaeus's as a pragmatic artificial system in the absence of anything better. But what could an eighteenth-century naturalist possibly have meant by speaking of a "natural" classification? The answer to this question illustrates fundamental assumptions that natural philosophers in that period made about the conceptual structure of the world and how people could come to know it.

Natural Classification and Its Problems

Nowadays, after Darwin's work, we can easily say what we mean by a natural classification: one that groups species that are closely related, literally, through a common ancestor, so that successively higher-order divisions of natural classification represent a genealogical tree. But before Darwin, that interpretation was unavailable; the prevalent idea was that species were fixed, natural kinds that did not change or evolve over generations. One of the basic understandings of the organic world was the axiom "like breeds like": cows are born to cows; they don't change over the generations. Without a genealogical view of classification, then, there had to be a different understanding of what natural classification could mean. The difference is clear from considering eighteenth-century taxonomy in areas other than botany and zoology.

Physiology, for instance, developed in large part around tissue types, which focused on the idea of different basic tissues in the human body.

Albrecht von Haller, in the 1730s, established two basic types, "irritable" and "sensitive," plus a vaguer category called "cellular" tissue, but that was only the start.[7] By the beginning of the nineteenth century, the French physiologist Xavier Bichat had extended the classification to twenty-one basic tissue types. The act of distinguishing and categorizing things was itself a form of knowledge.[8]

In chemistry, Antoine Lavoisier, at the end of the century, invented the system of nomenclature that, in most respects, has been the standard ever since. But he developed his chemical system in conscious imitation of natural history's. So, like Linnaeus, Lavoisier had a binomial nomenclature: Linnaeus named species with a generic name followed by the specific name; similarly, Lavoisier named substances by the type of compound, like a genus, preceded by a specific name—copper carbonate, sodium carbonate, calcium carbonate, and so forth (see chapter 6). Linnaeus himself, besides producing taxonomic schemes for plants and animals, also produced one for minerals, so as to have one each for all three kingdoms of nature: animal, vegetable, and mineral.

Clearly, in these cases there was no possibility that classifications represented genealogies. The whole idea of taxonomy in the eighteenth century had no necessary connection to laying out literal family relationships, not in zoology any more than in mineralogy. And although there was a practical imperative for classification in botany, that did not apply to eighteenth-century classification mania in general. In many areas where formal classification was done, it was not a practical imperative, as with tissue types or classifying diseases (nosology).

In fact, the ultimate meaning of the classification of nature in the eighteenth century (and the point holds for the most part until Darwin) was the idea that a true natural classification would somehow reveal the plan according to which God Himself had organized the world. In effect, a natural classification would mirror God's blueprint for Creation.[9] The obvious affinities between things in the world—between different kinds of cats, or fir trees—served as evidence that such a blueprint existed.

Classification was a matter of looking for order in nature, and looking for order, including mathematical order, is something that scientists do as much today as their forebears did in the eighteenth century.

Taxonomic classification in natural history made perfect sense to most naturalists and natural philosophers even before Darwin and the acceptance of a genealogical interpretation. At the same time, it was not universally regarded as unproblematic. There were active contestations and criticisms of some foundational assumptions underlying classification that called into question the very possibility of natural classification.

The Englishman John Ray was the author of one of the most influential and widely used classification schemes in botany before Linnaeus. Ray did most of his work in the late seventeenth century (he died in 1705), and in a booklet of 1696 he got into a controversy about the correct way to go about botanical classification. This took the form of criticisms of two continental botanists: a German called Rivinus and the Frenchman Joseph Tournefort, both of whom had come up with their own taxonomic schemes. This is what Ray said to Rivinus: "The correct and philosophical division of any genus into species is by essential differences. But the essences of things are unknown to us. Therefore, in place of these essential characters, some characteristic accidents [in other words, superficial characters of some kind typically associated with the kind of thing in question] should be used, which are present in only some species and in all the individuals contained in these."[10] The *essence* of a thing—a plant or anything else—was whatever made it the kind of thing that it was, and an essential character was supposed to express that essence. For example, a piece of ice must be solid or it does not count as ice. An *accident*, in contrast, was some character that did not express the essence of a thing. An oak tree can be ten feet tall or fifty feet tall and still be an oak tree, so height is an accidental rather than essential character, as is, say, the exact number of the tree's branches or leaves. But a tree that propagated by means of seeds embedded in apple-like fruits rather than through acorns would certainly not be an oak tree, and that characteristic would be counted as an essential character.

Ray's point was that there is no way of telling with certainty in all cases what ought to be counted as essential and what as accidental characters when looking at a plant or animal. To modify one of his own examples: How can one decide whether it is essential or accidental that whales have lungs and live births? If those things are essential, then whales might be grouped with cows and horses. On the other hand, if it were decided that essential characters of a whale included absence of legs and inhabiting the sea, that might make it a fish. (Notice, too, that all those attributes, whether potentially accidental or essential, are found in all whales.) From Ray's perspective, deciding which characteristics of a whale are truly essential would amount to knowing *what God has in mind when he thinks of a whale*. And Ray's point was that we cannot read God's mind.

His conclusion was that the only way to classify in practice is to select whatever characteristics of the individuals being studied yield groupings that accord with what seems reasonable. There are no particular traits, such as individual features of flowers, that will always provide classifications that agree with the overall similarities found in a group of plant species. The only thing to do is to pick and choose among plant characteristics to find ones that give the most appropriate arrangements. Ray was careful not to portray his system as entirely arbitrary; he claimed that his approach likely did a better job of grouping together natural kinds than other, less flexible schemes. (He grouped whales with quadrupeds because of shared basic features, including but not restricted to their mode of reproduction.) But he abandoned the possibility of proving it. Truth could be discovered, but not always demonstrated.[11]

Most other naturalists, however, especially Ray's main rival Tournefort, rejected that pragmatic attitude because it ran counter to the prevailing view of classification. Tournefort rather dismissively described Ray's arguments:

> Nature, so it is said, is wholly alien to systematic arrangement. But even if it is conceded that men can comprehend nature only in complete obscurity, without a method, in things so numerous and diverse as in Botany, nevertheless it is necessary to allow the conviction of experience.... The Author of things, who gave us the faculty for giving names to plants, placed in the plants themselves signifying marks from which should be sought that similarity which is required of species of the same genus. We can neither change these signifying marks,

nor refrain from the use and contemplation of them, if we would eliminate error.[12]

This was an unfair characterization of Ray's position insofar as Ray had not claimed that "Nature is wholly alien to systematic arrangement"; he had simply said that we cannot come to know that systematic arrangement with certainty—a point that Tournefort, and most other taxonomists, simply ignored.

In France it was Tournefort's system, and in the German lands to a lesser extent Rivinus's system, that became the standard for botany in the first few decades of the century, whereas Ray's was used more in England—all three for basically nationalistic reasons. Then Linnaeus's work overtook them all in popularity. Linnaeus's basic philosophical sympathies might be seen as being with Tournefort in that Linnaeus thought that a natural system of classification was not impossible to develop. But he was sufficient of a pragmatist to focus on a useful working system that he thought unlikely to be truly natural.

That left the situation nicely resolved with the best of both worlds: a pragmatically devised taxonomic system, recognized as basically conventional or artificial, for actual classificatory work, while retaining the idea that taxonomy aimed at, and could ultimately achieve, a natural system.

Buffon: Natural History as Natural Philosophy

There was one major exception to the eighteenth century's acceptance of Linnaeus and the assumptions behind the idea of the natural system. The most eminent naturalist of the entire century, the zoologist Georges-Louis Leclerc, comte de Buffon, was an important member of the French Academy of Sciences and dominated French natural history from the 1740s to his death in 1788. In the earlier part of his career he was a vocal advocate of Newtonianism, and related aspects of physical and mathematical philosophy remained lifelong features of his thinking.[13]

The first edition of Linnaeus's book *Systema naturae* had appeared in 1735, with subsequent editions and other more specialized works following it. In 1744 Buffon delivered a talk criticizing Linnaeus's approach at a session of the academy. The criticisms were published in 1749 as an introduction to volume 1 of Buffon's greatest work, the vast *Histoire naturelle* (natural history).[14]

Buffon did not restrict himself to a criticism of aspects or details of Linnaeus's classification. He wanted to criticize the whole philosophical basis of contemporary taxonomy, using Linnaeus's work as the most prominent example. One of Buffon's central points was similar to Ray's: that divisions of the natural world into subdivisions, with hierarchies of classes and orders and families and genera, was a conventional approach that had no real scientific justification. There simply was no basis for claiming that these groupings reflected the true, essential natures and relationships of things.

In particular, the only reasons for picking some characters rather than others as the basis for classificatory distinctions were pragmatic, or even completely arbitrary. That was necessarily the case; there was no way around that fact, so Linnaeus's and other people's searches for natural classifications were futile. Classification was usually done on the basis of the idea that, even if a truly natural taxonomy had not yet been achieved, it was still a realizable goal. Buffon's criticism was, at root, that this is no way to do natural philosophy, because it lacks conceptual coherence. How do you justify the use of one grouping of individuals rather than another?[15]

This might sound as though Buffon had given up on any way of making sense of the living world, as if he were saying that all we have is a plethora of diverse organisms with no defensible way of ordering them. In fact, this was not Buffon's position, but the character of his work in *Histoire naturelle* almost makes it seem like it. This is because the vast treatise comprises detailed descriptions of animals: pictures, accompanied by accounts of their appearance, behavior, habitat, and uses to human beings.

Buffon talked about how natural history as a true science ought to be done in the opening discourse to the first volume of *Histoire naturelle*:

> The history ought to follow the description, and it ought to treat only the relations which the things of nature have among themselves and

FIGURE 5.1. Buffon's naturalistic and contextual picture of a rat from *Histoire Naturelle* (1749–). *Source*: Terence Kerr/Alamy Stock Photo.

with us. The history of an animal ought to be not only the history of the individual, but that of the entire species. It ought to include their conception, the time of gestation, their birth, the number of young, the care shown by the parents, their sort of education, their instinct, the places where they live, their nourishment and their manner of procuring it . . . and finally, the services which they can render us and all the uses which we can make of them.[16]

Buffon did not deny that there is an intelligible order to nature. His point was simply that its order is different from the abstract one assumed by taxonomists like Linnaeus. The correct kind of order in nature lies in concrete, observable, physical truths about how nature behaves. An anachronistic way of putting it would be to say that Buffon recommends an *ecological* mode of understanding.

Buffon went so far as to defend a classification system based on the relationship of an animal to human uses. Although such an approach would be, in a sense, perfectly arbitrary, he argued that it would be no more arbitrary than existing classificatory schemes. It would in fact be more justifiable because it would relate directly to our experiences of animals and of the world. Such a perspective reflects Buffon's commitment to the epistemological stance of Newton and his philosophical underlaborer John Locke, who stressed the role of the senses in creating natural knowledge.

In practice, Buffon treated species individually, but he grouped them for treatment in *Histoire naturelle* according to overall similarities rather than formal criteria of distinction: horses, donkeys, and zebras appear alongside each other, but without taxonomic comment. Accordingly, during his earlier career (the 1740s and '50s), Buffon refused to allow the category *genus* as anything but a pragmatic term or to recognize that there was necessarily any kindred relationship between even apparently quite similar animals. The very concept of species was somewhat suspect, since the usual approach treated species in the same kind of way as it did any of the higher taxa like genus or family: as a transcendent, idealist category serving to link things together intellectually. Buffon instead used a simple, operational definition of species. Individual animals—individuals being the only realities—would be counted as being of the same species if they belonged to a common breeding community. Looking for formal morphological similarities is then utterly irrelevant: if a community of individuals can produce fertile offspring, by definition, those individuals count as being of the same species.[17]

Consequently, for Buffon, species as a whole were defined by their existence and interrelatedness through time. Since an entire group of individuals will count as the same species if they form an ongoing breeding population, Buffon had effectively introduced something that looks a lot like a genealogical definition of a species. But he had no basis for extending that idea to higher classificatory levels. If it could be shown, he said, that two apparently distinct, although similar, species were literally related, that they stemmed from the same original stock, then it would make sense to say they were truly of the same genus as opposed to being put there for mere classificatory convenience. By the 1760s Buffon claimed that it is possible to point to examples of such true generic relations between species—but only in domestic animals, where human selective breeding has clearly, in historical time, created a distinct breed from a wild prototype (as with goats or sheep).

Subsequently, in the 1770s, Buffon began to speculate that maybe one could conjecture, by analogy, the existence of real genera or families in the wild. But that was very tentative speculation on his part, and in any case he thought that the emergence of new forms would be a process of degeneration from a limited initial number of species, not genuine evolutionary development of brand-new species.[18]

Despite Buffon's dominance in French natural history, Linnaean classification became standard in France during his lifetime, especially among botanists (Buffon was chiefly a zoologist). The reason that happened was that Linnaeus's system—which he himself designated "artificial"—was *useful*. But the ambition to develop a natural classification, shared by Linnaeus, remained. This was probably in part because Linnaeus's system and variations on it were so successful in solving most practical problems of classification. Naturalists could afford the luxury of working on a natural system because the practical problems were under control. The most prominent botanists to pursue this goal in the latter decades of the century were the French botanists Michel Adanson and Antoine Laurent de Jussieu. The latter's approach to the subordination of characteristics in evaluating the classificatory priorities of plant features was to shape Georges Cuvier's functional approach to zoological classification. But Buffon had raised serious philosophical questions, and these new attempts at natural classification at the end of the century begin to look much more, to our eyes, modern. At the same time, living organisms were not the only things inviting classification and the imposition of order.

6

Cleaning up Chemistry

THE CLASSIFICATION OF MATTER

PLANTS AND ANIMALS were not the only things whose sheer variety challenged natural philosophers in the eighteenth century. Linnaeus proposed a classificatory scheme for minerals, too; wherever diversity appeared in nature, an organizational scheme to contain it was sure to be proposed. Nowhere was this more true than in matter theory. At the end of the seventeenth century there were a number of closely related mechanistic views of matter as being made up of small particles of different shapes arranged in different ways. These ideas, because they were really pictures of what everything material is supposedly made of, were intended to lay the foundations for an understanding of the behavior of all substances under all conditions. On its face, chemistry looks like the ideal subject for developing the details of matter theories; it might appear a possible way of getting at the fundamental constituents of matter and how they interact. After all, Newton's ideas on the corpuscular structure of matter in the Queries, especially Query 31, were routinely illustrated and motivated by chemical phenomena.

Elements

Newton was especially interested in reactions in solution: precipitation reactions or the effervescence produced by putting certain metals into acid. He thought these were the most promising ways of learning about the supposed attractions and repulsions between the corpuscles of matter—short-range action at a distance, analogous to gravity. But when others tried to follow Newton's lead, they found it difficult to come to grips with matters as obscure as the interaction of invisible corpuscles. The difficulty is quite basic: it was very hard to tell, with any given chemical substance plucked from an apothecary shelf, be it sulfur or gypsum or various acids, whether what you had was simple or compound at the corpuscular level.[1] Something that seemed simple might just be very difficult to decompose. So although for most practical purposes it behaves as if it were simple, it might in reality be a compound made from more than one basic ingredient, corresponding to different kinds of particles. It was even difficult to tell, at the level of the supposed corpuscles themselves, whether different substances were made up of different kinds of corpuscles or different arrangements of the same kind of corpuscles.

Prior to Newton's ideas about particles and distance forces, the dominant view in England had been of particles combining mechanically by hooking onto or meshing with one another. One of the chief exponents of that picture had been Newton's older contemporary in the Royal Society, Robert Boyle. Like Newton, Boyle's interest in matter theory involved him in writing about chemistry, and this point about the difficulty of distinguishing simple from compound substances was one of the chief arguments Boyle tried to make in his book *The Skeptical Chymist* in 1661.

Popular accounts of the history of chemistry long held that Boyle was the first person to define the modern concept of a chemical element. While there is a small amount of truth in that claim, it is fundamentally misleading: Boyle insists that the proper definition of an element is that it is a simple substance that cannot be decomposed into other, more basic substances. The definition is at root operational: if a substance cannot be analyzed or decomposed, then it is an element. But this is where it would be misleading to credit Boyle with inventing the modern notion of chemical elements. The reason he goes to the trouble of saying all this in the first place is to argue that there are no known elements at all.

Boyle's discussion of elements focused on the traditional seventeenth-century versions, of which the most common were earth, air, fire, and water—Aristotle's elements. There were other versions, most notably (in regard to chemistry) the tripartite principles proposed by the sixteenth-century mystic and physician Paracelsus: salt, sulfur, and mercury. Boyle's central point was that people who think they know what underlying elements compose the world cannot really know any such thing, because you can never tell whether any particular substance claimed to be an element might not be able to be decomposed further. Just because some substance had not yet successfully been decomposed did not imply that such decomposition was not possible; it might be the case that the proper technique for doing so had not yet been found.

Newton had a similar idea. Both Boyle and Newton thought that the particles of matter were probably all ultimately the same, with different substances made up of higher-order aggregate particles in different arrangements. On that basis, there was no reason in principle why any substance could not be transmuted into any other—like lead into gold, for instance. The problem was finding a way of doing it in practice.[2]

Although speculations on matter theory encouraged the study of chemistry as a way of understanding matter, people like Newton and Boyle were of little help in providing additional understanding of chemical phenomena. Matter theory supplied a way of talking about phenomena only after they had become known through chemical practice. It was little help in developing new phenomena or predictions; it just rationalized things previously known.

When chemistry really took off as an independent discipline in the eighteenth century, it did so by becoming a true part of natural philosophy rather than a basically practical art used for various technical processes like dying cloth, handling metal alloys, or making medical remedies, as it had traditionally been. The change was accomplished by finding ways of dealing with the phenomena of chemistry in their own terms, rather than using them to get at the underlying structure of matter. In the article on "Chymie" in volume 3 of Diderot's great *Encyclopédie*, the chemist Gabriel François Venel drew distinctions between physics (*la physique*) and chemistry (*la chymie*) that attempted to place chemistry in the higher position despite what Venel decried as a general ignorance of its true character. He wrote, "Chemistry is little cultivated among us; this science is only very modestly familiar even among scholars, despite the pretension to universal knowledge that is nowadays the dominant taste.... Now, since it has come about that people are mistaken and even have formed a prejudice on the nature and extent of chemical knowledge, it will not be an easy or light matter to determine precisely and incontestably what 'chemistry' is."[3]

Many people, he continued, think of chemistry only in relation to practical concerns such as medical remedies, or associate it with the work of philosophers who focus on the structure of matter (generally corpuscularian), such as Boyle or the Dutch Newtonian Boerhaave. But the "superior views" of Stahl are not widely known.[4]

The German physician Georg Ernst Stahl (1660–1734) is perhaps the most important figure in early eighteenth-century chemistry in regard to the development of specifically chemical concepts that were not reducible to physical principles. Stahl had a big impact on chemical thought because he provided a set of concepts that helped to make sense of observed phenomena and also to direct further practical research. His most influential book was called (in English) *Fundamentals of Chemistry*, which first appeared in 1723, in Latin. It sets out at the beginning the same kind of speculative matter theory as Newton and Boyle had suggested, involving a hierarchy of aggregate particles of differing arrangements. In fact, Stahl's book was translated into English from Latin in 1730 probably because of the English interest in chemistry inspired by Newton's discussions of it in the *Opticks*. But where Stahl differs from Newton and Boyle is in the way he discarded matter theory ideas in favor of different, properly chemical concepts.

He did so as an explicit recognition that ideas about matter theory were of little use in actually *doing chemistry*. He advised avoiding talk of corpuscles and their organization. Ultimately, chemical phenomena would be reducible to these sorts of physical terms, but it was necessary to leave those issues aside temporarily. This was his point of view:

A difference, at present, prevails between the *physical* and *chemical Principles* of mix'd Bodies.

Those are called *physical Principles* whereof a Mixt [in other words, a compound] is really composed; but they are not hitherto settled. . . . And those are usually term'd *chemical Principles*, into which all Bodies are found reducible by the chemical operations hitherto known.[5]

Stahl's approach, then, was a pragmatic, operational one. *Chemical principles*, here meaning starting points, refer not to ideas but to the basic substances that are handled in the course of chemical experiments; neither are they abstract structures of corpuscles based on some kind of mechanistic matter theory. The latter are what he calls *physical principles* and are what will ultimately explain chemical properties, but it is futile to talk about them because they elude direct investigation. Stahl's position was that, although all chemical substances might be composed of the same kind of matter, that was of no help in understanding chemistry. Chemical phenomena had to be understood in terms of chemical concepts. Notice, too, how Stahl talks about chemical *principles* rather than chemical *elements*. The word *element* implied an ultimate constituent in some absolute sense, so his avoidance of it is rather like Boyle's criticism of it.

During the course of the eighteenth century the idea of there being just three or four elements—such as earth, air, fire, and water—in terms of which the chemist could conceptualize all chemical phenomena gradually got pushed off to one side. The idea of "the elements" wasn't rejected so much as it ceased to be interesting because it did not help in comprehending the diversity and complexity of actual chemical substances and reactions. The Aristotelian elements were sometimes still used as section headings in books on chemistry, as in the Dutch chemist Herman Boerhaave's 1732 book *Elements of Chemistry*, which presents the role of the Aristotelian elements as "instruments" of material transformation rather than as constituents of bodies.[6]

Combustion and Phlogiston

The old idea of a handful of basic elements making up all the substances met in the laboratory steadily gave way, in the first half of the eighteenth century, to concepts centered on understanding specific laboratory chemical materials and their behavior—the kinds of things the chemist actually handled and observed. Stahl developed probably the most famous of eighteenth-century chemical concepts, phlogiston (although the basic idea, not under that name, was invented in the seventeenth century). The word *phlogiston* was based on the Greek word *phlogistos*, meaning inflammable, and was designed chiefly to explain phenomena of combustion. The idea of phlogiston was so important because combustion, the phenomenon it dealt with, was traditionally central to chemistry as the most obvious case of chemical change. More broadly, alchemists, who developed a lot of the practical techniques that came to make up chemistry, had always done their work by heating substances rather than by performing reactions in solution.[7]

The central cases of combustion in the chemistry of this period, the ones that were most important for chemists, were those where metals were burnt to produce what was called a calx (*calx* is a Latin term for chalk). The reason that combustion of metals was treated as the model for combustion generally was that it was the most controllable. In the kinds of combustion that are familiar in everyday experience, things literally go up in smoke. That's not a very informative or useful procedure for understanding what's happening to the substances involved—above all, what's being produced in combustion. Burning most metals was much more controllable. Metal would be burnt by concentrating heat on it, often using lenses that would concentrate sunlight, leaving behind dull, powdery, lumpy stuff: the calx.

Stahl's conceptualization of combustion was quite straightforward. All combustible substances are taken to contain some material called phlogiston, and when a substance of that kind burns, the heat causing the combustion drives off the substance's phlogiston. This view implied that the calx of a metal is simpler than the metal itself: a metal consisted of its particular calx plus phlogiston. Combustion released the phlogiston into the air, leaving the calx behind. The idea of phlogiston was significant in part because it made all combustion phenomena into a single class of reactions governed by the same basic chemical process: the release of phlogiston. In addition, phlogiston explained the appearance of flames when something burns; the flames are escaping phlogiston. In fact, it was sometimes identified with an idea of Boerhaave's found in his *Elements of Chemistry* concerning fire contained in ordinary matter.

Another major example of this eighteenth-century concern with specifically chemical concepts, rather than reduction to universal matter theory, is the classificatory approach of what came to be called elective affinities. This idea didn't help in understanding combustion, and arguably led to something of a dead end in the long run, but it did succeed in making sense of certain kinds of displacement reactions in solution to the extent that it allowed theoretical prediction of experimental outcomes.

Elective Affinities and Their Tables

Newton's name plays its inevitable role in the career of elective affinities, too. Newton had described a version of the idea in Query 31 of the *Opticks*, although subsequent versions seem to have been at least partly independent (though only partly, because all natural philosophers in the eighteenth century knew Newton's *Opticks*).

Newton described the simple experiment of showing that iron displaces copper from solution in acid (the equivalent of our nitric acid) so that the iron dissolves and the previously dissolved copper precipitates out; similarly, copper displaces silver, and so on for six different metals (the others being tin, lead, and mercury). Newton interpreted this in terms of differing strengths of attractive force between the particles of the acid and those of the various metals, with iron being most strongly attracted, copper next strongest, and so on down the line. That was part of Newton's overall approach to chemistry in the Queries to the *Opticks*, which was all about understanding reactions in terms of both attractive and repulsive forces between the particles of different chemical substances. But however this series of displacement reactions was interpreted, the result in practice was a table of metals arranged according to which of any given pair would displace the other.[8]

In 1718 a French chemist, Étienne-François Geoffroy, published in the *Memoirs* of the Academy of Sciences a set of such tables. Geoffroy's tables consisted of a series of parallel columns headed by the symbol of

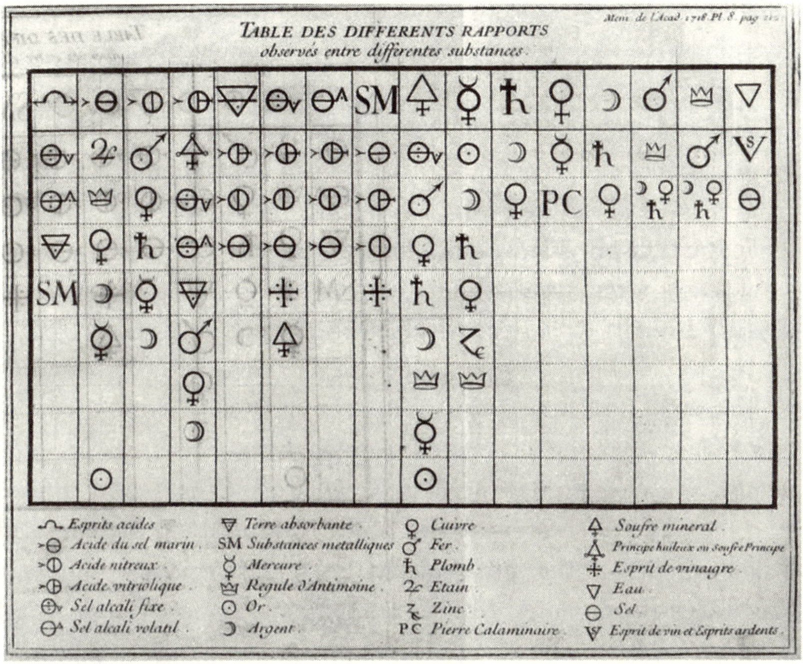

FIGURE 6.1. Geoffroy's affinity table. From E.-F. Geoffroy, "Des différentes Rapports observés en Chymie entre différentes substances," *Mémoires de l'Académie royale des sciences* (1718), pp. 256–69.

a particular substance, typically an acid or salt solution, with the reactants that combined with it listed below it. Each reactant would displace any of those below it from combination with the substance at the head of its column.[9]

It remains unclear whether Geoffroy owed anything to Newton's work, although he did have contacts with the Royal Society. But unlike Newton, he was not concerned with using the tables to support particular theoretical ideas about matter. He said nothing about corpuscles or attractions; he used the words *rapport* or *relation*, committing himself to no interpretation of the tables on the basic level of matter theory. Instead, the phenomenological concept of *rapport* coordinates and lends order to a whole set of chemical reactions. Geoffroy presented a formalized digest of chemical experience comparable, perhaps, to Mendeleev's periodic table in the nineteenth century.

In the eighteenth century many chemists subsequently took up Geoffroy's tables as an apparently promising way of gaining reliable chemical knowledge that would be capable of reducing many different reactions to the same basic order. It went beyond the simple listing of individual chemical behaviors, because any claim to an understanding of those phenomena required the prediction of differences between them. This Geoffroy's table provided. Notice too how the tabular arrangement resembles taxonomic systems in natural history, which are also ways of reducing diversity to comprehensible order.

Work on developing and expanding affinity tables was pursued most assiduously by Guillaume-François Rouelle and his private pupils in the 1750s and '60s. Among those pupils were such subsequently illustrious chemists as Lavoisier, Macquer, and Venel. Rouelle's circle adopted the same practical attitude toward the tables as Geoffroy: they dismissed speculations about the causes of the different reactivity of different substances; they wanted to interrelate phenomena. Some natural philosophers, especially in England, did put forward ideas about powers and affinities among particles to explain the tables, but the chemists who actually worked with them tended not to.[10] The historian Wolfgang Lefèvre has recently clarified a crucial point about affinity tables and other parts of chemistry that carries such work through the culminating achievements of Lavoisier: "In the new system, the simplicity of the building blocks and the compounded nature of chemical substances were defined on the basis of experimental knowledge about the various possibilities of their decomposition, hence in a relative way and not as an absolute property of these entities."[11] The relationships represented in an affinity table were meant to express chemical experience, not reveal some underlying essence.

Airs: Hales and Black

Antoine Lavoisier, one of Rouelle's pupils, was to become the central figure in the transformation of chemistry at the end of the eighteenth century. But Lavoisier's work was more broadly that of a natural philosopher than a traditional chemist (focused on practical recipes, often part apothecary), and he did not pursue an approach focused on affinity tables. Instead, he focused on physical aspects of matter in different states, while also pursuing a Stahlian approach to chemical principles that nonetheless rejected Stahl's ideas about phlogiston.

Lavoisier's concern for physical states of matter appeared in his interest in gases and combustion and derived from his interest in the production of gases from various kinds of dense matter. In the early eighteenth century there had been no general concept of gases; instead, natural philosophers used the term *airs*. Airs were not regarded as a group of chemically distinct substances in the way that we think of different gases like oxygen or nitrogen. In the view of most people, including natural philosophers, there was one basic kind of air that existed in different degrees of purity.

Poisonous gases in mines were impure airs, for instance, whereas bracing mountain air would be pure air. The focus was always on air as a physical state; it wasn't thought of in connection with specifically chemical processes. Having impure air simply meant that there were certain "earthy" admixtures dissolved in it, as there might be in impure water. The atmosphere was thought of as being a kind of sea of air with impurities dissolved in it, just like a body of water. There was little interest in the production of gases in chemical reactions, even in effervescent reactions where the process is particularly clear, because it was regarded as nothing more than an associated, accidental physical process without any chemical significance.

When serious investigation of gases began, it too was basically out of interest in the physical nature of the process. Newton had suggested, in Query 31, ideas about how the particles of matter might exert attractive and repulsive forces on each other. He thought that in effervescent reactions, particles contained in the solid or liquid bodies involved might be so agitated as to be thrown far apart, when longer-range repulsive forces between them would overcome the shorter-range attractive forces that had held them together. Because of this mutual repulsion at greater distances, the particles become an "aerial fluid" and escape from the solid or liquid state. In the 1720s Newton's idea encouraged Stephen Hales, a fellow of the Royal Society, to carry out experimental work collecting evolved airs and examining the properties of their generation.[12]

Hales used what he called a "pneumatic trough," now a familiar sight in chemistry laboratories: an upturned glass vessel filled with water, sitting neck-down in another water-filled vessel—in Hales' case, a bucket—with the gas evolved from the examined materials bubbling up into it through a tube. Because Hales, like Newton, was primarily interested in the physical nature of gas production, he looked at a diverse group of phenomena, not just effervescent chemical reactions. He found that air could be produced by heating substances, including organic things like peas or tobacco, as well as inorganic things such as alkaline earths, now called carbonates. He also collected air from fermentation, which was regarded as an inorganic process until Louis Pasteur's work on yeast in the middle of the nineteenth century.

The purpose of Hales's work was really physical rather than chemical, in that Hales wanted to examine Newton's conjecture about the generation of air from solids. Accordingly, Hales concluded that there must exist something he called "fixed air," which was air bound up in dense matter and released from certain substances when they were heated. Hales made no suggestion that there might be distinct *kinds* of gases; he still adhered to the standard idea that there was just air, more or less pure depending on whether it had earthy admixtures dissolved in it, that could, sometimes, be "fixed."

Hales's experiments on airs generated by heating substances, and the apparatus that he developed for collecting those airs, enabled work by someone who really was a chemist. Joseph Black was a Scot who held professorships at the universities of Glasgow and then Edinburgh. Although an exception to the rule that most natural philosophers in the eighteenth century were not university professors, Black's role was as a medical man, not a natural philosopher. He did a lot of important work, including inventing the concepts of specific and latent heat, but his study of magnesia alba—what we know as magnesium carbonate—which was published in 1756 as his MD dissertation, reframed the whole question of airs and their nature.[13]

In the course of investigating magnesia alba as a possible treatment for kidney stones, Black found that when it was heated, an air was released. Afterwards, the alkalinity of the original substance had disappeared; when mixed with acid, a salt solution resulted. He found that the same thing happened with alkaline earths generally, and that he could get what seemed to be air with identical properties bubbling off by putting the starting material directly into the acid, producing effervescence and leaving behind that same salt.

Black interpreted these results in terms of reactions with acid. Either the air was released by heating, after which the result could be put into the acid, or the air could be released by putting magnesia alba directly into the acid—hence the fizzing. In each case, the same salt was produced in the solution (in modern terminology, Black used hydrochloric acid, producing the salt magnesium chloride). Knowing of Hales's earlier work (thanks to the well-established publication practices of eighteenth-century natural philosophy), Black also called the kind of air he was now dealing with "fixed air." But for Black that name had a much more specific meaning. His fixed air was the air generated from alkaline earths, explicitly distinguished from ordinary air on the basis of simple tests such as whether it precipitated lime water or extinguished a flame. In effect, Black had, for the first time, identified a particular species of air, instead of thinking in terms of one kind of air that could be more or less pure. It also implied the possible existence of other distinct species.

This new conception of different kinds of airs pointed toward seeing them as chemical constituents of substances, instead of being a special, physical form of generic air that could sometimes be trapped inside ordinary dense matter.

Lavoisier and States of Matter

The idea that airs could be fixed in solids became crucial to Antoine Lavoisier's early work on combustion. Lavoisier had started his career in natural philosophy in the 1760s after completing a law degree in Paris. He had become interested in a wide range of natural-philosophical subjects, especially subjects associated with natural history: geology, mineralogy, botany, meteorology, and chemistry, together with the taxonomic impulse found in all of them.

In 1766–67 Lavoisier went on an expedition to eastern France for mineralogical and geological surveying purposes with a number of established geologists headed by the naturalist Jean-Étienne Guettard. His presence was due chiefly to the fact that he came from a well-connected family who knew these people already. Partly because of his connections, although by no means solely that, it took him an unusually short time to become a member of the Academy of Sciences—a restricted, quasi-professional group, unlike the Royal Society in London—being elected in 1768 at the age of twenty-five. By this time he was starting to concentrate his attention on physical and chemical questions about the nature of matter, although throughout his life he also pursued interests in physiology. One of his interests was the question of the elements in the traditional sense of earth, air, fire, and water.

As we saw with Boerhaave, the Aristotelian elements had remained standard topical headings in chemistry even though they were being used less and less in chemical practice. Part of the reason they were retained is that they still seemed to relate to one very evident aspect of experience to do with matter and its behavior. Earth, air, fire, and water were convenient ways of designating what we would call the *states* of matter, with fire being a sort of "energized" form of matter associated with heat.

Lavoisier knew about Hales's work on the release of air when substances were heated, but he was apparently unaware of Black's work. Consequently, he thought in terms of generic *air* and not chemical *airs*. During the later 1760s and especially in a manuscript he wrote in 1772, he played around with some ideas that were deliberately designed to make sense of the production of what, in elemental terms, would be called air out of earth. In other words, Lavoisier was interested in the physical process involved in the release of fixed air and how it was possible for air to be "fixed" in the first place. He developed the idea (not entirely original) that air in its usual state consisted of a special kind of particle combined with the matter of fire, which formed a kind of atmospheric envelope around each air particle. When the air particles were not combined with fire, they were fixed in solid matter—just as with Hales's fixed air.

Lavoisier assumed that heat was a manifestation of the presence of subtle fire particles. When something was heated, what was really happening was the introduction of the matter of fire into the body being heated. In the case of heating solid matter that contained fixed air particles—like alkaline earths—these heat or fire particles were able to combine with fixed air particles to produce the expansion and creation of air in its gaseous form.

This speculation on the four elements in their traditional Aristotelian sense doesn't mean that Lavoisier routinely used them in place of properly chemical principles. Many chemists, including Stahl, sometimes talked about the elements to describe or account for apparent physical properties of substances. And this is the point: Lavoisier was not doing chemistry; he was thinking about the nature of different physical states of matter. In fact, Lavoisier soon moved away from the four-element idea altogether as he became more involved in chemistry itself. That meant using chemical principles—handleable chemical substances that resisted decomposition—as his basic concepts, again very much like Stahl.

States of Matter and Chemical Species

For combustion, in the guise of the formation of metallic calxes, Lavoisier's approach involved examining differences among the metals. It had been known for some time that the weight of calx left after burning lead or tin was greater than the weight of the original metal. But these had usually been regarded as anomalous cases, since for other metals there seemed to be always less calx than there had been metal. The temptation to ignore lead and tin came from the dominance of Stahl's phlogiston theory. If combustion meant the loss of phlogiston, then the calx ought to weigh less than the original metal—or if phlogiston were weightless, the calx should not weigh more.

This comfortable state of affairs was shaken up in early 1772, when a book appeared by the French chemist Louis-Bernard Guyton de Morveau. It presented results that showed on the basis of careful weighing experiments that most metals, not just lead and tin, become heavier when burned to form a calx. Perhaps that had not been noticed before because material was usually driven off during the burning. Now, however, it seemed that if the experimenter were careful about not losing material, all increased in weight. A few months later another French chemist, Pierre-François Mitouard, showed that phosphorus, too, gained weight in combustion.

The idea of phlogiston now began to be regarded as troubled, although not necessarily refuted. As a last resort it was always possible to say that phlogiston had *negative* weight. This was a perfectly viable option at the time, when Newtonian ideas about repulsive as well as attractive forces and ideas of completely weightless subtle fluids (like electricity) were common. Weight, or gravity, wasn't regarded as a necessary, essential property of every sort of matter. But here, clearly, was an anomaly.

The ambitious Lavoisier immediately took an interest in this work. It seemed to him that it could have relevance to the ideas about fixed air that he had been considering. What struck him was the possibility that these calcination reactions—the formation of calxes by combustion—might be cases of fixing air in solid matter, a sort of inverse of the phenomenon he had considered previously where air was released from solid matter in Hales's fashion.

The first thing Lavoisier did was to perform his own weighing experiments. He knew about the recent finding that phosphorus appeared to gain quite a lot of weight when burned. In fact, Mitouard, the chemist who had reported this, had suggested that absorption of air might be involved. So Lavoisier began with phosphorus. He found, using careful weighing techniques, that the product of burning it was indeed heavier than the original quantity of phosphorus.

He tried the same thing with sulfur (chosen because of its similarity to phosphorus—both produce an acid when the combustion product is mixed with water). Again taking the same precautions to avoid loss of material, Lavoisier found a weight increase there, too. In both cases it seemed to him that a lot of air was being used up in the closed reaction vessel. What he did then was to go at the problem from the opposite direction: producing air from a calx.

Lavoisier used Hales's pneumatic trough apparatus on one of the calxes of lead (litharge, PbO), which we now know as one of the two main lead oxides. He heated it in a closed container in the presence of charcoal—the standard way to produce metal from its ore, as with smelting iron. He could then observe the large quantity of air being produced, bubbling up from the tube leading from the reaction vessel. Air stopped coming off when all that was left was metallic lead, which supported the idea that when metals were burned, that's to say, turned into calxes, they used up air, which was released when the calxes were reduced back to the metal.[14]

This was at the end of 1772. Only in 1774 and '75 did Lavoisier come up with what became his famous oxygen theory of combustion, made possible by work being done at the same time in England by Joseph Priestley.

By 1774 Lavoisier had decided that combustion, specifically the calcination of metals, involved the "fixing" of air into the metals. He now knew about Joseph Black's special form of fixed air, which precipitated lime water and put out candle flames, but he didn't know how that fitted in to the things he was looking at, or what it really was. All he was able to say was that air was in some way combining with ordinary dense matter during calcination. Lavoisier made no distinction between different kinds of air, despite Black's work; he was still thinking of air as one kind of thing that could be more or less pure depending on whether it contained vaporous or earthy admixtures.

Also, Lavoisier still had not entirely abandoned the phlogiston theory. Although he was sure that air became fixed in calxes when a metal was burnt, he remained unsure whether calcination also involved the release of phlogiston. There was still the matter of explaining the production of fire and flame during calcination, which the idea of phlogiston clearly accommodated.

In 1774 Joseph Priestley announced that he had found a new kind of air to put alongside Black's fixed air. It was distinguished from Black's air in that it supported combustion and respiration even better than ordinary atmospheric air, whereas Black's fixed air was terrible for both of those things. Priestley generated the new air by using a pneumatic trough and heating red calx of mercury, (mercuric oxide, HgO), which can be reduced by heating alone—no charcoal needed—to yield mercury and this new kind of air. On learning of the discovery (perhaps hearing of Priestley's talk of it during the latter's recent visit to Paris), Lavoisier set to work on red calx of mercury in 1775. He soon decided that only one portion of common air was involved in calcination. This was a crucial idea, because Lavoisier was thinking for the first time that common air might be a mixture of different kinds of air, which themselves could be treated as distinct chemical species.[15]

He performed experiments to find out what proportion of common air was involved in combustion: about 20 percent by volume. After a variety of precise and carefully designed experiments, he concluded that combustion was not a matter of the release of phlogiston, which Priestley and many other people had continued to believe it was. Instead it was the chemical combination of the burning substance with this stuff that made up 20 percent of atmospheric air. This was the same as Priestley's new kind of air, and Lavoisier later invented for it the name *oxygen*.

But there is also a sense in which Lavoisier never entirely abandoned the idea of phlogiston. Part of the strength of the phlogiston theory was that it made sense of the appearance of fire and flame during combustion, where heat and light are clearly being produced—it is not simply a combustible sucking up oxygen. Lavoisier wanted to be able to explain that. This is where his earlier speculations on the physical nature of air and how it could be fixed in solid matter reenter the story. Lavoisier had been thinking of fixed air (in Hales's generic sense) as being released from ordinary matter on heating by virtue of the matter of fire or heat (taken as the same thing) uniting with the air particles by surrounding each of them as a kind of mantle or atmosphere. That was how air became an expansive fluid, escaping out of the matter in which it had been fixed. This view remained Lavoisier's basic physical understanding even after he had developed the oxygen theory of combustion and started thinking of the atmosphere as a mixture of different gases, not just one air.

Combustion, in Lavoisier's new interpretation, was now quite simple: it was the opposite of that process. Oxygen particles in their aerial state are surrounded by clouds of heat particles. When a metal (or anything else) burns, the oxygen particles become fixed in the metal to form the calx, and therefore must lose, or shrug off, their associated mantle of heat particles.

So combustion involved more than the combination of the combustible substance and oxygen; it also involved the *separation* of the oxygen and the matter of heat. The release of heat causes the appearance of flames. Lavoisier called his version of the matter of heat or fire, with its family resemblance to phlogiston, *caloric*, and he regarded it as a chemical substance on a par with other chemical substances like iron or sulfur or oxygen.

Chemistry and the Naming of Experience

One new result that emerged from Lavoisier's combustion theory was a demonstration that water is an oxide of hydrogen. In 1766 Henry Cavendish in England had identified a new kind of air made from sulfuric acid and iron filings, which was very inflammable. In 1783 Cavendish reported that the product of the combustion of this gas was ordinary water. Lavoisier used other arguments and experiments to confirm that identification, including analyzing or cracking water back into its constituents. Lavoisier gave the new gas the name *hydrogen* because it generated water.

As for *oxygen*, Lavoisier chose that name because it means "producer of acids." That may be surprising at first—one might expect that he would have characterized it by its role in combustion, that being the way by which it had been identified in the first place. But before the naming of oxygen, Lavoisier had followed his theory of combustion with a theory of acids that characterized them as containing oxygen, and he chose to make that rather than combustion the defining feature of the new gas. The basic argument for this was that commonly known acids like sulfuric acid and phosphoric acid could be produced by combustion plus water. Burn sulfur, add water, and get sulfuric acid. The combustion is, of course, a combination with oxygen. (Nitric acid, HNO_3, apparently produced with no combustion, was obtained by treating saltpeter with sulfuric acid, which was understood to contain oxygen already.)

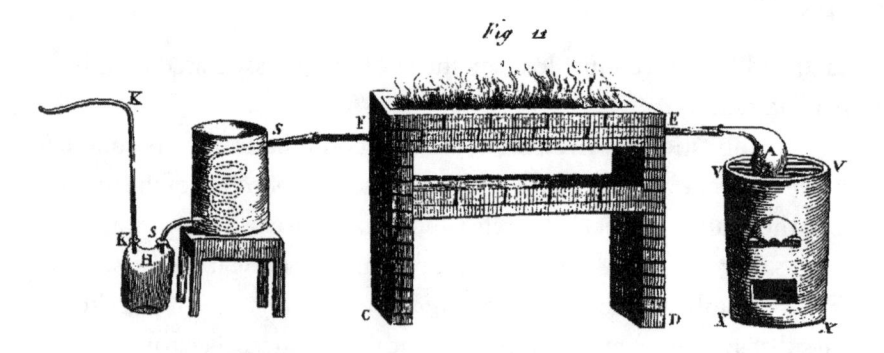

FIGURE 6.2. Lavoisier's experiment to crack water into hydrogen and oxygen, conducted at the Paris Arsenal before witnesses. From Lavoisier, *Elements of Chemistry*, trans. Robert Kerr (1790).

So Lavoisier decided that acids were formed by the chemical addition of oxygen. This ran into a difficulty, however, in that Lavoisier was unable to analyze or synthesize so-called marine acid, which is our hydrochloric acid, to show the presence of oxygen. But since there was nothing else in all other acids that might account for their acidity—sulfuric acid was simply sulfur, oxygen and water—there seemed to be no other candidates for the principle of acidity.[16] Lavoisier had to leave marine acid as an anomaly to be resolved in the future.

Why did Lavoisier choose the supposed acidifying property of oxygen to characterize it rather than the more straightforward property relating to combustion? It seems likely that he did it in part to give his ideas a greater scope. The original oxygen theory applied only to combustion phenomena, but by making oxygen central to acidity he could extend it to another major class of chemical reactions, namely reactions in solution involving acids, alkalis, and salts. He thereby placed the stamp of Lavoisierian chemistry on the field as a whole. This was truly, as he said, a revolution in chemistry.

The revolution was consolidated by Lavoisier's classificatory approach to chemistry, an initiative that would transform chemical practice through analytical naming procedures. Lavoisier was systematic and ambitious in his chemical taxonomy in ways that resonate with Stahlian notions of properly chemical concepts related to observed chemical phenomena, as opposed to viewing chemical reactions as evidence for or against theories about the nature of matter (as Newton did). Lavoisier formalized and codified the practice of doing chemistry on the basis of its own principles, following Stahl, and in fact distanced it even further from a reduction to fundamental matter theory despite his earlier interest in the physical nature of airs and fixation. He did so by inventing a new system of chemical terminology that is much the same as what chemists use today. It was laid out in a book that Lavoisier and three collaborators published in 1787 called the *Method of Chemical Nomenclature*, which incorporated central aspects of Lavoisier's own approach. It also confirmed the selected materials earlier developed in affinity tables and elsewhere in eighteenth-century chemistry as the privileged chemical principles that directly foreshadow the nineteenth-century chemical elements that we know today.[17]

In 1789 Lavoisier's own book appeared, the *Traité élémentaire de chimie*, or *Elements of Chemistry* in the 1790 English translation. It summarizes Lavoisier's chemical ideas and uses the new nomenclature. This was effectively a new language for talking about chemistry, designed to be an analytical description of chemical experience—in other words, the terminology was supposedly rooted in chemical operations, as Lavoisier explained in the book's preface, quoting from the eighteenth-century French philosopher Condillac's *System of Logic*. Condillac wrote: "We think only through the medium of words.—Languages are true analytical methods.—Algebra, which is adapted to its purpose in every species of expression, in the most simple, most exact, and best manner possible, is at the same time a language and an analytical method.—The art of reasoning is nothing more than a language well arranged."[18] As a direct consequence of developing a nomenclature for chemistry along these lines, Lavoisier says he found that "while I thought myself employed only in forming a Nomenclature . . . my work transformed itself by degrees, without my being able to prevent it, into a treatise upon the Elements of Chemistry."[19] Properly forming a nomenclature for chemistry for the purpose of *doing* chemistry inevitably created this chemical textbook.[20]

Lavoisier in effect noticed that nomenclature, mediated by pedagogy, encapsulated formal empirical knowledge: science. "However just the ideas we may have formed of these facts, we can only communicate false impressions to others, while we want words by which these may be properly expressed."[21] All substances that were, from a chemical point of view, simple—that could not be chemically decomposed, like oxygen or hydrogen or sulfur—were given single names. In the case of the new gases, the names reflected their properties. These were in effect chemical principles as discussed by Stahl. Compound substances, composed of more than one chemical principle, were named by adding a word indicating their constituent chemical principles and sometimes marking degrees of oxygenation, as with ferric oxide and ferrous oxide.

> The acids, for example, are compounded of two substances, of the order of those which we consider as simple; the one constitutes acidity, and is common to all acids, and, from this substance, the name of the class or the genus ought to be taken; the other is peculiar to each acid, and distinguishes it from the rest, and from this substance is to be taken the name of the species. But, in the greatest number of acids, the two constituent elements, the acidifying principle, and that which it acidifies, may exist in different proportions, constituting all the possible points of equilibrium or of saturation. This is the case in the sulphuric and the sulphurous acids; and these two states of the same acid we have marked by varying the termination of the specific name.[22]

Lavoisier's binomial nomenclature closely resembles Linnaeus's codification of genus and species, seen in chapter 5.[23]

Lavoisier's terminology was supposed to embody chemical knowledge of how substances are made or what they do, a linguistic summary of certain chemical knowledge attained through unquestionable direct experience. In practice, that meant that it embodied Lavoisier's theories. It was a powerful weapon in getting his ideas accepted when many other chemists (most notably Joseph Priestley) continued to defend phlogiston against the oxygen view of combustion. The new nomenclature, as much as Lavoisier's theoretical ideas, stands for what the so-called chemical revolution was all about.

The eighteenth century achieved vastly greater control over chemical experience by abandoning attempts to understand the ultimate structure of matter via chemistry. Instead, chemists stuck to chemical concepts and principles like oxygen or sulfur—things that they were identifying and handling in the laboratory, that had operational significance. Lavoisier was the capstone on this process. But there is profound irony attached to his achievement. At the beginning of the next century, almost immediately after the triumph of Lavoisier's chemistry across Europe, John Dalton in England introduced an atomic theory of chemistry that turned these practical chemical principles—Lavoisier's so-called simple substances—back into physical elements.

7

Laplace, Revolutionary Order, and the Invention of Mathematical Physics

A NEW STYLE of physical science was invented in France in the years around 1800. It was pioneered by Pierre-Simon de Laplace, a mathematician whose interests moved toward mathematical-physical sciences in the 1780s and included collaboration with Lavoisier in precision experimental work (calorimetry). At the beginning of the nineteenth century, Laplace used his powerful political position to promote work by others that similarly combined precision experimental measurement with theoretical mathematical models descended from the rational mechanics of eighteenth-century mathematics.

Physical Sciences and Education in France

The experimental fields of electricity and chemistry in the eighteenth century show an increasing focus on their own specific subject matters and the development of uniquely electrical or chemical concepts, such as electrical induction or chemical principles. But despite the importance of specialization in these subjects, the ultimate goal for natural philosophers in the eighteenth century remained the unification of various areas of study. Everything—mechanics, electricity, light, heat, even chemistry—ought ultimately to be reducible to the same underlying laws of nature; remember Stahl and the inaccessible physical principles underlying chemical processes. Newton's matter theory speculations in the *Opticks* had been directed toward that end. By the close of the eighteenth century, however, French natural philosophers began to address physical processes in ways that promised to create a common language in which to explain all aspects of the physical world. This program for physical science combined a Newtonian forces-and-particles view of matter with the mathematical techniques that had been developed in eighteenth-century rational mechanics to deal with gravitational forces and masses. In effect, they laid a cornerstone for the modern discipline of physics.

All this happened in the context of one of the most important events in European history: the French Revolution. In 1789 the French absolutist monarchy went into crisis after the bankrupt government found itself forced to call the Estates General (the French equivalent of a national parliament) for the first time since the early seventeenth century. It had to do it as a last resort to raise more money because it had dug itself into a financial hole by supporting the United States in the Revolutionary War (War of Independence) and had failed to climb out again.

The Estates General provided a powerful forum for people who were dissatisfied with the autocratic government of France. The criticisms of the previous decades by *philosophes*, people like Voltaire and D'Alembert, now had an effective institutional channel. Reformism turned into revolution, and in 1793 Louis XVI, who had handled attempts at compromise rather badly, was executed by guillotine, and the French Republic was declared.

One of the first bastions of the ancient régime hit in 1789 was the established church. The French branch of the Roman Catholic Church was effectively an arm of the French state, privileged and wealthy, with vast landholdings targeted by reformers for confiscation. This anticlericalism tied in with the *philosophes*' dislike for the church and religion generally.

But there was a practical problem that followed from dispossessing the church of property and income, one that had direct implications for the sciences: the church ran most of the secondary schools (*lycées*) that prepared students for advanced studies. The sudden disappearance of much of the French secondary educational system generated a marked crisis, but threats of foreign invasion and disagreement over what a new school system ought to look like meant that it was not until late 1795 that new state-run schools began to be established.[1]

Out of this mess came the introduction to a central place in French secondary education of mathematics, chemistry, and physical sciences in general, which included work in areas like heat, electricity, and the behavior of gases. These subjects overtook Latin and Greek, previously emphasized as part of the traditional humanistic education that had marked the properly educated.[2]

By 1794 a college had already been established in Paris that soon became a centrally important institution for higher education in the physical sciences: the École polytechnique. It was designed to produce military and civil engineers of the highest quality for the service of the state. Its curriculum consisted above all of mathematical training up to the very highest standards, as well as training in the physical sciences. The foremost French mathematicians and mathematical scientists were drafted to teach courses and write textbooks. The entry standards were extremely high, and so was the prestige of getting in, which meant that teaching at the secondary-school level in these subjects improved to meet the entry requirements.[3]

The result was that quite soon a new generation of well-trained mathematicians and mathematical scientists was appearing from the École. Some of them, rather than becoming engineers, moved into research, with the best taking up teaching posts at the École polytechnique and other reformed institutions. That process worked in large part through personal patronage on the part of already established figures, especially those at the École polytechnique.[4]

This system really began to take effect after Napoleon took power in France in a coup d'état in 1799. Napoleon regarded the sciences as the epitome of rationality and the key to effective political and administrative organization. Two of France's foremost physical scientists benefited directly: in 1803 Napoleon appointed the mathematician and physical scientist Laplace and the chemist Claude Berthollet as senators. Although the senate's only real power was to agree with Napoleon, senators received an annual salary somewhere in excess of fifty thousand francs. This was at a time when a professor at the École polytechnique, an exalted position in itself, only earned six thousand. It was a great deal of money.[5]

Napoleon was so enamored of scientific expertise that one of his first acts in 1799 was to make Laplace Minister of the Interior, a job that Laplace did so badly that he had to be replaced after six weeks.[6] But all the money given to Laplace and Berthollet as senators enabled them to pay for their own research facilities and to provide financial support for promising young natural philosophers who kept in their good graces.

Although these are senses in which the aftermath of the Revolution turned out to be favorable to science (the life sciences benefited, too), it had not looked that way at first. Lavoisier, who was about to start on a program of physiological research particularly connected with respiration (oxygen again), was executed by the guillotine in 1794. This was during the days of the Terror, under Robespierre, when anyone deemed an enemy of the people was in grave danger. Lavoisier's death had little to do with his scientific activities; the trouble was that he had been a tax farmer before the Revolution. This was a sort of private contracting job collecting taxes on behalf of the government in return for a commission. Not surprisingly, the whole system was thoroughly corrupt, and tax farmers were obvious targets after the Revolution. Lavoisier's scientific standing was insufficient to save him.[7]

Even before this blow, in 1793 the Academy of Sciences was abolished. It went along with other nonscientific state-funded academies on the grounds that it was an elitist holdover from the monarchy. In practice, however, the damage was slight: the most important of these supposedly elitist academies were soon incorporated into the Institute of France. The Academy of Sciences simply became the so-called First Class of the Institute, and things carried on largely as before. After Napoleon's final downfall in 1815 it was simply a matter of turning the organization back into the Academy of Sciences by changing the name again.[8]

"Laplacian Physics"

The First Class of the Institute, as the continuation of the academy, remained the central forum for French science during the Napoleonic era. But the personal patronage of Laplace and Berthollet was perhaps the most important factor in shaping the French approach to the physical sciences. They had money to provide research facilities, and they also had the political power to secure teaching posts at the École polytechnique and places in the First Class of the Institute for favored protégés. That put them in a very strong position to determine what kind of research should be done.[9]

The group that centered on them came to be called the Society of Arcueil—Arcueil being the village then just outside Paris where Laplace and Berthollet lived and held court from around 1801 onward (they were next-door neighbors for several years).[10] In 1807 the group even started its own journal. The research program that Laplace and Berthollet laid out and encouraged other people to follow was an attempt to bring all physical sciences, including chemistry, into the same form as mechanics. The sophisticated mathematical techniques developed in the eighteenth century to deal with problems of Newtonian masses interacting through gravitational action at a distance were, in principle, applicable to *any* system of particles that exerted attractive or repulsive forces. Laplace and Berthollet wanted to treat electricity, heat, light, and chemistry in exactly that way: in other words, to mathematize them. In order for that project to make sense, the whole of nature had to be seen as made up of particles exerting attractive and repulsive forces on each other.

This was not a new picture, of course; it was really just the Newtonian one. The difference was that throughout the eighteenth century this worldview had been basically qualitative, except for the treatment of gravitation. Many people subscribed to the idea of other types of distance forces besides gravity, but no one had successfully treated any of them mathematically, with the partial exception of electricity. Laplace was the first important mathematician to give it a serious try. As he put it in a paper published in 1810:

In general, all the attractive and repulsive forces in nature can be reduced, ultimately, to forces of this kind exerted by one molecule on another. Thus, in my Theory of Capillary Action, I have shown that the attractions and repulsions between small objects floating on a liquid, and generally all capillary phenomena, depend on intermolecular attractions which are negligible except at insensible distances. Similarly an attempt has been made to reduce the phenomena of electricity and magnetism to intermolecular action. The behavior of elastic bodies also may be treated in the same way.[11]

Laplace had become interested in these possibilities in the 1780s, after having been a "pure" mathematician; however, he only laid it out as a formal program of research in 1805. Berthollet, as a chemist, wanted to develop chemistry on a similar basis of forces and particles and was completely in sympathy with Laplace's program. The route Berthollet followed was of elective affinities, looking at substances' tendencies to combine with each other. Berthollet tried to account for these phenomena by using the idea of forces of differing strengths acting between particles of varying masses. In the end the complexity of the phenomena turned out to be too much and the project failed, but for some time it was a very serious one.[12]

Laplace's more general program dominated French physical sciences between about 1805 and 1815, both through his own work and through the work of others encouraged and supported by him. Laplace made some progress in providing general mathematical descriptions of the refraction of light and with capillary action, but the most significant work was done by other members of the Arcueil circle.

One of the advantages of being in a position of power in the institute was that Laplace could promote his own research program through setting the questions for the institute's prize competitions. These were regularly offered financial prizes designed to encourage people to work on specific problems that the members of the prize committee thought were important.

A product of the École polytechnique, Étienne Malus, in work done for one of those prize competitions, discovered the polarization of light by reflection in 1808 while doing some experimental research designed to push forward the Laplacian program of treating refraction as the consequence of short-range forces acting between light particles and particles of ordinary matter. Malus invented the term *polarization* because he explained the phenomena connected with it in terms of light particles having two sets of poles: north-south and east-west. These poles determined the directions in which the light particles interacted with particles of ordinary matter, whereby they were filtered into two orthogonal components.

The chief respect in which the Laplacian approach to physical sciences was genuinely new is that for the first time people were trying to turn areas that had previously been the province of purely qualitative experimental natural philosophy into fully (mixed) mathematical sciences along the lines of Newton's celestial mechanics. In this enterprise, the experimental component remained crucial.

The mathematicians who engaged in rational mechanics in the eighteenth century had usually managed to get along quite well without having to worry too much about experiments and observations. They were basically mathematicians exploring the ramifications of Newton's laws, including the inverse square law of attraction, in analyzing increasingly complicated systems of interacting masses. But for the phenomena in which Laplace was now taking an interest, there were no preexisting laws of force. In consequence, Laplace's program had to be as much experimental as mathematical in order to get the quantitative data necessary for determining force laws. Of course, the properties of invisible particles were more difficult to get at than the properties of planetary motion. So Laplace and his younger coworkers had to devise and carefully conduct experiments with the idea of using precise measurements as the basis of equally precise forces-and-particles explanations.

In practice, their ambition tended to outrun the results. But even when they were less successful than they wanted to be, the Laplacians at least established highly precise experimental measurements. Laplace had pioneered such work in the 1780s, when he had collaborated with the ill-fated Lavoisier using an ice calorimeter to measure specific heats (understood in terms of caloric, one of Lavoisier's so-called simple substances).[13] It is notable that they stressed, in presenting this work, the methodological significance of their quantitative experimental procedures:

> It is with the greatest caution that we present these results concerning the amounts of heat produced where carbon is burned in an ounce of oxygen gas. We were only able to make one experiment on the heat evolved in this combustion and, although it was carried out in quite favorable circumstances, we will be really convinced of its accuracy only after having repeated it a number of times. We have already said, and we cannot stress this fact too much, that it is less the result of our experiments than the method we have used that we offer to scientists [*physiciens*], inviting them, if this method seems to offer some advantage, to check these experiments which we ourselves propose to repeat with the greatest care.[14]

This style of scrupulous measurement became characteristic of Laplace's circle during the Napoleonic period. An indication of what that work could look like can be seen in one of the more outstanding, yet typical, investigations carried out by members of the Arcueil group, as detailed by the historian Eugene Frankel.[15] This was a study in 1805 of the refraction of light (the bending of light rays) in different gaseous media performed by Jean Baptiste Biot with the assistance of François Arago. Biot was one of Laplace's earliest protégés and students, and he was strongly committed to the Laplacian program. Laplace himself, via the First Class of the Institute, had commissioned the study.

Laplace's interest in this particular refraction problem came from his attempts to calculate the refraction of light in the atmosphere to correct astronomical positional observations. He had set up a differential equation to describe the refraction of a light particle as it passed through the atmosphere, based on the sine law of refraction and including the variables of air temperature and density. To obtain quantitative results, he then had to use hypotheses about the variation of atmospheric temperature and density with altitude. Laplace's problem was that he assumed that the refractive index of air was a simple function of its density, with no regard for the possibility of proportions between the various gaseous constituents of atmospheric air changing with altitude.

So the problem that Biot faced was to see whether different transparent gases have different refractive powers at the same temperature and pressure. He needed a very high degree of accuracy to make these determinations, because the absolute refractive powers were going to be very small to begin with and the differences between them even smaller. Biot used an astronomical instrument designed for measuring the angles between celestial bodies. It employed two small telescopic sights, and the angle between them would be read off when their respective images of two different objects coincided, providing the angular separation of the objects themselves.

Biot used a hollow glass prism filled with the gas he was investigating and arranged the astronomical instrument with one of its sights looking directly at a distant object—a lightning rod on the Paris Observatory—and the other at the same object as seen through the prism. Making the images coincide gave him the angles for the refraction the light had undergone in passing through the gas prism. He did all this in Laplace's senate chambers with the equipment paid for by the First Class of the Institute (2700 francs).

A striking feature of this work is the almost unprecedented level of precision employed in these experimental investigations. Biot corrected his results by a factor (determined experimentally) to take account of the slight lack of parallelism of the glass walls of the prism. He also took into account the dilation of the glass depending on temperature, as well as the temperature, pressure, and humidity of the gas inside the prism and of the air outside. He even invented a new kind of stopcock for the prism to prevent any air from leaking in during the experiment.

It is symbolically appropriate that Biot used an astronomical instrument in this work, because observational astronomy had long stood as a model of scientific accuracy. Perhaps the only person to come close to this level of experimental precision was Lavoisier, with his weight measurements; recall that before the Revolution he and Laplace had collaborated on accurate determinations of specific heats using their calorimeter. This kind of experimentation, aimed at producing mathematical science, was something new, and was specifically a product of French physical sciences in the late eighteenth and early nineteenth centuries. Notice that the crucial aspect here is the high level of precision rather than simply a concern for measurement.

The experiments that Biot carried out with Arago measured the refractive powers of air, oxygen, nitrogen, hydrogen, ammonia, and carbon dioxide (carbonic oxide). Something then had to be done to make use of these raw figures for Laplace's problem. Biot and Arago had found that the refractive power of air seemed to be rigorously proportional to density, which made matters a bit simpler. But Biot wanted also to be able to say something about what to expect when mixing gases in different proportions. He needed some kind of theoretical model to do so.

The model he chose was perfectly Laplacian—based on forces and particles—and it aimed to be more than a rule of thumb to deal with straightforward gas mixtures. Biot thought that perhaps he and Arago had found a way to determine the chemical composition of compounds by means of measuring refractive powers (this also aligned well with Berthollet's approach to chemistry). The central idea was a version of Newton's view of the nature of refraction. A Newtonian static gas model was made up of mutually repelling particles. Biot postulated that the particles also exert short-range forces on light particles so as to divert them from the direct path they would otherwise follow. This is how Biot and Arago set it up in their 1806 paper:

> Since the action of bodies on light is exercised in a sensible manner only at very small distances, the intensity of this action is necessarily linked to the nature of the particles of bodies and to their arrangement; that is to say, to their most intimate properties. Thus, the physicist who observes the refractive powers of substances to compare them one to another acts exactly like the chemist who presents successively to a given base all the acids, or to a given acid all the alkalis, to determine their respective (combining) powers and their degree of saturation. In our experiments the substance we present to all the bodies is light [one of Lavoisier's simple substances], and we evaluate their actions upon it by their refractive power, that is, by the

increase in live force [*force vive*, i.e., *vis viva*] that the action of their molecules tends to impress upon it.... Since each substance brings to its combinations its own character and seems even to conserve, up to a point, the degree of force with which it acts upon light, let us try to calculate the influence of the constituent elements in a given mixture or compound.[16]

In other words, the effect of a mixture or compound on light might be calculated by assuming that the constituent substances retain their individual influences. This will yield the overall effect by summing the individual contributions. Biot thought that this might apply to compounds as well as to the simpler case of mixtures by analogy with gravitational forces: if you lump several masses together, the gravitational effect of each part is unchanged by the presence of the others; they just add together.

The expression that Biot derived is very simple. He was interested in the so-called refractive power: the degree to which the medium bends light rays. This is defined as (n^2-1), where n is the medium's refractive index. According to Biot's assumptions, the value is calculated for a compound substance like this:

P is the refractive power of the compound.

P', P'', P''', ... are the refractive powers of its constituents.

x', x'', x''', ... are the fractional weights of those constituents in the compound; hence all the x's together sum to *one*.

Then

$$P = P'x' + P''x'' + P'''x''' + \ldots$$

So for a compound with just two constituents:

$$x' = (P-P'')/(P'-P'') \text{ and } x'' = 1-x'$$
$$= (P'-P)/(P'-P''),$$

which is all very neat.

On this basis, the refractive power of a gas mixture from knowledge of its constituents, or, much more impressively, the proportions in a compound gas's chemical composition, could be determined simply through measurements of the relevant refractive powers.

Biot applied these formulae to various compounds and mixtures. On comparing the results with experimental data from direct measurements, he found good agreement. For air—just a mixture—the prediction of its refractive power showed an error of only 0.49 percent. With ammonia, this time a chemical compound, Biot had to reverse the procedure and calculate the fractional weights of hydrogen and nitrogen in ammonia from measured refractive powers of all three gases. The calculated fractional weights came out to 0.203 and 0.797 respectively, compared to 0.200 and 0.800, the figures recently determined by chemists including Berthollet (this all preceded Dalton's work, it should be noted). So the result for ammonia in particular was an excellent fit.

Biot and Arago's work, like Malus's on the polarization of light and some of Laplace's on capillary action, looked quite good for the assumptions underlying the Laplacian program. Forces-and-particles explanations could apparently be provided to account for precisely measured experimental phenomena. Those explanations could be put in a mathematical form (usually much more complicated than the one outlined above) that provided predictions often fitting experimental data quite closely. These were the beginnings of characteristically modern experimental physics as an arm of theoretical, mathematical physics.

Yet this vigorous and successful research program fell apart in the years after 1815.[17] A major factor was Laplace's loss of political power after the fall of Napoleon. As a result, Laplace had less control over the dispensing of patronage to people following his ideas and less power over the prize committees. Also, it so happened that at the same time two important and powerful new mathematical theories were developed by French physical scientists who were not members of Laplace's circle.

These two theories were Joseph Fourier's on heat and Augustin-Jean Fresnel's on light, and neither of them had anything to do with Laplace's forces-and-particles approach. Fourier's analysis of heat flow, developed in the 1810s, was independent of any particular model of the nature of heat, unlike Laplacian analyses. Fourier said that the question of the underlying nature of heat made no difference to the results his theory produced regarding the phenomena themselves. His work certainly provided no support for Laplace's view of heat as a subtle material fluid made of repelling particles. Fresnel's wave theory of light, also from the 'teens, postulated light as an effect of transverse waves in a fluid medium, which ran contrary to the Laplacian view of light as particles. Nevertheless, and despite the presence on the prize committee of Laplace, Biot, and Arago, among others, Fresnel was awarded the prize in 1819 (there was no serious competition).

Although Laplace and his followers were still important arbiters of French science in the years immediately after 1815, they no longer held sufficient power to assimilate these new theories and maybe modify them to suit their own program. Laplace had not been convinced by Fresnel's work. But there ensued a sort of anti-Laplacian bandwagon effect, and by 1825 Laplacian physics in the sense we have been considering was finished.

But in an important sense it did have a lasting effect: Laplace's promotion of a program for mathematizing experimental science in the Napoleonic period brought together most of the areas of study that became physics. It set up a style of science, with precise measurements and often complex mathematical theories to account for those measurements, that turned many areas of imprecise natural philosophy into mathematical physical sciences. Even though Fourier and Fresnel rejected the forces-and-particles assumptions about nature that underlay the Laplacian approach, they had already inherited this French style of physical science embodied in the new institutions of higher learning. And the twin threads of quantitative experimentalism and mathematization, which had never quite joined since the seventeenth century, were now being tied firmly together to form a direct antecedent of our modern physics.

Entr'acte

INSTITUTIONS AND PEDAGOGY

OUR MODERN WORLDVIEW, or cosmology, relies on the authority of scientists, a special class of people who act as "priests of nature," presenting and explaining the world to us.[1] Scientists receive rigorous training in their vocation, emerging at the end of a lengthy process of education and apprenticeship as priests qualified to carry forward a long tradition of knowledge and practice in the interpretation of nature. Scientists are also magicians; part of their authority resides in their ability to manipulate nature.[2] This typically takes the form of technological capabilities premised on theoretical discourses that are derived from, or related to, scientific models of understanding, these models not obviously connected to the technical capacities themselves. Professionalization of science has meant above all the establishment of pedagogical regimes: scientific disciplines are bodies of knowledge, skill, and practice that can be taught. The historian Owen Hannaway long ago made the case, for example, that chemistry became a true scientific discipline in the seventeenth century when it became *teachable*, as represented by the appearance of chemistry textbooks.[3]

By the second half of the nineteenth century the mass production of science was fully underway, taking the form of the large-scale training of scientists working in laboratory settings for experimental sciences, as well as in textual competitive settings for theoretical and mathematical sciences. Starting in the 1820s, Justus Liebig's organic chemistry laboratory was designed to generate doctoral research, both results and people (see chapter 14), to which can be added examples such as the physics laboratory at Göttingen; for theory, the training for the mathematics tripos at Cambridge that fed Maxwellian physics there in the later nineteenth century.[4] While this last is a non-German example, Germany was first to establish this form of training beginning in the 1820s and '30s.

Before the nineteenth century, university teachers were not expected to do research; they were just that: teachers. If a professor chose to do research, as some did, that was his own private affair. The concept of a university as a place of research was invented in Germany in the first half of the nineteenth century, and it was not restricted to scientific research. It developed from the standard view of the university as an institution devoted to teaching (and as the source of professional accreditation), now augmented by the idea that learning to do research was the best way to become truly educated.[5]

This process of change began on a large scale after the defeat of Napoleon in 1814, when the German lands began to organize themselves in the wake of Napoleon's subjugation. Germany was still a collection of small principalities and city-states, dominated by one fairly large state, Prussia, with its capital in Berlin. But although Germany's political unification would not occur until the Franco-Prussian war of 1870–71, cultural development during the century followed broadly similar lines in all the states, largely because of the common language. One of the results of the end of French domination was a renewed sense of German nationhood. Because Germany lacked strong political unity, it compensated with a form of nationalism that stressed German cultural identity instead. Universities played a central role.

German universities came to be seen as a repository of German culture and nationhood. The more prominent among them tended to draw students from all over Germany, not just from their local states. Theories of education fleshed out a view of the cultural function of universities. Two basic ideas came to dominate German higher education in this period, both first applied in Prussia. The first centered on the word *Bildung*, referring to the development of students' cultural and ethical character as the proper goal of education.

The central discipline initially involved in the ideal of *Bildung* was the study of classical antiquity, the world of ancient Roman and especially Greek culture. Classical culture was regarded as the highest level that human civilization had ever attained. Studying ancient writings, Latin, and Greek, was a way of imbibing ancient virtues—turning everyone into idealized Athenians by reading Homer in the original Greek. One of the foremost proponents of this concept of advanced education, Wilhelm von Humboldt, was the Prussian minister of education for a while. He was centrally involved in regenerating the Prussian universities and secondary school (*Gymnasium*) system to raise standards, and particularly to encourage *Bildung* as the goal of advanced education. The University of Berlin was founded in 1810 to embody these values.[6]

The other central educational concept, which overarched all the individual subjects being taught, had as its key word *Wissenschaft*. This is a very rough equivalent of the English word *science*, but has a much broader meaning. It applies to any organized and rigorously structured academic discipline with its own special skills and techniques—in other words, any academic discipline that is *disciplined*. Besides the natural sciences, the term also covered law, history, and the study of classical civilization: *Altertumswissenschaft*, the science of antiquity.

The idea of *Wissenschaft* involved a stress on training in research techniques as the best way to acquire the sort of cultural elevation that *Bildung* expressed. A student would not simply acquire learning about particular subjects; he would learn a discipline by learning to practice that discipline, that is, becoming a practitioner of that discipline: a researcher. This represented a high level of accomplishment. The candidates for this kind of education in the universities were provided by the secondary school or *Gymnasium* system as it was established in Prussia. Over the course of time this Prussian model was adopted, to some degree or another, over most of Germany.

The teachers in these prestigious secondary schools were required to have received doctoral training at a university. Thus, while producing new university teachers and state bureaucrats, the universities were also turning out schoolteachers who were themselves excellent scholars. This system really took off after about 1820, when the philosophical faculties of the universities began to expand rapidly in the wake of Humboldt's reforms in Prussia. (The philosophical faculty comprised all subjects except the professional doctoral faculties of law, theology, and medicine—more or less the same as what are called "arts and sciences" in a modern American university). Indeed, Humboldt's new University of Berlin had been established with nothing but a philosophical faculty, offering its own kind of doctorate: the PhD, or doctorate of philosophy.

The most important development in German university teaching in this period was the research seminar, in which research techniques were taught and students performed original research under professorial supervision. The seminar format had emerged at the end of the eighteenth century, but really grew in the post-Napoleonic reform period. It was initially used exclusively for *Altertumswissenschaft* but was subsequently adopted for modern history as well. Other subjects, including the natural sciences, stuck at first to the lecture format, but the growing influence of the *Wissenschaft* ideal encouraged the notion of the natural sciences as intellectual disciplines in their own right (as opposed to the eighteenth-century German view of sciences in the university as being allied to practical subjects, especially medicine). This newer, *Wissenschaft* view of the sciences soon encouraged the adoption in university science teaching of the same research ideal that had been developed in the study of classical culture.[7]

The development of research seminars in the natural sciences occurred in the 1830s. The first was a mathematical physics seminar at Königsberg (now Kaliningrad, part of the Russian Federation); significantly, it was started by C.G.J. Jacobi, who had originally intended to become a classicist. He had studied in the Berlin classics seminar before deciding to move into mathematics, and when he went to Königsberg to teach he transferred those research seminar techniques to the study of mathematical physics.

Over the next couple of decades this approach was adopted in universities throughout Germany, along with practical research work in laboratory sciences, including chemistry. Because these universities were all state-supported, often generously, university-based scientific research did very well indeed. And because of the seminar system and laboratory research training, they were soon turning out large numbers of rigorously trained scientists. The German system also encouraged competition between universities to hire successful researchers (because of the prestige and students they generated), and between individuals in the same university for promotions, which reinforced the research orientation even more. The state, especially Prussia, provided money for research facilities, and the university system produced plenty of researchers to make use of them.[8]

As a consequence, by the middle of the nineteenth century the concept of the university as a place of research, and as a well-funded location for scientific research in particular, was firmly established in Germany. Scientific research had become for the first time a professional activity, in which a career could be carved out in a highly competitive academic environment with the ultimate aim of being awarded a professorial chair. Being a professor no longer meant being a teacher who might also do research, but to be above all a researcher.

This structure for organized scientific research produced enormous strength in German science, especially physics, well into the twentieth century, from Kirchhoff, Helmholtz, and Hertz to Planck, Einstein, and Heisenberg, to name only a few. The system was particularly effective in turning out large numbers of scientists and an enormous volume of research; the big names are just part of a crowd.

In the twentieth century the United States replaced Germany as the foremost producer of scientific work, but the German model itself played an important role in bringing that about. The modern American university system, particularly scientific research in the American universities, was self-consciously based on (what was taken to be) the German system.

The modern American university was created in the last three decades or so of the nineteenth century, following the Civil War. Before the war, there had been many colleges throughout the United States focused on a broad "liberal arts" curriculum without specializations and without graduate schools for training researchers. By around the middle of the century, Americans who wanted (and could afford) more advanced training in a specialized subject after their undergraduate college education would often attend a European university, frequently one in Germany. A consequent American admiration for German universities came to focus especially on *Wissenschaft* and the German research ideal. The American version used the term *pure science* in a comparable way, as much to describe its effects in developing character as for the knowledge it produced.[9]

A major landmark in the reform of American higher education was the establishment of the Johns Hopkins University in Baltimore in 1876. Hopkins had originally been planned as a purely graduate institution for doctoral study. Local civic pressure resulted in the admission of undergraduates as well, but graduate study still predominated. The structure of teaching was explicitly based on the German model, especially with the use of the research seminar. Hopkins quickly turned into a leading scientific research center, especially in physiology, certain aspects of experimental physics, and chemistry. Other universities adopted versions of the German model too, typically in the form of establishing a graduate school in a preexisting liberal arts college (as Columbia and Harvard did).[10] There was also a spate of new universities: Cornell, founded in 1865; the University of Chicago, founded in 1892 with Rockefeller money (and, characteristically, with a very high emphasis on research); Stanford, founded in 1885; and major state universities such as the University of Michigan.

American universities never adopted all the overtones of the German *Wissenschaft* ideal; scientific subjects tended to benefit more than others, although even today the standard form of American graduate education involves research seminars even in subjects like history. The principal American adoptions from Germany were the idea of the university as a place of research rather than just teaching and a system for producing large numbers of research scientists.

It would be tedious to track the various national paths taken in the nineteenth century toward what amounted to a similar overall picture by the year 1900. But it is worth noting that in France and the United Kingdom, two of the most important and productive scientific countries, the organizational pattern deviated significantly from the German model and yet resulted in comparable outcomes: the establishment of university training regimes in the sciences that helped to define who counted as a "real" scientist, and industrial as well as governmental interest (both military and economic) in more obviously practical scientific endeavors to be pursued by trained scientists. France pursued strong centralization, with the state apparatus in Paris and the Academy of Sciences exerting considerable control over academic science. The aftereffects of the large-scale reforms in education that were made after the French Revolution, with their particular emphasis on the sciences, resulted in nineteenth-century French science being primarily university-based, with the exception of a few research institutions outside of universities such as the Museum of Natural History.

In the United Kingdom, organized science was slower to integrate with the state, although those two magisteria had long overlapped (think of the Royal Observatory at Greenwich, or the Royal Society of London, both founded in the seventeenth century). The British Empire and the Royal Navy were important actors in the efflorescence of British scientific endeavors in the nineteenth century, to the extent that the distinction between imperial and scientific projects is hard to draw. Nonetheless, it was only in the second half of the century that training in the sciences became fully established in the leading universities of the United Kingdom, being particularly behindhand at Cambridge and Oxford despite their long traditions of astronomy and mathematics. By the 1870s the celebrated Cavendish Laboratory at Cambridge, under the direction of the first Cavendish professor of physics, James Clerk Maxwell, and various professorships in other natural sciences joined the example of other universities in England and Scotland (at Manchester, London, Edinburgh, Glasgow and elsewhere) as leading centers of scientific training in Britain, providing serious competition to German scientific excellence.

One notable feature of earlier British science had been the coordination of the work of individuals through scientific societies. The years around 1800 had seen the creation of a new set of specialized scientific societies separate from the Royal Society, which had tended in the later eighteenth century to become less and less serious, until by the early nineteenth century it was seen by many scientific people as not much more than a gentleman's club. The *Philosophical Transactions* was still a place to publish research, but the society itself was losing its status as a forum for serious scientific discussion. But by that time other groups were emerging.

The Linnean Society, to do with natural history, especially botany (named after Linnaeus, of course), was founded in London in 1788, and became important in the nineteenth century with the increasing British interest in those studies. The Geological Society of London was founded in 1807 as a conscious rejection by geologists of the Royal Society; it was central in providing a focus for British geology, which led the world in the first half of the nineteenth century.[11]

But the Linnean and Geological Societies were just two in a spate of specialized societies in the early nineteenth century. By 1830 there was quite a list of them, including the Zoological Society, founded in 1826, and the Astronomical Society, founded in 1820. These societies, all based in London, published their own journals, and they coordinated British scientific research and gave it a strong institutional base. They constituted serious fora in which the foremost people in their fields met, presented papers, published, and argued about things. Like the Royal Society, they were not research institutions in the sense of places where people did their scientific work. The research was still conducted by individuals rather than in some communal laboratory or university department; there was no German production line.

Thus Charles Darwin, for example, a man of private means obtained in part by marrying into wealth, spent most of his life devoted to full-time research in his country house in Kent, while keeping in touch with the London geological and natural historical communities. James Prescott Joule, one of the formulators of energy conservation, was also independently wealthy and did his work in his own private laboratory; no university position was needed to establish him as a serious natural philosopher. Even James Clerk Maxwell, who spent most of his life in universities, first at Aberdeen, then King's College London, and finally at Cambridge, illustrates the point: he spent five years as a private gentleman with no formal academic position before he accepted the post of first Cavendish Professor of Physics at Cambridge in 1871. Those five years hurt neither his research work nor his scientific reputation.

Besides specialist societies, the British Association for the Advancement of Science (BAAS), founded in 1831, shows the growing self-consciousness of research scientists. It represents the growing idea that science could be a profession, with its main purpose being to publicize the importance of science to the nation (the word *scientist*, which encapsulates this idea, seems to have been first used by William Whewell at the group's inaugural meeting in 1831). The BAAS acted as a sort of "super-society" that brought together the leading lights of the specialist societies. It provided a platform for the public proclamation of the value of science and has always held its annual meetings in various provincial cities, never London.[12] An American twin, the American Association for the Advancement of Science (AAAS), was founded in 1848.

By the end of the nineteenth century, despite differences in national patterns, the overall picture had become fairly uniform. Scientific research was now an integral part of a university's function, even though the majority of researchers in the twentieth and twenty-first centuries have worked in industrial or governmental settings of some sort.[13] Above all, universities had become indispensable as the institutions that trained new scientists and perpetuated the traditions of scientific disciplines. The establishment of science as a profession associated with university degrees and positions now clearly marked the distinction between the scientist who could be taken seriously by the scientific community and the amateur enthusiast. The days of private gentlemen with their own laboratories were gone, and science had taken on the social and institutional structure with which we are nowadays familiar—a prominent feature of the world as we know it. Science and the state were by then firmly intertwined in an iron clasp that would only grow tighter.[14]

8

Classification and Extinction

CUVIER AND NATURAL HISTORY IN THE EARLY NINETEENTH CENTURY

THE ORGANIZATION of science in France in the decades around 1800 had consequences beyond the physical sciences. The institutional upheavals of the French Revolution brought to prominence newly invigorated practices in the life sciences, and the natural history sciences more generally, that gave practical significance to preexisting themes in taxonomy and earth history. The central figure in this story, the life sciences' equivalent of Laplace, was Georges Cuvier.

Charles Darwin's work would have been impossible without Cuvier and his conceptual restructuring of taxonomy. Although Darwin is arguably the dominating figure in nineteenth-century life sciences, he owed a great deal to other naturalists, and Cuvier was the most important of all. Darwin's achievement was to draw together problems that had become increasingly central to natural history in the first half of the nineteenth century and to provide unified solutions for them. The issues of classification that ran through most sciences in the eighteenth century played a central role.

Darwin's most famous book is called *On the Origin of Species*, clearly designating the problem area that it addresses. There are many areas that it does not directly address: fields such as physiology, microbiology, or embryology, for example. The area of the life sciences that talked about species, and the area from which Darwin's work emerged, was natural history. And the essence of natural history was the classification of species: grouping some together, or separating them, or linking them at different taxonomic levels.

Natural history had already undergone some major changes at the beginning of the century at the hands of the zoologist Georges Cuvier. What Cuvier did in zoology—his work on anatomy, classification, and paleontology—radically changed the image of natural history as a whole.

Georges Cuvier

Cuvier (1769–1832) was a native of eastern France, born in Alsace, a semi-independent principality at the time, as much German as French. In the late 1780s, after the end of his formal schooling at a German military academy, he got a job as a tutor to a family in Normandy, where he spent seven years. He had received some instruction in natural history during his schooldays, and he spent most of his spare time in Normandy (he apparently had quite a lot of it) in reading, fieldwork, and dissection, particularly of marine animals. After the Revolution he became a local organizer and volunteer bureaucrat in the neighboring town, and in 1795 he was taken up by a visiting agronomist from Paris.

Cuvier was already becoming known in Parisian natural history circles from research papers he had been sending, but now he was encouraged to go to Paris. Not long after he arrived, Cuvier was given a junior post at the Paris Museum of Natural History, the leading natural history establishment in France (Buffon had been its head for decades, under its prerevolutionary name). He was also elected in 1795 to the renamed Academy of Sciences, the First Class of the Institute.[1]

Cuvier soon became the dominant figure at the museum and was awarded a chair there in 1802. The museum was the perfect place for Cuvier, because it gave him a superb place in which to work. It should not be thought of as a static collection with specimens in cases. It provided some teaching, mostly to medical students, but it was primarily a research establishment, funded by the French state, with laboratories and workrooms and assistants for leading figures like Cuvier. Unsurprisingly, Cuvier spent the rest of his career there. He had considerable administrative abilities, which, added to Napoleon's fondness for the sciences, got him into high government positions, particularly to do with educational reform, in the first decade of the nineteenth century. He was so good at administration that he remained an important civil servant even after Napoleon's downfall. All this was in addition to his leading involvement in Parisian scientific circles, particularly the First Class of the Institute.

But Cuvier's importance stemmed not only from his administrative abilities or his aggressively ambitious personality. Cuvier was recognized as the best zoologist of his time, and he worked very hard at it. Above all, his work changed the character of natural history by developing new concepts and techniques that changed the image of classification.

Recall that Linnaeus, in developing the basic taxonomic system that later naturalists employed, had characterized his systems as artificial: he was unable to justify them as natural systems, although that was always his ideal. The attitude was shared by most naturalists following Linnaeus, although most did not go as far as Buffon, who actively opposed the adoption of Linnaean classification even for practical purposes. Cuvier effectively demolished that image of classification by creating new grounds for regarding it (or his version of it) as a system that provided information on genuine structural and functional relationships within the animal kingdom and, by implication, the plant kingdom as well. Talking about the mind of God was no longer necessary to justify the idea of natural classification, it seemed.[2] The concepts that Cuvier developed to justify his own "natural" system of classification depended crucially on the concept of the immutability, or fixity, of species. Evolution is antithetical to Cuvier's approach, and yet Darwin's work would have been impossible without the ideas that Cuvier incorporated into natural history.

Cuvier's major technical innovation was in virtually inventing systematic comparative anatomy, the description of an animal organism by comparison with other species. His first major publication on the subject was a five-volume version of his lectures at the museum, the *Lessons in Comparative Anatomy*, published between 1800 and 1805. Comparison of the structures of different species had always been part of anatomy at a more informal level, but Cuvier turned it into what he claimed was a precise technical tool. He saw comparative anatomy as a way of discovering laws governing the animal organism. And his aim was to raise natural history to the status of a true science.

Models of Science in Early Nineteenth-Century France

The accepted model for a true science in Cuvier's cultural environment was a mathematical, physical science modeled on Newton's celestial mechanics. Cuvier wanted to do for natural history what his scientific colleagues in Paris, preeminently Laplace, were trying to do for the physical sciences: he wanted to formulate what could pass for rigorous Newtonian-style laws that relied on experimentation. That was what carried the real prestige in post-Revolutionary French science, as it had in one way or another throughout the eighteenth century. Real science was Newtonian science, according to this perspective, and Cuvier wanted to be the Newton of natural history. This was a typical goal for ambitious students of the life sciences in this time and place.[3]

At the beginning of the century the leading French physiologist Xavier Bichat (1771–1802) formulated an influential perspective on the goals of physiology. His work was characterized by his conception of the body as being made up of various distinct kinds of tissue, each of which was understood to have its own properties and to serve particular functions. From a methodological point of view, Bichat notably made no real attempt to investigate how the animal organism worked with respect to the physical and chemical features of the behaviors and properties that he examined; instead, he insisted on sticking to concepts specific to physiology as an autonomous science. Bichat held that phenomena of life were radically and categorically different from physico-chemical phenomena; animals don't work through chemistry. Consequently, performing chemical analyses of physiological processes, fluids, or tissues would be irrelevant to understanding those processes as manifestations of life. This position is usually known as *vitalism*, as opposed to *materialism*, where life *is* taken to be reducible to the physico-chemical behavior of ordinary matter.

Bichat set out his basic position in a book in 1801, later published in English as *General Anatomy Applied to Physiology and Medicine*. As much as anything, the work is a call for the autonomy of his discipline, occurring in the same cultural environment as that of Cuvier. Cuvier wanted to develop "laws" in comparative anatomy and natural history to put his science on a par with physics and its Newtonian laws. Bichat wanted to do the same for physiology, making explicit reference to Newton. He wanted to distinguish the sciences of life from the various physical sciences, and did so by making a sharp distinction between their objects of study.

He explained himself in *General Anatomy*: "Let us leave to chemistry its affinity, to physics [*la physique*] its elasticity and gravity. For physiology let us employ only *sensibility* and *contractility*."[4] So these were to be the basic, irreducible concepts: *sensibility*, the property of an organ or type of tissue to detect that a response is appropriate, and *contractility*, the response itself (Bichat wanted to reduce all physiological responses to contractions).

Classifying and investigating the behaviors of different organs and their constituent tissues on that basis meant that Bichat's practical research often differed from that of a modern physiologist or of physiologists later in the century. Because he regarded his work as examining properly physiological, and therefore nonchemical, properties, there was no reason for him to investigate the chemical properties of living tissues or fluids. He would apply experimental tests, such as treating tissues with acids or boiling them, to see how they responded and as a basis for classifying different types of tissues. But he could not by definition do chemical physiology.[5]

One thing that Bichat could do was vivisection. Using the well-established techniques of the medical anatomist and surgeon, Bichat carried out physiological experiments on living animals in ways that set the dominant approach to French physiology in the nineteenth century. Just a few years after Bichat's untimely death, François Magendie carried forward Bichat's strongly experimental approach, and adopted similar methodological ambitions. Like Bichat, Magendie had trained as a physician, and after receiving his doctorate in 1808 he went into teaching and research as well as becoming a medical practitioner.[6] (Physiologists in this period, whether in France or Britain or Germany, typically entered the field through medicine, a connection that also enabled them to deflect criticism for the cruelty of vivisection.)

During and after the Napoleonic period Magendie had the support of Laplace, which got him a good academic post and a timely exemption from military service. He also earned kind words from Cuvier in referee reports on papers that he submitted for publication; as a result, Magendie soon became well established in the Paris scientific community. Since a feature of that local culture, exemplified by Laplace as well as Bichat, was the formulation of programmatic statements about the nature of their own sciences and how these ought to be pursued, it is unsurprising that Magendie published a paper in 1809 containing a programmatic statement of what physiological science should be and how it should be pursued.

> The character of a certain and perfected science is to have for its foundation only a small number of principles to which a great number of facts are connected. The physical sciences, properly speaking, generally have this character; chemistry, physics, and above all astronomy, are remarkable examples of it. The physiological sciences, having many more difficulties to overcome, cannot yet at all be placed in that same pattern [*la même ligne*]; however, the progress that they have made in the last fifty years, the direction that they currently follow, give room for hope that one day they will be able to follow a similar path.[7]

Comparing physiology with the most prestigious science of the day, astronomy (in the sense of Newtonian celestial mechanics), represented the ideal of a precise science governed by a small number of general laws—the same model that Bichat and Cuvier, following Laplace, had invoked.[8]

The Newton of Natural History

Cuvier's adoption of similar ideas in anatomy had involved the development of laws governing animal organisms that would have the same fundamental status as Newton's laws of motion. The approach he took was to determine how the parts of an animal relate to each other, and how those interrelationships in turn relate to the animal's way of life.

Proper scientific laws for a good Newtonian were supposed to derive their legitimacy from experiments. But while experiments can be performed in physiology, matters seem otherwise in natural history or even in anatomy; such work seems on its face to be restricted to description. But the importance of comparative anatomy for Cuvier lay in the way that it could, he thought, serve as the natural historical equivalent of experimentation. He described different species as nature's experiments with animal form: comparing the anatomical structures of different species enabled the discovery of general laws governing *all* species. Just as a regular experimental science was supposed to draw its laws from experiments carried out under varying conditions, so natural history would draw its laws from nature's experiments: different species, properly and systematically compared with each other.

Cuvier theorized his approach in the terms of what he called "anatomical rules." Cuvier's approach centered on explaining the form or morphology of an animal, particularly the individual parts of an animal, by referring to the functions those parts were supposed to serve—for example, explaining why ducks have webbed feet by reference to the way that this makes them more effective as paddles. In some respects, this approach mirrors the work of the ancient Greek philosopher Aristotle, whom Cuvier admired very much. This dimension of functional explanation is what gives force to Cuvier's claim that his anatomical rules are not only generalizations or descriptions; for him, they were true laws of nature because they have explanatory force.[9]

Cuvier formulated the two basic anatomical rules early on in his career, and they originally appeared in the *Lessons in Comparative Anatomy*. The most basic is called "correlation of parts." It says that every part of an animal is functionally linked in some way to every other part, so that an animal as a whole forms an integrated organism. This is how Cuvier once explained it:

> Every organized being forms a whole, a unique and closed system, in which all the parts correspond mutually, and contribute to the same

definitive action by a reciprocal reaction [a nod to Newton's third law of motion]. None of its parts can change without the others changing too; and consequently each of them, taken separately, indicates and gives all the others.[10]

For any particular animal, the appropriate correlation of parts by which all parts would be able to work together to render the animal viable depended on what Cuvier called the animal's "conditions of existence." This expansive concept refers both to the kind of animal under consideration (a whale is large and heavy, for example) and the way it relates to its environment (as a whale lives in the sea). It refers to everything affecting the organism's existence, both external and internal.

As an imaginary instance of conditions of existence determining an animal's correlation of parts, consider an animal with powerful jaws and teeth suitable for tearing raw meat. Functional considerations then reveal that one of the animal's conditions of existence is that it must be a carnivore. On that basis (and assuming that it is a hunter rather than a scavenger), correlation of parts then tells us that for this animal to be viable, it will need in addition to have powerful legs to chase its prey, sharp claws for grappling with its prey when caught, and a digestive system appropriate for raw meat.

Cuvier claimed that the rule of the correlation of parts, if applied correctly, would make it possible to reconstruct an entire animal of a species never before studied on the basis of a single jawbone. On one occasion he played the showman by predicting to colleagues that a fossiliferous rock would reveal a hidden feature of the fossilized bones that he specified before ever a chisel was taken to the specimen. His prediction was confirmed—a characteristically marsupial (opossum) bone was revealed in the already identifiably marsupial fossil.[11]

Of course, making such a prediction would have owed more to anatomical experience and analogies with already known species than to the application of an abstract principle. Cuvier freely acknowledged the importance of analogy in comparative anatomy, and had himself probably the broadest experience and knowledge of any anatomist around. But correlation of parts goes beyond empirical generalization or the summing up of practical experience. It acts as more than a guideline to what features usually go together in an animal, such as sharp claws and powerful jaws. It also tells us why we do not find certain combinations of features.

For instance, it says that there will never be found an animal with, say, the body of an antelope and the head of a dog, even though that might be conceivable from a purely physiological point of view. Such a combination would simply not serve a viable function: the dog's jaws are adapted to chewing meat, but an antelope's digestive system cannot process meat. These considerations, dealing with the functional interrelatedness and interdependency of the parts of an organism, is what the correlation of parts is all about: making sure that everything works together.[12]

Another way of expressing the point is to say that for Cuvier, an animal organism is perfectly adapted to its environment and its way of life. The long-term significance of that idea for the role of adaptation in Darwinian natural selection is apparent, which makes it all the more interesting that Cuvier used the correlation of parts as his central argument against the possibility of organic evolution. Cuvier's main opponent here was Jean-Baptiste de Lamarck, a colleague at the Museum of Natural History. Lamarck, on little evidence beyond the general appearance of the fossil record, supported an idea of evolution whereby continual, spontaneous ascent always occurred from generation to generation, from the lowest organism to the highest, which was man.

Cuvier argued against the possibility of evolution regardless of what driver Lamarck or anyone else might propose for it. His argument came straight out of his fundamental conception of the laws of the animal organism as expressed in correlation of parts: if all the parts of an animal were functionally interrelated, then changing any feature to a significant extent would throw the whole organism out of balance, and the organism would therefore cease to be viable.[13] A good analogy would be with a mechanical watch, where, if the shape or size or arrangement of teeth on just one cogwheel is altered even slightly, the watch will stop working. Hence species could not transmute into categorically new forms, because an animal was a functionally discrete, integrated unit where the proper working of each part depended on all the others, just like a watch. Organisms, according to Cuvier, were perfectly adapted to their conditions of existence, which meant there was no slack in the system for significant variation: change something appreciably and the animal will simply die.

Comparative Anatomy and Anatomical Rules

Because Cuvier's view of natural history disallowed any kind of evolution, animal species were discrete, separate kinds that fitted into an essentially static classification-scheme. The second of Cuvier's anatomical rules acted to justify taxonomic classification (his version, anyway) as a natural, not arbitrary system.

This rule was called "subordination of characters." It says that for any kind of animal, some functions are more important than others, and the more fundamental functions help to determine and constrain the less fundamental ones. These functions are represented by characters of the organism that are consequent on the functions or that facilitate them. The hierarchy of functions and their characters for any given kind of organism is determined by its conditions of existence—its external environment and internal constraints.[14]

Cuvier said that individual groups of organisms, such as reptiles, have similar conditions of existence. Cuvier justified classifying reptiles together as a subgroup of vertebrate air-breathers on the basis of what he claimed was their overriding condition of existence, the thing that constrained all their other characters: the fact that they are all cold-blooded. That's what makes reptiles count as reptiles. Because being cold relates to inefficient respiration that fails to burn oxygen very effectively, Cuvier then used respiratory function to justify the four orders into which he divided the class of reptiles. The particular characters that express the respiratory function, and therefore the ones that he focused on at this level of classification, concern capability of rapid movement. If a reptile has a more efficient respiratory system, it will be able to move quickly when necessary, which implies the existence of legs adapted to rapid movement. The structure of the heart is also relevant: more efficient breathing implies a more effective circulatory system.

Cuvier used such arguments to justify his selection of particular classificatory characters as his criteria for setting up these orders (subdivisions of the class) of reptiles and assigning reptiles to each one. The idea was that the finer levels of distinction—family, genus and species—would be defined through characters that express successively less fundamental functions. The hierarchical structure of taxonomic boxes simply reflects a hierarchy of vital functions.

In practice, Cuvier's classificatory criteria tended to accommodate existing practice. Teeth were a useful classificatory character in Linnaean-style zoological taxonomy, for mammals as well as reptiles, and Cuvier justified their use by asserting their obvious relation to the nutritive function. To a large extent, Cuvier arranged his practical taxonomy using exactly the same sorts of rules of thumb and pragmatic choices of criteria as any naturalist; however, the significance of Cuvier's anatomical rules is that they appeared to make natural history as a whole, including classification, a genuine *science*—which was Cuvier's aim all along.[15] Prior to Cuvier's work, Buffon's skepticism regarding the reality of divisions such as family and order still carried weight, but after Cuvier there was at last a justification for taxonomy that claimed to be based on scientific principles.

The success of Cuvier's ideas owed a great deal to his enormous reputation, which was based not on his theoretical ideas but on his practical work, embodied in standard reference works like his massive *Animal Kingdom*, which went through countless editions from 1817 onwards. His work in descriptive and functional anatomy covered everything from elephants to fish. Because everyone knew he was a superb anatomist (and neither should we forget his institutional power within the French scientific community), they took his ideas, and his approach to anatomy, very seriously.[16]

Following Cuvier's work in the early nineteenth century, it was taken for granted that the divisions of practical taxonomies corresponded to the true order of nature. Later on, all Darwin had to do was to change that sense of "natural" from Cuvier's functional meaning to his own genealogical meaning.

Paleontology and Extinction

The significance of Cuvier's natural classification went beyond the organization of animals existing in his time. By using his rules of comparative anatomy, Cuvier was able to show, in ways that convinced most of his contemporaries, that the fossilized bones of large animals that had been turning up intermittently in Europe and North America throughout the eighteenth century belonged to extinct species. Previously the matter had been unclear; usually all that was found was a jumble of assorted bones, tusks, and teeth, often from different kinds of animals, rather than a complete skeleton. These were mammals from the Tertiary Period, including saber-toothed cats and forerunners of elephants (a mastodon was found in the Ohio Valley in 1739), rhinoceroses, antelopes, and so forth.

But incomplete skeletons were no problem for Cuvier, because correlation of parts was tailored to allow the reconstruction of a complete animal from pieces of it, such as a jawbone. And the advantage of Tertiary mammals in particular was that they were fairly similar to living species that Cuvier knew about already. That made analogical argument easier, justified theoretically by the correlation of parts. From work he did between the late 1790s and early 1800s, mostly on fossils recovered from gravel beds around Paris, Cuvier established an entire extinct fauna of mammals. These were similar to, but anatomically and hence taxonomically distinct from, existing species, none of them very old in geological terms (because they were from superficial, and therefore presumably recent, gravel deposits).[17]

The importance of this work was that it was the first time that anyone had shown convincingly that there had been large-scale extinctions in the past. Cuvier was able to do it not just because he had a lot of these fossils coming to light—he and a geologist collaborator, Alexandre Brongniart, collected many of them—but also because of his anatomical methods. These made him much more capable of determining (on grounds that other people found persuasive) that a specimen was from a separate *species* rather than a variety of one already known. That was the kind of determination needed to make the idea of extinction meaningful.

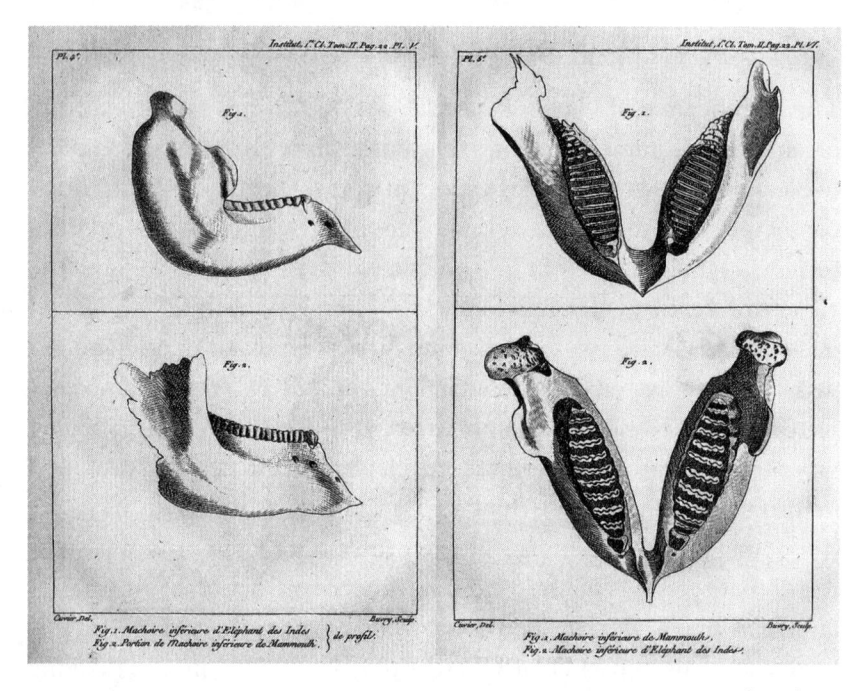

FIGURE 8.1. Comparative anatomy: Georges Cuvier's elephant jaws (1798).

The best example of Cuvier making such a distinction is probably his work on elephants in 1796. He argued that Indian and African elephants were separate species on the basis of the structure of their teeth, which was the clinching criterion in his classificatory scheme, backed up by the functional argument for why teeth ought to be taken as important. This criterion established, Cuvier could use the same kind of argument in 1799 to show that frozen mammoths that had turned up from time to time in Siberia and Northern Europe were not simply a hairy variety of elephants but a distinct species, similar to modern elephants (in the same genus) but nonetheless separate, and therefore, as a separate and clearly vanished species—it helped that they were so large—extinct.[18]

Arguments of this sort, derived from Cuvier's approach to zoological classification, enabled a new view of earth history in which many or most animals no longer existed at the present day; without the subordination of characters rule, the concept of past extinctions had no real scientific meaning. But Cuvier now had a new problem. Having identified an entire extinct fauna, he had to face the question of what had happened to it.

Because he was committed to the fixity of species, Cuvier could not say, as Lamarck promptly did, that these extinct animals had evolved into present-day species. But the fact that the bones had been found in gravel beds suggested to Cuvier that there must have been a catastrophic extinction caused by a flood, presumably the sea inundating the land and then withdrawing. The evidence remained inconclusive, but it looked as though there had recently been what Cuvier called a global revolution that had destroyed the species of a large region of the earth (revolutions were much in vogue in this period in France).

Further research on fossils from older deposits uncovered even more collections of extinct species, and Cuvier started to speculate on the possibility that the earth's history had been punctuated by not just one flood—which he carefully avoided identifying with the Biblical flood—but by a series of major catastrophes. The full account, published in 1812 (and dedicated to Laplace), represented a succession of catastrophes, each one of which would have wiped out the population of a large area of the globe. After the catastrophe was over and the waters had receded, the affected area would have been repopulated by other species migrating from unaffected regions.[19]

Cuvier's own work provided evidence that this picture could not be satisfactory, however. His paleontological work indicated that the older the rocks, the less developed the organisms fossilized within it. Old rocks had shells, later ones added fish and reptiles, still later ones mammals, in the progressionist sequence that Lamarck relied on for his evolutionary ideas. But the lack of apparent continuity between the fauna of different ages—between successive fossil-bearing strata—was what encouraged Cuvier to assume sharp catastrophes between them and steered him clear of evolution. One might imagine that it would have been hard to avoid the conclusion that, over time, new species had appeared that had not existed before. Nonetheless, Cuvier resisted giving way, largely by relying on ambiguities and uncertainties.[20]

Cuvier's arguments necessarily implicated geology. The catastrophes that he posited as having punctuated earth history must have involved, he thought, massive seismic events to cause the devastating floods; these he associated with the rise of mountain ranges. The view was pursued by others, especially the French geologist Jean-Baptiste Élie de Beaumont, in the 1820s, and was soon to be dubbed *catastrophism*. These events were destructive, not constructive, and they created a geological past that lacked long-term continuity. But the perspective incorporated the massive extinctions that Cuvier required.[21]

Consequently, the natural history that Darwin inherited owed enormous amounts to Cuvier, who was arguably more important for British naturalists and anatomists in the first half of the nineteenth century than he was for the French.[22] Cuvier's approach to anatomy stressed the functions of animal organisms and how the form of an animal was adapted to its whole way of life. He established classification as a natural system so that Darwin was in a position to reinterpret it as a family tree. And he established that there had been vast extinctions in the past, leaving the problem of explaining where later species had come from. Cuvier in large measure set up the problems that Darwin tackled and the terms in which he tried to solve them. Without Cuvier, there could have been no Darwin.

9

Darwin's Taxonomy

GEOLOGY AND THE ORGANIZATION
OF LIFE

DARWIN'S IMPORTANCE in framing the modern world cannot be overstressed. Although he was not single-handedly responsible for all the beliefs and attitudes that came to be associated with his name, his career and cultural productions both informed and were appropriated by the ambitious rising scientific priesthood of Victorian Britain. The success of British natural philosophers, or "scientists," in inserting themselves into the life of the state and the intellectual formation of its leaders owed much to the worldview that Darwin's new natural history provided.

Darwin's Decision

Charles Robert Darwin was born in 1809, and most of his life was spent under the reign of Queen Victoria, who came to the throne in 1837. He was straitlaced and serious-minded; in that sense he was a stereotypical Victorian. He displayed his character perhaps most acutely in an episode in July of 1838, when he had reached a crucial point in his career.

Darwin had already acquired a good reputation in the British natural historical and geological communities. He was living in London, belonged to the appropriate professional societies (the Geological and the Linnean), and although he was still a bit of a scientific newcomer, he was sufficiently highly regarded as to have been elected to the council of the central forum for British geology, the Geological Society. In private he had been considering a very touchy issue: the possibility of the transmutation of species, including speculation about human evolution. In public, however, he had been cautious, publishing on much less controversial geological subjects. Now he was preparing himself to make a decision that he knew would affect his life profoundly. In order to help himself make it, he employed a methodical, Darwinian procedure: he drew up a two-column list of pros and cons. This piece of paper still exists in the Darwin papers at Cambridge, with the words "This is the Question" written across the top.

The question was "Marry" versus "Not Marry." Darwin's list provides a remarkable insight into his character and thought habits. He listed more points against the idea than in favor of it, although the objections are somewhat repetitive. He was worried about the restrictions on his freedom if he were to burden himself with a wife and family, so he wrote of not being able to go on geological expeditions in Wales, or visit America, or learn French, or make balloon ascents. (He never had flown in a balloon, but he seems to have thought it would be too risky once he had family responsibilities.)

The main problem, though, was *"loss of time"* for working; he was a very ambitious man. The things he jotted down in favor of marriage, on the other hand, show Darwin's finer feelings. He wrote, in the "for" column, "Only picture to yourself a nice soft wife on a sofa with good fire, and books and music perhaps ... constant companion, friend in old age ... better than a dog, anyhow."[1]

Private Thoughts and Public Pronouncements

As a result of Darwin's methodical approach to problems, his tendency to write everything down, and above all the fact that he kept all his papers—that memorandum on marriage was filed away systematically, like everything else—modern historians of science have been able to follow in detail the step-by-step development of his ideas, especially his intensive consideration of the problem of species in 1837 and '38. Darwin had returned from his famous five-year voyage around the world in late

1836. In the summer of 1837, he began recording his ideas in what are known as the species notebooks. They open at the point at which he had already become convinced that new species originate by transmutation from other species, but he still had no real idea what the causes might be. By late 1838, at the end of a long trail of notebook entries, he had come up with a causal explanation that he found convincing. And he filed the notebooks away carefully, just like his speculations on marriage, and thought about what to do next.

Darwin was much too careful and much too concerned about his professional reputation to rush into print with nothing but a bright idea, particularly when almost everyone else was convinced that species were fixed kinds (not to mention the implications about human evolution that Darwin was drawing from his theory). On the other hand, he had decided to get married, at the beginning of 1839, to one of his first cousins, Emma Wedgwood (a member of the Wedgwood potteries family). Emma also brought a lot of money with her. So Darwin was able to carry on the life of a gentleman of private means, concentrating on research and writing; "loss of time" proved not to be a big problem.

With that difficulty out of the way, Darwin still had to decide what to do with his ideas on species. There was no point in having what he understood to be a spectacularly important idea if he never had the courage to tell anyone; it was a matter, therefore, of deciding on the best way of doing it. Things just kept dragging on—he was busy writing papers on geology, and pursuing various other projects, in the next few years, and evolution remained something that he worked on privately in addition to his public scientific work.

In 1842 he finally wrote a 35 page "sketch" of his theory, and two years later a 230 page "essay."[2] Neither was intended for publication; they were basically drafts to see how the argument would work. But soon after he had written the essay, Darwin let somebody else see it in confidence: the botanist Joseph Hooker.

Hooker thought Darwin's ideas interesting, although he remained unconvinced by the argument as it stood. He advised Darwin to do some serious, detailed work on zoological classification to get the practical knowledge that he would need to be able to argue

FIGURE 9.1. Darwin in his prime, ca. 1854. Photograph in University College London Digital Collection.

convincingly for his theory, which relied heavily on taxonomy. Hooker's advice suited Darwin; he was in no hurry, and he probably feared jeopardizing his career by publishing anything less than the best argument he could. In 1839 he had produced an ambitious paper proposing a radical theoretical interpretation of a topographical feature found in the Scottish Highlands, the correctness of which would have further confirmed a geophysical theory that was close to Darwin's heart (and to that of his great patron in the 1830s, the geologist Charles Lyell). The idea was soon defeated by a much more plausible alternative having to do with the new concept of an Ice Age and its associated glaciation, and Darwin's sense of humiliation in this experience may have played a role in dissuading him from casually or hastily publicizing his species theory.[3]

So Darwin continued to publish more work on geology, and in 1847 he embarked on an eight-year project classifying barnacles—his way of following Hooker's advice. In the 1850s he started breeding pigeons, which was related to his interest in variations and their inheritability; not until 1855 did he finally return to working seriously on his species theory. The following year he commenced a multivolume blockbuster presentation of his theory intended for publication, a treatment designed to convince by weight of references.[4]

Fortunately, he never finished it. Two years later, in 1858, the big book still a long way from completion, he received a letter from a fairly obscure young British naturalist who was writing to him from an expedition in the East Indies. The naturalist's name was Alfred Russel Wallace, who had come up with an idea about organic evolution while suffering from a fever. He had immediately written an account of the idea and sent it off to the well-known and respected naturalist Charles Darwin, back in England, asking him what he thought of it and whether he would arrange for its publication if he thought it worthy. Wallace's idea, of course, was what Darwin called natural selection, and the paper differed in few essential respects from Darwin's own elaboration of the theory.

That was probably the only occurrence that could have rushed Darwin into print. He was very ambitious in his scientific career and cared a lot about his reputation. Several years earlier he had made out a will that had the strict provision that if he died, his executors would ensure that the 1844 essay on natural selection would be published. Wallace's paper forced him to declare himself or lose priority. By this time a small number of Darwin's scientific friends in addition to Hooker knew something about his ideas on evolution. They arranged for a joint presentation of Wallace's paper and one put together for the occasion by Darwin himself, which took place at a meeting of the Linnean Society in London later in 1858. And then Darwin started scribbling.

In November 1859 there finally appeared, over twenty years after he had first come up with the idea of natural selection, Darwin's *On the Origin of Species by Means of Natural Selection; or, The Preservation of Favoured Races in the Struggle for Life*. Darwin described it as an "abstract" of his unfinished massive work, but it's a good-sized book in itself. Notably, in its basic layout, the structure of the argument, the kinds of evidence used, and particular ideas that develop it, the book is very similar to the 1844 essay. The main difference is that the *Origin* is a much fuller treatment, with more material to act as evidence fleshing out the argument.

Adaptation

Cuvier's approach to zoology dominated natural history, especially in Britain, in the first few decades of the nineteenth century, through his ideas and his achievements in anatomy and paleontology. Cuvier's approach to anatomy was functional; it centered on understanding the parts and the interrelations of the parts of an animal in terms of what they were *for*, what job they did. And it disallowed transmutation because that would destroy the correlation of parts that controlled the delicate balance between the animal's conditions of existence and the functions those conditions required.

This view of the perfect adaptedness of species found its way into British naturalists' thought not only through Cuvier's writings; British receptiveness to Cuvier also no doubt owed much to the British tradition of natural theology. That meant that it had rather touchy religious connotations. As we have seen, British natural theology had its roots in the later seventeenth century. Natural philosophers like Robert Boyle and the botanist John Ray saw one of their tasks as using natural philosophy to prove the existence and goodness of a Creator. The classic early nineteenth-century example was a book by the Reverend William Paley titled *Natural Theology*, first published in 1802.

Paley's book concentrated on the organic world; its basic strategy is the so-called argument from design. Paley's most famous argument runs as follows: If you find a watch on the ground, you can immediately infer that it had a maker. It has clearly been designed to perform a particular function; no one would believe that it got that way by chance. Now consider the eye: It's obvious that an eye is as much designed for its function as a watch is; the eye is a precision optical instrument. Therefore it is just as valid to infer that the eye was designed for its function, and has a maker, as it is in the case of a watch. Paley made this sort of argument for many apparently designful adaptations in nature (again following a well-trodden eighteenth-century path; Derham had made similar points). Paley was very good at making these arguments, and Darwin said in his autobiography that the book's logic gave him "as much delight as did Euclid."[5] The theme remained strong in the 1830s, when a series of books by various authors appeared along broadly similar lines, commissioned through the will of the Earl of Bridgewater. These are known as the Bridgewater treatises, and their focus was still primarily on adaptation and the argument from design as proofs of God. On both fronts—Cuvierian anatomy and British natural theology—Darwin existed in an environment that encouraged a conception of organic nature in which animals and plants were well-designed, both in themselves and in relation to their ways of life. Function was a fundamental element in understanding why a plant or animal was the way it was.

After Darwin graduated from Cambridge University he took a five-year voyage around the world, between 1831 and 1836, on the Royal Navy ship H.M.S. Beagle. The Beagle was sent on a surveying expedition along the coast of South America and in the Pacific, and Darwin came along as captain's companion and unofficial ship's naturalist, collecting specimens and making natural historical and geological observations. But at the time of his return to England, Darwin was not a proponent of the transmutation of species. In the following months, however, he convinced himself that the only way to explain his own observations, as well as those of others, was to imagine that existing species could give rise to new species. The so-called species notebooks were intended to consider possible ways in which that might happen—some kind of explanatory causal process to make it more plausible.

The major factor that converted him to the idea of transmutation was something that he had been in a privileged position to witness as a result of traveling round the world. This was geographical distribution, a topic that would cover two chapters in *Origin of Species*. Darwin had seen, in South America particularly, situations where similar but distinct species lived in areas that were close but had some kind of barrier between them. In Argentina he had noticed that on the pampas a particular species of South American ostrich, the rhea, occupied hundreds of miles of terrain. On the other side of a large river bordering this territory there was a different species of rhea, similar to but clearly distinct from the first.

Most famously, in the Galapagos islands, in the Pacific ocean, west of Ecuador, there were kinds of mockingbirds ("mocking-thrush") and finches very similar to ones on the mainland, but not the same. Darwin noticed that there were three clearly distinguishable kinds of "mocking-thrush," each living on different islands.[6] He was less clear on the finches, being not much of a taxonomist at that stage of his career; he assumed that they were just varieties of single species until an Oxford ornithologist, John Gould, examined Darwin's specimens back in England and convinced him that these variants should count as distinct species—classified, of course, on the basis of nonevolutionary taxonomy, which for birds in the same genus concentrated especially on beaks.

> A group of finches, of which Mr. Gould considers there are thirteen species; and these he has distributed into four new sub-genera. These birds are the most singular of any in the archipelago. They all agree in many points; namely, in a peculiar structure of their bill, short tails, general form, and in their plumage.... It is very remarkable that a nearly perfect gradation of structure in this one group can be traced in the form of the beak, from one exceeding in dimensions that of the

largest gros-beak [a finch characterized by a large beak], to another differing but little from that of a warbler.[7]

While in situ, Darwin had so little regarded such differences among the finches that he had failed to keep records of which of the various islands in the archipelago corresponded to which specimens.

Taken together, such considerations later persuaded Darwin, after his return to England, that species could change into new species—that the rheas, mockingbirds, and finches each had common ancestors that had produced divergent species due to geographical separation, whether by river or sea. Geological theorizing also came into play: when Darwin had set off on the Beagle voyage in 1831, he took with him the newly published first volume of Charles Lyell's *Principles of Geology*, and had the second of the eventual three volumes sent out to him the next year. Lyell's book, together with Darwin's observations over those years, converted him to Lyell's ideas, which were to be crucial in setting the terrestrial and temporal stage for organic evolution.

Charles Lyell's Systematic Geology

Lyell, a lawyer by training, was one of the leading lights of the Geological Society of London. Lyell's views on geology were what came to be called *uniformitarian*.[8] The basic feature of uniformitarianism was a steady-state view of earth history: Lyell thought that all geological changes and processes have always been of a similar kind and intensity to the ones we see in action today (a position more usefully labeled *actualism*).

From Lyell's point of view, it was a better theory than the contrarily named catastrophism of Cuvier and most other geologists of the time. Cuvier's view of earth history as having been punctuated by huge global cataclysms, or revolutions, that were responsible for massive extinctions had quickly become the dominant view. But Lyell preferred his theory because it utilized known processes to explain change, instead of unknown catastrophic ones. His main objection to catastrophism was that the intensity of the catastrophes had to be so great that the causal mechanism would be quite out of the ordinary; it certainly would have been of a completely different order from existing processes, as with the idea of rapid mountain formation.

So Lyell opted instead for a steady-state picture. Land gradually rose and fell over long periods, perhaps pushed up a little bit at a time by earthquakes, and sinking in a similar way; there were also effects of erosion and volcanic activity. Overall, these processes balanced out so that the broad features of the earth were always about the same. Continents might turn gradually into oceans, and vice versa, but there was no long-term qualitative development. That made a stark contrast to catastrophism, where the cumulative effect of the catastrophes meant that the earth today was qualitatively different from its state in previous epochs.

Lyell's commitment to actualism led him to believe that human activity would never leave long-lasting marks on the geological record. He insisted that, as far as any evidence could reveal, the processes that produce change in the fabric of the earth have always been the same as they are now. Like his fellow geologists, Lyell had no truck with a Biblical timescale for the earth's history, but where his peers allowed the earth's age to extend as far back as might be needed to accommodate geological change, Lyell in effect denied that *any* extension back in time to the earth's beginnings was relevant to geology, because he could not countenance a starting point for the earth at all; the earth's origins were, methodologically, inaccessible. This left the question of humanity: although not bound by a literalist account of human history as derived from Christian doctrine, Lyell accepted the special, and hence irregular, status of human beings in the earth's past. At the same time, he did not wish to allow human historicity to violate the orderliness of his steady-state picture of the earth.

Practically, Lyell wanted his geological cosmology to match the Newtonian cosmology of the solar system. Laplace, in his great work *Mécanique céleste* ("celestial mechanics," five volumes, 1798–1825), had demonstrated the stability of the solar system in the face of mutual gravitational interactions among its constituent parts, thereby rendering Newton's own imperfect system (intended to require periodic divine intervention) free of any tendencies to decay.[9] Lyell's geological system of the earth displayed a similar balance: secular tendencies canceled out one another, from an indiscernible past to an unbounded future. What, then, to do about humanity? Lyell had no interest in challenging Man's theological significance, but neither did he want to allow Man's presence to ruin his system. In the first volume of his great work *Principles of Geology*, which Darwin brought with him on the Beagle, Lyell addressed the problem directly:

> Not a single bone of a quadrumanous animal [animals with "four hands," like apes and monkeys] has ever yet been discovered in a fossil state, and their absence has appeared, to some geologists, to countenance the idea that the type of organization most nearly resembling the human came last in the order of creation, and was scarcely perhaps anterior to that of man.[10]

Lyell then noted the radically incomplete preservation of fossil remains in Tertiary formations, so as to make it clear that the absence of quadrumanous remains is not evidence that such animals did not then exist.

> It is, therefore, clear that there is no foundation in geological facts, for the popular theory of the successive development of the animal and vegetable world, from the simplest to the most perfect forms; and we shall now proceed to consider another question, whether the recent origin of man lends any support to the same doctrine, or how far the influence of man may be considered as such a deviation from the order of things previously established, as to weaken our confidence in the uniformity of the course of nature.[11]

Lyell proceeded by noting the unquestionably recent origin of Man, which "is not controverted by any geologist," and elaborating on the nature of the evidence from recent periods.[12] Above all, he wanted to argue that the appearance of Man nonetheless gives no credibility to a doctrine of progressive development of animal forms. Even if we grant that "the animal nature of man, even considered apart from the intellectual, is of higher dignity than that of any other species; still the introduction at a certain period of our race upon the earth, raises no presumption whatever that each former exertion of creative power was characterized by the successive development of *irrational* animals of higher orders."[13]

Lyell was concerned to show that this comparatively recent advent of Man did not violate the uniformity of nature because if it did, the idea that other such violations might have arisen in the past or could do so in the future became more credible. Accordingly, Lyell downplayed the possible impact of humanity on the processes that the earth underwent and on the expected evidence of human discontinuity that might result. He used the example of the British colonization of Sydney in Australia, describing it as a "revolution" that had introduced elaborate new arts as well as new plants and animals and began "rapidly to extirpate many of the indigenous species."[14] He claimed that there was nonetheless no impropriety in assuming that the system remained uniform under such circumstances, even though the recent changes there were much greater than those brought about in earlier ages by human agency. The physical effects of human intervention in nature were, he thought, much less than human intellectual capacity might suggest, since its exercise was always constrained by the properties of the things that we work upon. "Domestication and garden culture" could do much, for instance, but only by using the materials at hand: "assisting the development of certain instincts, or by availing ourselves of that mysterious law of their organization, by which individual peculiarities are transmissible from one generation to another."[15] Changes wrought by humanity were of a moral, not physical, nature.[16] For Lyell, the uniformity of nature expressed itself as a natural world that was uniform throughout accessible time. On that stage, all life and all terrestrial processes displayed unchanging activity, just like the celestial bodies in their endless circulation. This does not look like a promising venue for organic evolution, and yet it was as a Lyellian that Darwin would find his way.

Natural Selection

From Darwin's point of view, perhaps the crucial feature of Lyell's geology was that it implied that the earth was immensely old. There were no periodic geological and environmental catastrophes; instead, the world existed in a basically steady state. That allowed minor divergences like those between the mockingbirds to add up, over enormous periods of time, to yield widely different organisms. This was Darwin's genealogical solution to the problem of establishing a natural classification: the taxonomic boxes were generated over long periods of time by reproduction and descent with modification; common ancestors knitted the categories together. Now the problem was to determine how these transformations might actually occur.

There were various difficulties that Darwin knew he would have to take into account. By far the most important was the theme stressed by Cuvier and by British natural theology, namely adaptation. Random mutations would not in themselves produce functionally integrated organisms or account for the optical perfection of the eye. On the other hand, purposeful mutation implied divine interference, but Darwin wanted a purely natural process. He sought a scientific theory, not a mystery.

Darwin's goal was to find a natural way of producing changes that would be useful and advantageous to an organism. Inheritance of acquired characteristics, one of Lamarck's ideas, looked like a possibility— as with a blacksmith developing strong arms and passing that characteristic on to his children. Darwin always, to the end of his life, thought that such inheritance played some part in evolution, but the problem was that it was unable to explain very much. A Galapagos finch would not acquire a longer beak by trying to get at less accessible grubs in tree bark, so it could hardly pass that character on to its offspring.

Through the summer of 1838 Darwin kept reading and jotting down notes and observations in his notebooks. Because heredity was centrally implicated in this question, Darwin started reading up on domestic plant and animal breeding and how breeders selected particular features to breed from and accentuate. He was especially interested in animals like show pigeons, which comprise an enormous number of variants, all descended from common rock doves, but looking very different from each other.

The lesson he took away from his study of artificial selection was that small individual differences of the sort that distinguish one individual from another of the same species could be picked out, preserved, and accentuated over a number of generations. Darwin saw no reason why there should be any limit to that process—why differences could not continue to be selected to make one variety more and more different from another until they counted as different species. The trouble Darwin had experienced distinguishing the Galapagos finches and mockingbirds showed that the distinction between different varieties and different species was hazy.

These various ideas crystallized in Darwin's mind in September and October of 1838. He had been reading a well-known book by the Reverend Thomas Malthus, the sixth edition of *Essay on a Principle of Population*.[17] It had first appeared in 1798 and gone through an enormous number of subsequent editions; it was a popular book that Darwin thought it was about time he read, just for recreation. The impact of the book in early nineteenth-century Britain stemmed from its confrontation with one of the most pressing problems the country was facing throughout this period, a direct result of rapid social changes and displacement of population caused by the Industrial Revolution: the problem of the poor. Malthus provided an argument against the provision of state poor relief, maintaining that helping the poor only made matters worse.

Clearly this was a popular position in many circles in the period. Malthus's argument was pseudomathematical. He said that the human population tends to increase geometrically; the more people you have, the faster the population multiplies. Food supply, on the other hand, cannot realistically increase that fast. He gave this second point spurious exactitude by saying that at best food supply can increase only arithmetically, by adding each year to the amount of land under cultivation. Consequently, there will always be population pressure tending to increase the population beyond available resources.

Malthus argued on that basis that poor relief would only make things worse because it removes the incentive to avoid poverty. Poor people should be persuaded not to have too many children, and the best way of doing so is simply to allow this natural law to exert its full force. That sort of thinking had led to the establishment of workhouses for paupers in England in the 1830s, designed to ensure that people would try to avoid poverty at all costs.

Malthus's rather grim picture gave Darwin a crucial idea he had not previously considered. Malthus emphasized the so-called principle of superfecundity, which is the tendency of a population to multiply at a much greater rate than is required to offset the death rate. Darwin saw, as a consequence of that principle, competition for available resources between individuals of the same kind; this would apply not only to people, but to the individuals of any species.

That, at least, was how Darwin read Malthus, and although Malthus could not have given Darwin this idea if Darwin had not already been thinking about related problems and therefore been receptive to this aspect, nonetheless the book did trigger the central concept of natural selection, namely intraspecific competition.

Darwin's notebooks show that by November 1838 the mechanism of natural selection was fully formed in Darwin's mind. Superfecundity meant that there would continually be competition for resources between individuals of the same species, and that a large proportion of any given generation would therefore never survive to reproduce. Darwin also knew, particularly from his study of breeding domestic animals, that there were always small, random differences between individuals that acted as raw material for plant or animal breeders.

That meant that in the competition between individuals of the same species in the wild, an individual with a slight advantage would be just a little bit more likely to survive and pass its characteristics on to its offspring. Conversely, an individual with a random slight peculiarity that acted as a disadvantage would be less likely to survive and reproduce.

Here, finally, was the natural counterpart to artificial selection. Intense competition within a species would mean that apparently insignificant individual differences would, over the course of many generations, become cumulative. Favorable variations would be preserved, as would the ways of life that made them favorable, and unfavorable ones would be eliminated.

10

Evolution and Scientific Naturalism

Darwin's Diagram

This diagram from the *Origin of Species* incorporates several of the most important features of Darwin's reinterpretation of natural history in the middle of the nineteenth century. It can be understood in several ways. Most immediately, we have a set of family trees, or lines of descent, where divergence between lines ascending from bottom to top indicates increasing morphological divergence between organisms—the formation of new species, and in time even new genera, new families, and so on. Thus, as well as being a diagram of descent, the diagram also maps onto a taxonomic structure.

Before Darwin, the standard way of viewing a taxonomic scheme in natural history was as a static, timeless structure. Organisms were slotted into it according to various morphological criteria, typically justified in zoology by Cuvier's functional interpretations (the subordination of characters rule). Even when Cuvier had started applying his taxonomic scheme to *extinct* fauna, the scheme itself remained a timeless structure that applied to any period.

Darwin's taxonomic scheme, by contrast, is created by development over time: without the vertical dimension of time, there would be no taxonomic scheme. This means that Darwin's is a historical taxonomy. It is not justified by any theoretical conceptions of the nature of the

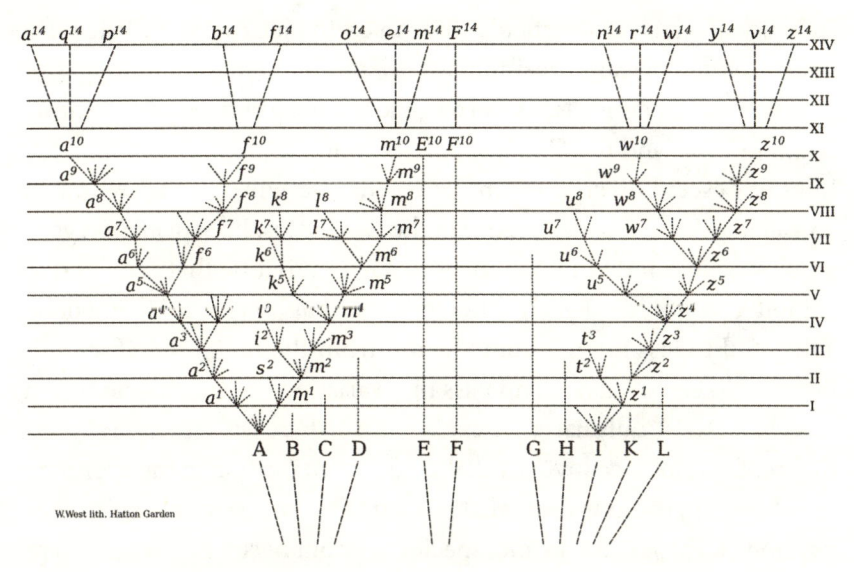

FIGURE 10.1. Foldout diagram of organic descent from the first edition of Darwin's *Origin of Species* (1859).

animal organism, or laws of nature, as Cuvier's was; it is justified on the basis that this is how existing species developed by transmutation and selection. Neither is there anything inevitable about any particular line of development; taxonomic groupings simply reflect how things happened to proceed throughout earth's history. One consequence is that Darwin's developing evolutionary taxonomy is not inherently progressive, in contrast to Lamarck's evolutionary ideas, among others. There is no upward development toward some predetermined end and no basis for saying that later organisms are qualitatively "higher" on an evolutionary ladder than earlier ones. In principle, there is no evolutionary ladder in Darwinian evolution, just change and diversification. In practice, Darwin always gave a special status to humanity's moral and intellectual qualities as the summit of evolutionary change, but this is not reflected in the diagram.[1]

The diagram is not simply an abstract graph of organic life against time; it also represents the most direct evidence of the history of life, namely the fossil record. Each of Darwin's horizontal lines represents the state of affairs at a particular time in earth's history as recorded in a particular geological deposition.

One crucial new idea that Darwin had come up with between writing the 1844 essay and the *Origin of Species* concerns the way in which the lines of descent diverge as time proceeds. Darwin came up with this so-called principle of divergence in the early 1850s. Its explanation goes as follows: if an individual happens to have a variation that enables it to exploit a way of life slightly different from that of its fellows, it will be subjected to less intense competition from them and therefore will be more likely to survive and pass that variation on. Consequently (in this characteristically narrative explanation), there will emerge a slightly different variant, occupying a slightly different "station in the economy of nature."[2] In the long term, that pressure for divergence to avoid competition can produce a distinct species, without necessitating the disappearance of the parent species. This was also a more generally applicable mechanism for separating emerging varieties than the geographical isolation of breeding populations that Darwin had earlier postulated for the rheas and Galapagos finches, where the groups were physically separated from each other. This version establishes social isolation instead.

A crucial early objection that Darwin had to face was why, if his transmutation idea were true, the fossil record does not seem to display gradual, continuous change but instead shows a succession of related but distinct species in strata separated by discontinuities that Cuvier had identified with global catastrophes. Darwin needed to come up with explanations about the causes of these apparent discontinuities, which he had also addressed to support his elimination of periodic geological catastrophes.

In his discussion of the fossil record in *Origin of Species*, Darwin was quite apologetic about the whole thing, which was appropriate given that it was one of the strongest objections to his theory. His basic solution was the same as Lyell's and focuses on saying that the fossil record is incomplete because sedimentary rocks are not laid down continuously; successive strata actually have enormous periods of time separating their respective depositions. In particular, Darwin used Lyell's picture of earth history to make it even more plausible that the individual strata that actually were formed would preserve little evidence of transmutation anyway.

For Lyell, the land in any particular region of the globe is either rising slowly or sinking slowly, and over the course of geological time any given region will rise and fall periodically. Darwin used this picture, which was an integral part of his own view of geology, to suggest that perhaps variation, and therefore speciation, occurs much faster when the land is rising. As that happens, more new habitats are formed, giving room for new ways of life and hence new species to use those ways of life. So if, as Darwin suggests, variation and speciation tend to occur chiefly when land is rising, that evolutionary process will not be visible in the fossil record because deposition does not occur when land rises. If anything, land will erode as it is pushed up; no rock will be formed to contain the fossils bearing witness to transmutation.

Darwin's horizontal lines on the diagram in the *Origin of Species* therefore also represent successive geological strata that are widely separated in time and therefore contain fossils that represent episodes in the development of life, not continuous records. In addition, these episodes have occurred at times when evolution is slow.

1840s, 1850s

Darwin's *Origin of Species* broadly resembles his essay of 1844, such that one can say that many of his basic ideas had not changed radically during the intervening period. The principal difference, pointed out fairly recently by Derek Partridge, is Darwin's focus in the essay on environmental alterations as the main triggers of transmutation through natural selection (so as to produce new forms suited to those new conditions), as contrasted with the much broader constant action of natural selection in adjusting the mutual relations of organisms at all times. Partridge's argument seems to account for occasional ambiguities in the *Origin of Species* relating to so-called perfect adaptation, which now appear as holdovers from the episodic adaptations highlighted in the essay.[3] Nonetheless, the many continuities between the two versions draw attention to the fact that the British natural history community was much readier to accept Darwin's ideas in 1859 than it would have been when he began to develop them in the late '30s and early '40s.[4]

For example, in the 1830s it was not established to everyone's satisfaction that the fossil record definitely showed a progression of forms (or increasing diversity) from the oldest to the youngest rocks, with whole new classes, like mammals, appearing at particular times where they had previously been absent.

Lyell denied progressionism because his steady-state view of earth history looked neater methodologically if the kinds of organisms in the world, although not necessarily the same species, had always been the same—if there had always been mammals, birds, and so on. He said that the apparent progressiveness in the fossil record must be an artifact of the differential preservation of fossils, again stressing the radical incompleteness of the record. But by the 1850s, progressionism was pretty well entrenched, in large part because of the enormous amount of paleontological work that had been done in the meantime—work that would have seemed severely problematic if progressionism were false.

There had also been a gradual shift in emphasis in work on anatomy. By the '50s, although Cuvier's strictly functional approach was still dominant in Britain, and function and adaptation to way of life were still the main things to be addressed in explaining the structure of an animal, there was also an increasing emphasis on form itself, that is, morphology.

Anatomy meant above all, after Cuvier's stress on it, *comparative* anatomy. But comparing animals showed not just similar functions and hierarchies of functions associated with similar anatomical structures. There were also purely morphological similarities—structural similarities that did not correspond to functional equivalences.

One of the prize examples at the time was the identical structure of bones found in the wing of a bat, a human hand, and the fin of a whale. This could not be explained on purely functional grounds, because the functions are quite different in each case. The same basic plan has been modified to fulfill diverse uses.

These structural similarities divorced from function were called *homologies*. They were investigated particularly by the foremost British anatomist of the period, Richard Owen.[5] (Owen produced the reports on the fossils Darwin had sent back from South America on the *Beagle* voyage in the '30s). In the 1840s Owen had come up with an idea to explain these basic structures: his theory of archetypes. This applied to the skeletal structures of all vertebrates, from fish to people. Owen saw them all as variations on a vertebrate archetype, which was the basic plan on which all the others were variants or modifications. The archetype was essentially a spine made up of the appropriate number of vertebrae and with the appropriate attached bones—a skeleton looking rather like a fish.[6]

Owen's vertebrate archetype was designed so that all vertebrate skeletons could be seen as modifications of the archetype, with particular bones lengthened, shortened, or fused to yield the skeleton of any animal. The justification of the idea was that the skeleton of any vertebrate could be derived from the archetype, bone for bone. It worked pretty well, although Owen had trouble with skulls; he tried to argue that they are made up of fused vertebral elements, but not everyone found his argument convincing. For Owen, the archetype was God's ground plan for vertebrates; there need never have been an animal with the structure of the archetype. In that sense it was far from an evolutionary idea.

In Britain the idea of archetypes was regarded as rather too metaphysical, but Owen had focused on a phenomenon in need of explanation. When Darwin's theory appeared, in the wake of Owen's, the idea of descent with modification made sense of anatomical phenomena, namely homologies, that were at the fore in natural history at the time. That had not been the case in the '30s or much of the '40s. After the *Origin of Species* was published, Owen immediately became one of Darwin's fiercest critics. Darwin's theory got rid of the need for archetypes, although at the same time Darwin could use Owen's work to argue in favor of his own idea. Worse, Darwin scarcely acknowledged Owen in the *Origin of Species*.

In the debate that followed its publication, Darwin's greatest champions were the botanist Joseph Hooker, who had read the essay version of Darwin's theory in 1844, and the zoologist Thomas Henry Huxley. The fact that Hooker now accepted Darwin's ideas enthusiastically, whereas he had not been convinced in 1844, is another indication that the time was now ripe for evolution in a way that had not been the case earlier. Wallace, having come up with a version of natural selection independently in 1858, was enthusiastically in favor of it, too, and gave Darwin full credit.

In the first few years, though, it was the more established figures of Huxley and Hooker who did most of the work of defending Darwin against critics, with Darwin himself on the sidelines at his country house in Kent. Part of Huxley's enthusiasm for Darwinism was his general interest in rocking the boat and presenting himself as a force of progress against the conservative establishment.

In the '50s Huxley had become one of the leading comparative anatomists in Britain. But he always had something of a chip on his shoulder about being of fairly humble social origins compared to other members of the natural history community, like Lyell or Darwin. There was also Huxley's personal antagonism and rivalry as a zoologist toward Owen; probably part of Owen's intransigence about evolution came from his dislike of Huxley. But by the mid-sixties there were few notable naturalists in Britain besides Owen who did not subscribe to evolution.

The Plausibility of Natural Selection

Soon the issue became not the reality of evolution, but the plausibility of natural selection as its principal engine. By the late 1850s, naturalists and others were ready for evolution, because it provided a way of making sense of a number of problems in natural history: the progression of forms found in the fossil record, homologies, and geographical distribution.

The explanation of geographical distribution was especially convincing to working naturalists, and it was what brought Lyell around to an acceptance of evolution in the mid-sixties. It was too much of a coincidence, if transmutation was ruled out, that taxonomically similar species would routinely be found in neighboring regions, whereas on a different continent, or on the other side of a mountain range, a very different group of species would be found occupying similar habitats and ways of life (think of marsupials in Australia, for instance). And there were plenty of more complex features of geographical distribution that Darwin, Wallace, Hooker, and others pointed to that they thought only transmutation could explain.

In effect, what Darwin had done was not to introduce a theory of evolution that naturalists found plausible—a lot of people did not find natural selection plausible as an evolutionary mechanism—but to make evolution scientifically respectable. Darwin provided a natural, causal process to explain evolution, which made the whole thing look much less mysterious. Even when people regarded natural selection as inadequate, the fact that Darwin showed that evolution could be approached with some degree of (even contestable) plausibility through a natural explanation lifted evolution out of the realm of crank ideas associated with books like the popular *Vestiges of the Natural History of Creation* that had created a sensation in the '40s.[7] It was possible to accept evolution and leave the process as something to be worked on, and Darwin opened that route. Once people are arguing primarily about the mechanism of evolution, evolution itself has been taken as a given.

Even those who accepted natural selection as the primary evolutionary driver had their doubts about Darwin's version of it—including Huxley. Darwin was convinced that evolution was a slow, long-term process, with natural selection operating on very slight variations between individuals. Huxley, and many others, thought that these kinds of differences were inadequate for the job. The problem resulted from the usual understanding of heredity.

For Darwin, random variations among individual organisms of the same species were inheritable, but he assumed, like everyone else at the time (except for Gregor Mendel in Moravia, who received no attention), that inheritance worked by the blending of characteristics. For any particular feature, the characteristics of the parents would be blended in the offspring—averaged out, rather than one or the other predominating—which meant that any abnormality in one parent would be diluted in the offspring. So an objection to Darwin's idea was that any slight favorable variations would quickly become swamped in a population and effectively disappear.[8]

The only solution that Darwin could think of was to imagine a population with such a high frequency of random variation that there was a significant probability of the same random variation occurring in many individuals at the same time, and those individuals could mate. Even that seemed inadequate, however, because if the variations, as well as being random, were supposed to be slight, then the extra survival value of the favorable ones would be marginal, in which case it might still seem unlikely that a swamping effect would not occur. Certainly, the conditions that would be required to prevent it looked contrived and artificial.

There were other problems, too, also to do with the sufficiency of small, individual variations as the raw material of natural selection. In the *Origin of Species*, Darwin used the analogy between artificial selection (in domestic plant and animal breeding) and natural selection as the central argument to establish the plausibility of his evolutionary mechanism. In effect, he said that the work of animal breeders can produce very different varieties, like of pigeons, which proves the power of selection as a mechanism for change; he then identified an analogous natural selective process. It was a good argumentative strategy, but Darwin still immediately found himself criticized on the same grounds as had previously been used to show that artificial breeding demonstrated the essential stability of species.

The argument was that domestic breeders had never been able to turn one species into another, which indicated that there was a practical limit to how much artificial selection could do. Particularly extreme varieties of pigeons, for instance, once obtained, tended to resist further accentuation of their peculiarities. Variability was not open-ended, according to this view; its effects seemed to cluster around a mean.

Huxley therefore used artificial breeding as further empirical grounds for doubting the sufficiency of small variations to promote evolution. Wallace took a different approach. He shared Darwin's belief that small variations in individuals are sufficient. But he could see no basis for questioning the apparent evidence of the limits of artificial selection, so he simply denied that it was a valid analogy. He said that artificial breeding focuses on abnormalities rather than ordinary variations, and therefore produces unstable freaks rather than proper varieties. Unlike Darwin, he believed that forms might revert to type in the wild.

Darwin's way around the problem was to refuse to be persuaded. He said that knowledge of the causes of variability was insufficient to make any firm judgement, and he stuck to the analogy. But it was a genuine difficulty for him.

Although Huxley had his doubts about the sufficiency of small variations, he still believed in evolution and natural selection, which meant that he had to rely on a different kind of variation. He argued that the principal raw material for natural selection consisted of so-called saltations. These were large, spontaneous jumps from one form to another, rather than small, gradual, cumulative changes. An individual of one species might spontaneously give birth to an individual sufficiently different to form the basis of a new species. The point of postulating jumps of that kind was to get sufficiently large variations to avoid the swamping effect that small variations would be subject to.

Huxley thought that there was good empirical evidence for saltations: he pointed to examples of notable, apparently spontaneous mutations, which also had the advantage of appearing to be stable, remaining undiluted in successive generations. Huxley's favorite case was of a human family recorded as having a long history of producing children with six fingers on each hand. He took this to show the possibility of spontaneous, fairly large jumps and the transgenerational tenacity of the forms produced. Another advantage in postulating evolution through jumps rather than by a gradual process was that it made the discontinuities found in the fossil record less of a problem for evolution.

Of course, occasional large jumps were not likely to produce organisms very highly adapted in all their characteristics to a particular way of life. Huxley's view of the matter necessarily placed less emphasis on adaptation than did Darwin's. Darwin's idea of constantly selected small variations presupposed a very high degree of adaptedness of organisms to their ways of life as part of the phenomena to be explained. The necessity of a high degree of adaptation, due to the "struggle for life," was one of the main dynamics driving natural selection and evolution for Darwin.

Huxley, on the other hand, laid much less stress on adaptation in any of his work. In his anatomy and paleontology, he was always much more concerned with form (morphology) than function; in other words, he had moved quite a long way from Cuvier's approach. In the '50s, before the *Origin of Species* appeared, Huxley caused a small scandal in zoological circles by claiming that anatomists were less Cuvierian than they pretended: for all their talk about function, they were really doing morphology. Huxley didn't think that species are inherently well adapted; instead, they make use of what they have.

Altogether, Huxley, while a ferocious defender of Darwin, was in some ways not much of a Darwinian. Or, perhaps, Darwin was more Cuvierian than Huxley. In any case, Huxley cared more about evolution than natural selection. Evolution made sense of morphological correspondences in the fossil record and homologies, the things with which Huxley, as an anatomist, was most concerned.

Perhaps the most serious challenge to Darwin's version of evolution came not from geology or biology, but from physics. That was a serious matter, because in the nineteenth century, physics became the most prestigious scientific discipline.[9] The physicist who raised it was William Thomson, later Lord Kelvin, after whom the absolute temperature scale is named. When he spoke, people tended to listen.

Thomson's contribution to the debate over evolution seems to have been a deliberate attempt to throw a spanner in the works. In the 1860s he produced a calculation of the age of the earth. This was based on assumptions about the rate of cooling of the earth from an initial incandescent state (though the operative thing here is the maximum temperature at which you could have any form of life). Given the earth's present temperature at the surface, Thomson got a rough order-of-magnitude figure for the age of the earth, putting it between twenty million and four hundred million years (the wide range had to do with ignorance of the internal composition of the earth).

Although a long time, this was far from what Darwin needed. It mattered little to Huxley, because his evolutionary jumps, or saltations, happened all at once, and might have happened comparatively frequently. But for Darwin's view of evolution as an extremely slow, gradual process it was something of an embarrassment. Correlating the absolute age with geological strata seemed to call for much too rapid a rate of evolution (and of deposition).

But Darwin was convinced by the evidence for his theory, especially in dealing with geographical distribution. He also believed in the explanatory power of Lyellian uniformitarian geology, which also needed a colossally old earth. These considerations were too convincing to warrant deferring to the calculations of a physicist. Kelvin might know about physics, but Darwin knew about geology and natural history. All Darwin could do was hope that Kelvin was wrong.[10]

None of this means that there were no real Darwinians—people who accepted not just evolution but also natural selection acting on small individual variations. There were some, and even though they were not in the majority, they were important. Apart from Darwin himself, there were, most notably, Wallace, Joseph Hooker, and the entomologist Henry Bates. All did work on geographical distribution, particularly variations in species of insects. They interpreted their observations in terms of varying degrees of struggle for existence, and hence differing selective pressures, on the same species in different regions.

Bates did some particularly striking work on protective coloration in insects. A species of butterfly that was toothsome to certain predatory birds could increase its chances of survival, Bates argued, by mimicking the coloration and markings of a different, inedible insect. The varying degrees of protective coloration among individuals of this species depended, he showed, on their geographical location, which indicated to Bates the presence of different degrees of selective pressure—in this case, birds—in the different areas. A virtue of this case was that, because the variation in protective coloration from region to region was gradual, it was clear that it was not a character that appeared all at once, as would have been the case for one of Huxley's saltations. After Bates published his findings, more examples were found with different insects elsewhere. This was among the best pieces of evidence for Darwin's kind of evolution; Darwin himself had a hand in publicizing Bates's work.[11]

God and Man

The impact of Darwin's ideas extended far beyond natural history or biology. It also made its mark on religion, partly because Darwin set as his main opponent in the *Origin of Species* the doctrine of "special creation," whereby each species had been made directly by God. Darwin clearly scorned religion's inability to explain. He liked to point out that, for many facts in natural history for which his theory provided an explanation, special creation could do nothing except accept the fact as a given, with no explanation at all: it was just the way God had chosen to do things.[12] But religious institutions also cared about Darwin's implication that people were a product of evolution as much as all other organisms; resistance to such a conclusion motivated much of the opposition to Darwin. In 1871 Darwin published *Descent of Man*, where he suggested ways in which human behavioral traits and moral feelings could have had survival value, so that natural selection could have produced human social characteristics. He dealt with the development of language and social behavior in a similar way, along lines called in the twentieth century *sociobiology*.

Darwin's approach to human evolution in *Descent of Man* followed a resolution of the controversies of the 1860s over both the arguments of the *Origin of Species* and its widely (and accurately) drawn implications of a doctrine of human evolution, which had added piquancy to evolutionary debates. By the time *Descent of Man* appeared, its approach was already anticipated, and Darwin's voice in the book quite assured. In the absence of fossil evidence bearing on human ancestors, Darwin resorted to what became known as the "comparative method" in order to trace, conjecturally, the development of human civilization from its earliest forms.[13]

The comparative method had been used by Darwinian anthropologists in the 1860s to infer the origins of various human social institutions from the present-day practices of human groups deemed "primitive," a designation already presupposing a lineal connection to modern "civilized" behaviors. By treating non-European human groups as if they were snapshots of some past era of developing European civilization (taken as the natural developmental terminus), Darwin and the Victorian anthropologists believed that they could reconstruct the evolutionary development of human society from its early beginnings (represented for Darwin by the indigenous inhabitants of Tierra del Fuego, witnessed on the Beagle voyage), to the pinnacle of Victorian achievement (unremarked and taken for granted). Human social evolution was spontaneous and, in the long run, inevitable; less developed societies might have stalled in their ascent of the ladder of civilization, but the ladder itself was clearly marked.

A related aspect of human evolution that Darwin felt obliged to address was that of race. Controversy among naturalists over the proper taxonomizing of human racial categories dated back centuries to biblical arguments over the sons of Noah and their descendants, an idiom that continued to dominate at the beginning of the nineteenth century. That approach was not exclusive, however, and Buffon had attempted to integrate his understanding of human diversity with his approach to natural history, concluding that human beings were all related and that the so-called races were not fundamentally different, all being descended from common ancestors. Cuvier's addressing of the question normalized it to his usual nontransformist approach, dividing humans into three fixed races. Darwin, in *Descent of Man*, took for granted the emergence over time of human racial groups from an original ancestral population (an idea by then known as *monogenesis*, contrasted with *polygenesis* for multiple origins of human races), but he fully accepted the taxonomic legitimacy of human racial divisions, which he argued probably ought to count as distinct subspecies on conventional taxonomic grounds, but conceded that the conventional designation "race" was likely too well-entrenched to be changed.[14]

The bulk of *Descent of Man* is concerned, however, not with human evolution but with sexual selection. Darwin used sexual selection to explain phenomena that he found inexplicable by natural selection. Natural selection could not explain the elaborate decorations found on male birds of many species, or the ostentatious developments found on others; what possible functional advantage could the display of a peacock bestow? Natural selection could account for adaptations that provided survival advantages, but not apparently for the bright colorations or elaborate antlers that seemed positively disadvantageous in the struggle for life. Darwin, noting the prevalence of such features in the males of certain species but not in the corresponding females, suggested that such sexual dimorphism was a result of female choice: the females of certain species preferred, aesthetically, particular features in the males with which they chose to mate. Alternatively, the males competed with one another to secure the most females, typically through combat, which might be advantaged by adaptive features such as larger antlers, although these too could have functions of an ornamental kind. Either way, some males succeeded in reproducing at the expense of others and passed on their successful characteristics.[15]

Darwin thought the ornamentation in males preferred by females of the species (often birds, and often taking the form of bright or elaborate coloration) amounted to a purely aesthetic preference not reducible to some functional adaptation to enhance the probability of survival. Female peahens preferred peacocks with beautiful markings and chose the best for their mates; hence those markings were selected for. Darwin found it only natural that, often, human notions of beauty corresponded to those of animals; we are all related.[16]

Sexual selection of this kind was usually driven by female, not male, choice. But there was a glaring exception: among human beings (and to some degree monkeys and apes), it was generally the males who chose their mates. *Descent of Man* enumerates many features of women that supposedly resulted from such choices by men in different societies. Besides sexual dimorphism (found across the animal kingdom), a consequence of sexual selection among human beings was what Darwin took to be marked differences between racial groups. Sexual selection was an important supplement to natural selection for Darwin, filling diverse explanatory gaps along with other auxiliary elements such as Lamarckian use and disuse.

Scientific Naturalism

After 1859 and the appearance of the *Origin of Species*, Darwin's arguments were promoted vigorously by various scientific luminaries in Britain—T. H. Huxley first among them—who promoted what is called "scientific naturalism."[17] They called themselves the X Club, and saw themselves as a kind of scientific priesthood, promoting truth in service of the state, much as the Anglican church had served as a prop and arm of the state. The group's members, including Huxley, John Tyndall, Joseph Dalton Hooker, the Darwinian anthropologist John Lubbock, and the social thinker Herbert Spencer, among others, opposed orthodox religion (Huxley coined the term *agnostic*), and advocated materialist explanations of the natural and human worlds. Evolution was central to their shared cosmology; they vigorously promoted both it and Darwin against theistic worldviews and religious dogma.[18]

All this was of a piece with Darwin's approach in the *Origin of Species*. He frequently drew attention to natural historical observations (such as details of geographical distribution or the structure of classificatory groupings) that his theory is capable of explaining, but about which a doctrine of special creation would have nothing whatever to say.[19] Thus, concerning classification of species, he wrote, "This grand fact of the grouping of all organic beings seems to me completely inexplicable on the theory of creation."[20] Setting up "creation" as the chief alternative to his own theory enabled Darwin to reduce alternative explanations to a single category, one that he found easy to eliminate, and thereby leave natural selection and descent with modification as the only reasonable account of organic diversity.

Although he soon found this argumentative strategy increasingly unnecessary, Darwin was still using it in 1863 when promoting Henry Bates's recent work on selective pressure as the cause of protective markings in South American butterflies.

> Some of the mimicked forms can be shown to be merely varieties; but the greater number must be ranked as distinct species. Hence the creationist will have to admit that some of these forms have become imitators, by means of the laws of variation, whilst others he must look at as separately created under their present guise; he will further have to admit that some have been created in imitation of forms not themselves created as we now see them, but due to the laws of variation![21]

But Darwin had a further argument to make on the point, one that relied on a rhetorical strategy that implicitly deprecated creationist assumptions in a style suited to the X Club:

> Prof. Agassiz, indeed, would think nothing of this difficulty; for he believes that not only each species and each variety, but that groups of individuals, though identically the same, when inhabiting distinct countries, have been all separately created in due proportional numbers to the wants of each land. Not many naturalists will be content thus to believe that varieties and individuals have been turned out all ready made, almost as a manufacturer turns out toys according to the temporary demand of the market.[22]

Darwin's analogy between organisms and manufactured items, proposed so as to ridicule the notion, evidently assumes that no serious person would imagine that the natural world operates according to the laws of commerce. This might seem surprising in someone who found Malthus's arguments regarding population dynamics in political economy fitting models for explaining the struggle for life at the heart of natural selection. But this perhaps is less remarkable in light of the implied parallel between the toy maker and the creator of organic beings. Darwin's point presumes that a God who responds to market forces is self-evidently ridiculous and not to be countenanced. Only his own naturalistic explanation of imitative variants was plausible.

By no means did all high Victorian scientists adopt such a position. The great physicist James Clerk Maxwell was a Christian who argued against determinism and for the doctrine of human free will.[23] He also took a very different view from Darwin's concerning the analogy between classes of individuals found in nature and the work of manufacturers. In his address to the British Association for the Advancement of Science in 1873, on molecules, published the same year in Huxley's journal *Nature*, Maxwell remarked on the identical properties of molecules of the same kind, with the characteristic features of their emission and absorption spectra.[24] His understanding of their significance departed radically, however, from the X Club's worldview.

No theory of evolution can be formed to account for the similarity of molecules, for evolution necessarily implies continuous change, and the molecule is incapable of growth or decay, of generation or destruction.

None of the processes of Nature, since the time when Nature began, have produced the slightest difference in the properties of any molecule. We are therefore unable to ascribe either the existence of the molecules or the identity of their properties to the operation of any of the causes which we call natural.[25]

The implication was clear: the unchangeable similarity of molecules bore witness to their supernatural origin. Isaac Newton would have approved.

But since this was the Britain of the nineteenth century and not the early eighteenth, Maxwell, like Darwin, invoked another familiar analogy: "The exact equality of each molecule to all others of the same kind gives it, as Sir John Herschel has well said, the essential character of a manufactured article, and precludes the idea of its being eternal and self existent."[26] For Darwin, a conception of a South American butterfly as a "manufactured article" had been an obvious absurdity. But for Maxwell, molecules were self-evidently like manufactured articles, and "because matter cannot be eternal and self-existent it must have been created."[27]

Other physicists in the Victorian era and after were, like Maxwell, opposed to materialism, and some gave credence to spiritualism and psychical research.[28] The members of the X Club and their like-minded colleagues, by contrast, roundly rejected attempts by some naturalists and members of the wider public at saving the world as God's creation by weakening the autonomy of natural selection and ascribing a directive role to God in its operations, as some opponents of materialism tried to do. Materialism was still a delicate issue, however. Huxley had little sympathy for doctrines of spontaneous generation as the origin of life, despite its apparent consistency with the materialism of Darwinian evolution. Unbroken lines of descent were the only acceptable elements for the promulgation of life, which left life's ultimate origins shrouded in obscurity.[29]

Perhaps the most uncompromising, and controversial, expression of the materialist wing of Victorian scientific ideology was John Tyndall's presidential address at the 1874 meeting of the BAAS, held that year in the Irish city of Belfast. Tyndall, a practiced and accomplished public lecturer who headed the Royal Institution in London, took the occasion as an opportunity to express an uncompromising scientific worldview whose perceived atheistic materialism alarmed some auditors, as Tyndall himself acknowledged in the version of the address published separately from the BAAS proceedings.[30] The address expresses an uncompromising version of scientific naturalism, tracing, in partly historical form, an account of the world developing from atomistic materialism in the physical realm to the development of life and, ultimately, thought. Tyndall applied the naturalists' principle of the continuity of nature, which Darwin had used in tracing the gradual evolutionary appearance of characters in organisms, to the appearance of life from nonlife, while carefully avoiding contradicting Huxley on the issue of whether living things must always have derived from other living things.[31] Tyndall turned the matter into a moral question of the methodological basis of scientific belief:

> If you ask me whether there exists the least evidence to prove that any form of life can be developed out of matter, without demonstrable antecedent life, my reply is that evidence considered perfectly conclusive by many has been adduced; and that were some of us who have pondered this question to follow a very common example, and accept testimony because it falls in with our belief, we also should eagerly close with the evidence referred to. But there is in the true man of science a wish stronger than the wish to have his beliefs upheld; namely, the wish to have them true.[32]

Believers in the doctrine that life can arise from inanimate matter, he goes on, "can justify scientifically their belief in its [i.e., matter's] potency, under the proper conditions, to produce organisms. But . . . they will frankly admit their inability to point to any satisfactory experimental proof that life can be developed save from demonstrable antecedent life.[33] None of this, and much more, would preserve Tyndall from religious condemnation, but it mapped out the route by which such issues would subsequently be carefully advanced: if in doubt, acknowledge doubt. That way, scientific naturalism did not need to yield ground on its fundamental assumptions.

11

Thermodynamics and Modern Physics

DARWIN'S WAS an attempt to bring order to a diverse group of phenomena in natural history, including fossils, geographical distribution, and anatomical homologies among different animals. Getting such a diverse group of problems to find solutions from a common set of ideas, a common hypothesis, was an extraordinary achievement; its sheer possibility was itself evidence of the truth of that hypothesis. Victorian philosophers had a name for such an explanatory confluence: *consilience*.[1] In important ways, the emergence of a recognizably modern discipline of physics in the nineteenth century resulted from a similar confluence, not so much in the postulated content of the explanatory hypotheses as in the techniques and skills that were brought to bear on the individual subjects of investigation. Thermodynamics was the central field in the incorporation of many of the elements of the new physics.

Carnot's Study of Heat Engines

In the early nineteenth century there existed a group of technical subjects usually referred to as the "physical sciences," including mechanics, optics, and electricity. They had a sort of family resemblance and were often investigated by the same people. But these different areas tended to be conceptually autonomous, with their own theories and concepts—optics had the law of refraction, for instance; mechanics had inertia;

electricity had charge. The ideal of uniting them, to show that they could all be dealt with using the same fundamental concepts, certainly existed; the problem was identifying such foundational concepts, whether ontological (such as a theory of matter) or methodological (rules of inference for hypotheses, for example). Not being self-evident, classifications of similarity and difference among physical phenomena had to be established, not simply observed.

The most important legacy of Laplace and his circle in early nineteenth-century France had been the idea of developing a unified treatment of physical phenomena using precise quantitative experimental techniques. These two things—unification and quantitative experimentation—played a central role in the emergence of a coherent discipline of physics by the middle of the century. But the crucial concept involved in bringing that about was *energy*.

Eighteenth-century rational mechanics, the mathematical treatment of motion and collision, had a concept of vis viva, or live force, measured as mv^2: mass multiplied by the square of the body's velocity.[2] But this was not really a version of the nineteenth century's concept of energy. Although live force was conserved in perfectly elastic collisions, it was not conserved in inelastic collisions (imagine two lumps of clay thumping together), and energy was characterized by being conserved through different manifestations (at least until Albert Einstein, who would make it convertible with mass).

The ideas leading to the nineteenth-century notion of energy came less from continuous development in this more or less Newtonian tradition of theoretical mechanics than from an engineering tradition.[3] The French engineer Sadi Carnot's work played a crucial role in the invention of thermodynamics. It was inspired and directed by an interest in steam engines, developed by the Englishman James Watt in the late eighteenth century, and in the production of mechanical *work* from *heat*.[4]

Carnot had been educated in the physical sciences and engineering at the École polytechnique during the 1810s; accordingly, his approach combined engineering interests with contemporary French, broadly Laplacian ideas about physical sciences. In particular, Carnot thought of heat as a fluid substance—caloric—as Lavoisier and Laplace had done.

The essence of Carnot's ideas about steam engines was that they were machines driven by the flow of heat, or caloric, from a hotter to a colder body. The engineering aspect of his approach involved applying that concept in the way engineers (like his own father, the engineer and politician Lazare Carnot) had done in analyzing machines driven by water and wind.[5] Sadi Carnot described the flow of caloric in a steam engine as a "fall" of caloric, and his analysis rests on conceiving of this fall of caloric as analogous to descending water. Just as falling water can be used to drive a wheel, so falling caloric could be used to drive a piston.

Carnot presented his argument in a short book titled, in English, *Reflections on the Motive Power of Fire*, in 1824.[6] Because he modeled his examination of steam engines on engineers' analyses of mechanical systems for moving weights, he immediately abstracted most of the actual features of a real steam engine out of the picture. Engineers often examined the relationship of a power source (wind or flowing water, for example) to work done in mechanical devices, independent of the mechanical linkages—for example, seeing how much weight could be raised how far by some applied force. Similarly, Carnot sought to use an analysis of input and output, regardless of the details of interlinkages, in his approach to heat engines.

Carnot considered an idealized heat engine rather than a real steam engine, assuming that the use of steam rather than any other material for transferring heat made no essential difference. Heat, or caloric, would pass through the engine and do mechanical work (which engineers measured as weight multiplied by the vertical distance raised against gravity). Carnot's basic conceptual tool was the idea of a reversible cycle.

Figure 11.1 represents a frictionless piston and cylinder with a body of gas confined below the head of the piston. This gas is imagined as being governed by the laws of a so-called ideal gas, that is, a gas obeying the relationships between pressure, volume, and temperature expressed by Boyle's and Charles's laws.[7] A and B are bodies maintained at a constant temperature, with A being hotter than B. The gas trapped in the cylinder can be expanded or compressed by moving the piston. Heat can be allowed to flow between the gas and bodies A and B by bringing

| 1–2 | 2–3 | 3–4 | 4–1 |

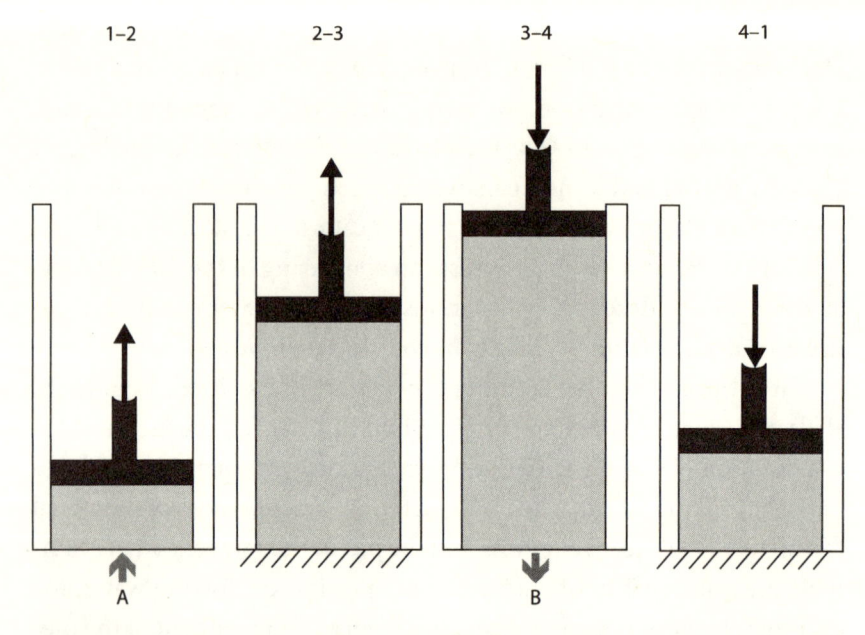

FIGURE 11.1. Piston positions illustrating the Carnot cycle to extract mechanical work from heat.

the cylinder into contact with one or the other. The laws governing an ideal gas, established by the end of the eighteenth century, indicated that if a body of gas is expanded without communication from a source of heat (caloric) to maintain its temperature, then that gas will cool down, whereas compressing a body of gas without allowing heat to flow away from it during the compression will make it warm up.

On that basis, Carnot provided a theoretical analysis of a process whereby caloric is transferred from body A to body B via the expansion and compression of gas in the cylinder. He envisaged an entire cycle of expansion and compression whereby the gas expands and, in effect, sucks up caloric from body A, and then is compressed again by pushing the piston back down, thereby squirting that caloric back out into the cooler body B. This is what became known as the Carnot cycle.[8] It allowed Carnot to point out that the temperature of the gas during the compression phase of the cycle is always lower than during the expansion phase for identical piston positions. Since a gas has greater "elastic force" the hotter it is, it will exert greater force against the piston during expansion than during the equivalent piston positions in compression. In other words, the gas *helps* as the piston is pulled out more than it *resists* as the piston is pushed in again. That difference is what provides the net excess of work from the process: more work comes out than is put in.

The novelty of Carnot's analysis lay in its establishment of a fundamental principle: to get work from a heat engine, heat, or caloric ("fire"), must flow from a hotter to a colder body. In Watt's steam engine, these are represented by the boiler and the condenser. In Carnot's terminology, there must be a fall of caloric from A to B, just as water must flow downhill to run a hydraulic machine. That is the only way in which heat can produce mechanical work.

The parallel with a hydraulic machine (a waterwheel) is closer still. Carnot regarded the total amount of caloric in the system as a constant. Caloric "falls" from a higher temperature to a lower one and produces work as it does so, but the process no more uses up caloric than a waterwheel uses up flowing water. There is no sense in which heat is a form of energy that can be moved around into different forms; there is in fact no general concept of energy here at all. When the concept of energy was invented, it was inseparable from the concept of the conservation of energy. It wasn't that people came up with the idea of energy and subsequently decided that it was conserved in processes like the one analyzed by Carnot. The idea of energy was invented so that there should be something to be conserved.

Joule

One of the central figures in the development of the energy conservation idea started out working on electric motors, a new discovery of the 1820s (see chapter 13). This was James Prescott Joule, a man of independent means (derived from the brewing trade) with his own laboratory near Manchester, at the heart of the Industrial Revolution (he studied for two years with the Mancunian John Dalton).[9] One of the most obvious effects of electrical currents is that the wires tend to heat up, and Joule soon turned to examining the relationship of this heating effect to the generated current itself. In 1843 he published a paper titled "On the Caloric Effects of Magneto-electricity, and on the Mechanical Value of Heat" in which he made the crucial claim that the heat produced by an electrical current is *created* rather than simply transferred from another part of the circuit. Pushing it back a further step, if the current is generated by a dynamo (a mechanical means of producing electricity recently developed by Faraday), then, according to Joule, we should think of the "mechanical power" of the dynamo being actually converted into heat via the intermediary of electricity.

That view had two important implications. First, it involved the idea that heat is not a conserved substance, contrary to Carnot's view in his treatment of heat engines. The other was that there might be a way of finding a common measure to compare qualitatively distinct physical phenomena. The importance of this step lay in its addressing of a fundamental problem: heat and electricity seem to be distinct things. Rendering them commensurable was like comparing apples and oranges. Joule's approach tried to avoid that difficulty.

In Joule's example, adding the dynamo yielded a series of such linkages—motion turns into electricity, which turns into heat—but these appear to be entirely different sorts of things, so much so that electricity and heat had often been regarded by people such as Laplace as distinct imponderable (weightless) fluid substances. Joule wanted to find a way to compare quantities of these categorically different things, and fixed on a good, solid, mechanical measurement: mechanical work, or weight multiplied by vertical displacement. Joule set out to determine the mechanical equivalent of heat. What he reported in 1843 involved using the intermediary of an electrical current generated from a dynamo run by descending weights to give an exact measure of what he called the mechanical "power" producing the electricity, which in turn produced the heat:

> The quantity of heat capable of increasing the temperature of a pound of water by one degree of Fahrenheit's scale is equal to, and may be converted into, a mechanical force capable of raising 838 lb. to the perpendicular height of one foot.[10]

Joule read the results of these experiments to the 1843 meeting of the British Association for the Advancement of Science. The BAAS brought together the leading lights of the various sciences with the idea of spreading support for science throughout the nation, to which end its rules require that it meet every year in a provincial city, never in London. In 1843 members gathered in Cork, in Ireland, then part of the United Kingdom. Joule presented his paper, but it made little impression. It was unclear what the consequences of Joule's results were, and in any case it seemed to cause problems for Carnot's analysis of the motive power of heat. That analysis had become widely known through the mathematical development of the Carnot cycle, complete with famous diagram, by Émile Clapeyron in 1834. Both Carnot and Clapeyron used the idea of a material caloric and its conservation, which Joule's analysis rejected.

In a further paper dated 1844, describing measurements on gas (heating or cooling air by the compressive or expansive movement of weights), Joule explicitly challenged Carnot's and Clapeyron's views of heat and heat engines. His criticism can be seen as a version of the conservation of energy principle.[11] Joule says that their conception of the work obtainable from heat implies that vis viva, live force, can be absolutely lost from the world. The analogy with running water makes Joule's criticism clear.

Carnot saw caloric as flowing from a higher to a lower temperature just as water flows from a higher to a lower level, with work being produced from the flow of caloric in a way analogous to its production from running water. The vis viva of rational mechanics was mv^2, a property of motion, which in the case of flowing water can be converted into work. If it is so converted, then the speed of the water will decrease depending on how much of its motion, its vis viva, is converted. If none is converted, the speed of flow will be unaffected.

In the case of the imagined fall of caloric, on the other hand, when the caloric, or heat, has flowed from a hotter to a colder body, there will be, in Carnot's picture, no difference between the case where that flow has passed through an engine and done work and the case where it flows directly *without* doing any work. This latter is what Joule had in mind when he said that Carnot's view entails the possibility of an absolute loss of vis viva, meaning in this case the vis viva "developed by the caloric contained in the vapour" traveling through a steam engine.[12] The potential work that the flowing heat could have done is lost absolutely if it is not run through a heat engine; it has simply vanished.

Joule's reason for seeing this as a problem is metaphysical, or really theological. Why shouldn't the heat's vis viva disappear? Joule explained, "Believing that the power to destroy belongs to the Creator alone, I entirely coincide . . . in the opinion that any theory which, when carried out, demands the annihilation of force, is necessarily erroneous."[13]

God conserves the sum total of all force in the universe. For Joule, instead of being a substance, heat was a dynamic phenomenon, a repository of what he called "force" just as an elevated weight or an electrical current or a moving body are repositories of "force." It seems fair to say that for all intents and purposes, Joule had formulated, in the mid-1840s, a principle of the conservation of energy in all but name.

There were some loose ends, however: the only hard figures or formulas concerned the conversion of mechanical work into heat, which is not the same as showing the convertibility of heat into work, as in heat engines. More conceptual, if not experimental, work was needed if a concept of energy (under whatever name) was to be made precise and generally applicable.

Conversions and Helmholtz

An important role had been played in all this by the discovery and investigation, during the first four decades of the century, of new physical processes, especially discoveries in electrochemistry and electromagnetism. These showed relationships between electricity and heat, chemical phenomena and electricity, motion and electricity, and so on (see chapter 13). Quite a few people besides Joule were now coming up with ideas pointing toward a conservation principle. Some were vague, talking qualitatively about the interconversion of "forces" in these processes, and some, like Joule, were a lot more precise.

What made Joule unusual in the late 1830s and '40s was his experimental work, which provided numbers and the techniques for generating them. A German theorist soon made use of Joule's results in what became the classic statement of the concept of energy and its conservation, an 1847 memoir called, in the English translation subsequently produced by John Tyndall, *On the Conservation of Force*.[14] The theorist was Hermann von Helmholtz, a young, Berlin-trained physiologist with an interest in physics.

Helmholtz, then working as an army surgeon, started with the then-widespread assumption that what he called "natural powers" in their various manifestations were somehow conserved. Like Joule, he had to find some kind of measure to link these manifestations. Helmholtz thought of the universe in a basically Newtonian way, as a mechanical system of forces and particles. What he developed was a new way of conceptualizing this standard picture in terms of what came to be called *energy*.[15]

Because of his mechanical worldview, and also because it was the most accessible example of a natural power, Helmholtz began with mechanics and the concept of vis viva. Like Joule and the engineers, he made use of the concept of *work* as measured by weight raised vertically against gravity. He equated this engineer's measure of work directly with the vis viva of mathematical rational mechanics. He imagined a weight suspended at a certain altitude being released and accelerating as it fell, and proposed that the work (as measured by the weight's height) that was lost as the weight descended should be understood as being steadily converted into the ever-faster downward motion of the weight.

> If we inquire after the mathematical expression of this principle, we shall find it in the known law of the conservation of *vis viva*. The quantity of work which is produced and consumed may, as is known, be expressed by a weight which is raised to a certain height h; it is then mgh, where g represents the force of gravity. To rise perpendicularly to the height h, the body m requires the velocity $v = \sqrt{2gh}$, and attains the same by falling through the same height. Hence we have $\frac{1}{2}mv^2 = mgh$; and hence we can set the half of the product mv^2, which

is known in mechanics under the name of the *vis viva* of the body *m*, in the place of the quantity of work. For the sake of better agreement with the customary manner of measuring the intensity of forces, I propose calling the quantity $½mv^2$ the quantity of *vis viva*, by which it is rendered identical with the quantity of work.[16]

Helmholtz went on to apply the same reasoning to Newtonian masses exerting gravitational forces on one another that vary with distance according to the inverse square law.

In this generalized form, Helmholtz called the quantities on either side of the equation *tensional force* and *live force*, formally identical to the slightly later terms *potential energy* and *kinetic energy*. When Helmholtz entitled his memoir "On the Conservation of Force," he was using the word *force* (*Kraft*) in a way exactly identical to the slightly later term *energy*. The modern terms became standard by the 1860s, from the usage of various British physicists such as W. J. Macquorn Rankine at the University of Glasgow. Helmholtz brought in the various conversion processes in his paper, mentioned Joule's work on heat, and tied everything to the mechanical basis of masses and Newtonian forces. Acceptance of Helmholtz's arguments meant that there was now something in the universe soon to be called *energy*, and just like matter, it was conserved.[17] Not only was it conserved, but it also circulated through its various forms. Helmholtz liked to play up a sublime vision of all activity and life on the earth as having its ultimate origin in the heat and light of the sun. In a lecture of 1854 (which was also published, in English translation, by Tyndall in 1856), Helmholtz concluded, "If this view should prove correct, we derive from it the flattering result, that all force, by means of which our bodies live and move, finds its source in the purest sunlight; and hence we are all, in point of nobility, not behind the race of the great monarch of China, who heretofore alone called himself Son of the Sun."[18]

This sort of thing played particularly well in Britain. The translator of that Helmholtz lecture, the great Victorian scientific popularizer John Tyndall, promoter of scientific naturalism and member of the pro-Darwinian X Club, played a major role in spreading the word about thermodynamics in relation to a mechanistic vision of the world like

Helmholtz's. Tyndall's public lectures at the Royal Institution in London in 1862 were published and republished many times under the title *Heat Considered as a Mode of Motion*. Tyndall was especially keen on the conservation of energy as a basic structuring principle of the universe.

After talking about the various links in a chain of conversions of energy, in the last chapter of the book Tyndall, echoing Helmholtz, discusses where the energy originated.

> Leaving out of account the eruptions of volcanoes, and the ebb and flow of the tides, every mechanical action on the earth's surface, every manifestation of power, organic and inorganic, vital and physical, is produced by the sun. His warmth keeps the sea liquid, and the atmosphere a gas, and all the storms which agitate both are blown by the mechanical force of the sun. He lifts the rivers and the glaciers up to the mountains; and thus the cataract and the avalanche shoot with an energy derived immediately from him. Thunder and lightning are also his transmuted strength. . . .
>
> He rears, as I have said, the whole vegetable world, and through it the animal; the lilies of the field are his workmanship, the verdure of the meadows, and the cattle upon a thousand hills. He forms the muscle, he urges the blood, he builds the brain. His fleetness is in the lion's foot; he springs in the panther, he soars in the eagle, he slides in the snake.[19]

H. G. Wells's novel *The Time Machine*, from 1895, is likely the best-known echo of this Victorian, almost religious, obsession with the sun as the source of all life. Wells's account of the distant future, with a large, dull red sun hanging in the sky over a dying earth, displays the inexorability of the laws of nature.[20]

In the early 1850s British physicists including Rankine and William Thomson (alias Lord Kelvin), both at the University of Glasgow, used the new approaches of Joule and Helmholtz as the basis for a unified characterization not just of the universe itself, but also of the specific discipline of physics. The theory of energy was applicable to all the physical sciences, because each individual branch of physics simply represented different states of matter, all governed by energy considerations. Physics was seen as a single coherent scientific discipline by British physicists in the second half of the nineteenth century because, for them, it was at some fundamental level *energetics*—it was concerned with energy transformations.[21]

Rankine and Thomson in particular made a point of emphasizing one central aspect of energetics: the law of "conservation of energy" (Rankine's coinage, dating from 1853) held independently of any particular hypothesis about the ultimate nature of heat, electricity, or anything else. It was precisely the claimed independence of energetics from specific hypotheses about particular phenomena that justified energy as the unifying concept for physics as a discipline.

Clausius and Entropy

The interconvertibility of mechanical work and heat, strongly indicated by Joule's work in the 1840s, was in flat contradiction to Carnot's idea of the conservation of caloric in the operation of heat engines. Regardless of whether heat was thought of as a substance or as a consequence of the motion of molecules, it was still necessary, given the conservation of energy, to admit that heat was lost, or used up, when producing work in a heat engine. This implication—that Carnot was wrong—resolved itself into a concrete theoretical problem in the later 1840s for William Thomson that closely resembles Joule's criticism of Carnot in 1844.

Even if one accepted Joule's position on the equivalence of heat and mechanical work and rejected the concept of caloric, one was still left with Carnot's basic idea that to obtain work from heat, heat must flow from a higher to a lower temperature. The same thing happens when heat flows through a metal bar as in a heat engine: the heat goes from a higher to a lower temperature (from the temperature of the heat source to that of the now uniformly heated bar) as it dissipates through the bar. We have an irreversible thermal process, the dissipation of heat, whereby the temperature that manifests it decreases. But in this case, no work is done.

According to Carnot's analysis, the work produced from a heat engine is a function of the upper and lower temperatures between which the heat flows. That means that any given temperature drop should in principle be capable of producing a particular amount of mechanical work. But in the case of heat dissipating through a metal bar, the temperature drop occurs without any work being done. Since the heat present is now at a lower temperature, it should be less capable of producing work than it was before. What has happened to the work that could have been done by the heat flow but was not?

That was the problem as Thomson saw it. It centers on the irreversibility of heat dissipation: heat does not spontaneously flow from a colder to a hotter body, only from hotter to colder. What became the generally accepted solution was provided by a theoretician at Berlin (and colleague of Helmholtz), Rudolf Clausius, in his 1850 paper "On the Motive Power of Heat." The secret lay in turning Thomson's problem into an axiom. Axioms, after all, don't need to be solved.[22]

Clausius's paper states as its first principle the equivalence of heat and work (soon to be generalized as the conservation of energy). The second principle is that in a heat engine, even a perfect cyclical one of maximum efficiency, when heat passes from a higher to a lower temperature, some of the heat is converted into work, but some always flows to the lower temperature. The relative proportions of heat energy turning into work and heat energy passing to the lower temperature are a function of the upper and lower absolute temperatures.

So Clausius's solution to the problem that bothered Kelvin was quite simple. When heat flows from a higher to a lower temperature without doing any work, the amount of heat energy remains the same; it just passes from heat energy at a higher temperature to the same energy at a lower temperature. What changes is the availability of that heat energy to do work, because availability is a function of temperature: as the temperature drops, so does availability. Thomson's problem is therefore solved by making a clear distinction between heat as a form of energy and temperature as a manifestation or mode of heat.[23]

Clausius's second principle turned into the second law of thermodynamics, and its definitive mathematical form required one extra step: quantifying the decreasing availability of heat energy at decreasing temperatures. The transformation of work into heat or heat into work was expressed in terms of the increment of heat divided by the absolute temperature at which the transformation occurs.

In an 1854 paper Clausius called this the *equivalence value* because it expressed the variable equivalence of heat and work. In 1865 he coined the less transparent word *entropy* by analogy with the term *energy*, which he had accepted from the British.[24] In principle, both quantities, energy and entropy, are rather abstract. But because of their level of abstraction—their supposed independence of particular physical hypotheses—they allowed energetics to unify the discipline of physics.

Still, there were other ways to understand energetics, or thermodynamics, than the approach focused on phenomena and the measurement of manifest properties of material bodies. During the 1850s Clausius was among a few people who developed a so-called kinetic theory of gases, whereby the properties of gases, such as pressure, were interpreted by thinking of gases as collections of molecules flying about in all directions and interacting only by collision. Pressure, for example against the walls of a container, would be due to the mechanical impact of molecules. This contrasted with the old Newtonian and Laplacian static gas model, which involved more or less stationary particles exerting repulsive action-at-a-distance forces on each other. The kinetic theory of gases was more consonant with a kinetic theory of heat, which the principle of the conservation of energy made the obvious choice as an alternative to the old idea of caloric.[25]

In an 1857 paper Clausius applied ideas from probability theory to the kinetic theory of gases as simplifying assumptions. But that amounted only to justifying the use of averages, not looking at velocity distributions among gas molecules. The Scottish physicist James Clerk Maxwell read Clausius's paper and set to work on his own treatment of the kinetic theory in 1859. He did it basically as a mathematical exercise, a kind of rational mechanics of gases that he was doubtful would square with experiments. The central idea was to use the bell curve, the astronomer's law of errors, to govern the assignment of molecular velocity distributions. These were techniques, and an attitude to scientific theorizing, that had emerged slowly since the eighteenth century from mixed mathematical approaches to the management of aggregate data of all kinds, as we will see in chapter 12. The late eighteenth and nineteenth centuries had seen those approaches elaborated into powerful instruments that Maxwell would now use to create a new kind of physics.

12

Chance and Determinism

NEW MODELS OF SCIENCE

IN BOTH the eighteenth and nineteenth centuries, scientific explanation and scientific understanding took two principal forms. One, as we have seen repeatedly, was taxonomic, evident especially in the natural history of living things, but also in chemistry, astronomy, and other fields; classification became an explanatory resource, not just a descriptive one. In aspiring to the status of natural philosophy, natural history infringed on the domain of causal explanation or causal understanding. Think of the explanatory use of classification in Darwin's work, where the process of descent with modification gave causally linked explanations of already accepted classificatory relationships.[1] Assumptions about scientific explanation in this period generally focused on causal relationships. Few people suggested alternatives even during most of the nineteenth century, and these tended to be a bit vague when offered.[2]

Rather than witnessing the full-fledged creation of a new scientific ideal that departed radically from a causal notion of science, the final three or four decades saw the development of a vision of a different way of understanding events in the natural world. The new scientific problems and procedures that this new vision addressed centered on statistics and the application of mathematical probability to empirical data.

Laplace and the Binomial Distribution

Mathematical probability in the eighteenth century was, as we saw in chapter 3, usually understood to be a mathematical modeling of the behavior of the so-called reasonable man. It was a branch of mixed mathematics that would codify human judgments in situations of uncertainty. It usually centered on idealized gambling situations where the judgment involved was usually how much to wager.

Applying probability theory to large amounts of empirical data was first developed on a large scale by Laplace in the late eighteenth-century, in connection with error theory. Laplace's main tool was the normal distribution, or bell curve. It had been invented in 1733 by Abraham de Moivre in England to deal with the numbers of possible outcomes of something like a thousand coin tosses (for example, to compute the chances of any particular result, such as 450 heads and 550 tails, actually occurring).

His solution to the coin tossing problem received little attention until the 1770s, when Laplace picked it up. He applied the curve to what he called "error analysis." Because there are always unknown causal factors disturbing individual measurements of some empirical value, the actual data fluctuate around a mean in a calculable way, and the mean is taken to be the underlying true value. Laplace argued that the binomial distribution represented the likelihood of error away from the central, true value due to such confounding factors. Laplace himself later applied the idea to astronomical observations to cope with observational error in positional measurements. This became the most important early use of error analysis, together with the associated method of least squares for fitting curves to data, which was developed by the French mathematician Adrien-Marie Legendre. Related analyses were produced by the German mathematician Carl Friedrich Gauss, who represented the binomial distribution by a continuous function, the so-called Gaussian distribution (what we often call the "normal" distribution, or bell curve). Laplace subsequently wrote a landmark book on probability theory and the use of the error curve: *Analytical Theory of Probabilities,*

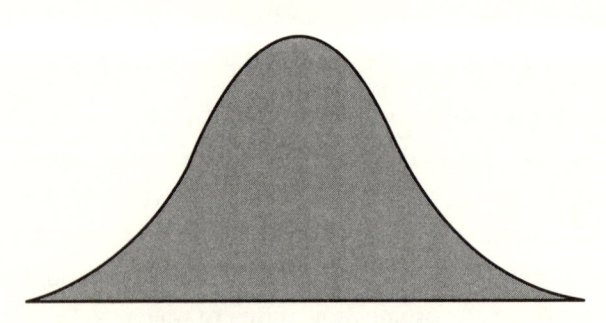

FIGURE 12.1. The binomial distribution as an error curve.

published in 1814, which remained the bible of probability theory for decades.[3]

There is immense significance in Laplace's view of the binomial distribution as fundamentally an *error* curve. Laplace thought of the universe as an entirely deterministic system of Newtonian forces and particles, where everything that happened was brought about by the immediately antecedent conditions. For Laplace, and everyone else in this period, true science meant determinate causal explanation. The scientific ideal was Newtonian mechanics, particularly as applied to planetary astronomy, where everything is determined and regular: forces cause motions and motions reveal forces with absolute mathematical necessity.

Laplace indicated what this meant by arguing that if there were a superintelligent being that knew in perfect detail the state of everything in the universe at some particular instant—the locations and motions of, and forces applying to, every single material body—then it could calculate the entire development of the universe, past and future, by extrapolating the system forward or running it backward in time. Everything is determined according to rigid causal connections by the immediately antecedent state of affairs. Of course, that makes the universe completely preprogrammed, and seems to rule out human free will. Worries about such determinism were constant themes throughout nineteenth-century thought, causing, in their way, at least as many theological problems as would Darwin.

For Laplace, mathematical probability was not an alternative to determinism, but a tool that yielded explanations of fundamentally the same kind as deterministic causal accounts did. To illustrate the idea, consider how Laplace regarded the classic case of drawing colored balls from an urn. Imagine an urn with an equal number of (otherwise) identical white and black balls. Proceed by drawing a ball (without looking), then note its color, mix it back in, draw out another, note its color, and so on. In the standard view, following Bernoulli's law of large numbers, this procedure should yield a ratio of black to white balls drawn that gets closer and closer, as the drawing proceeds, to the underlying 50:50 ratio. Laplace said that this is what happens with all regularities involving large numbers of events, including natural and social events. The reason it happens, he said, is that "in a series of events indefinitely prolonged the action of regular and constant causes ought to prevail in the long run over that of irregular causes."[4] In the case of the urn, the irregular causes are the accidents to do with exactly how the balls move in the urn when churned up and where the hand happens to go when it picks out a ball—the physical details of exactly what happens in each individual drawing. The single "regular and constant" cause in this case is the structural fact that the numerical ratio of white to black balls in the urn is 50:50.

In Laplace's deterministic view of the universe, every step of this process, including every detail of each drawing, would be fully determined by physical cause and effect starting from initial conditions. (Imagine, for neatness, having it all done by a robot.) From that point of view it becomes clear why probability, for Laplace, was connected with the idea of error and the error curve: only our ignorance of all the causes that Laplace called "irregular" forces us to resort to probability in the first place. If we were like Laplace's imaginary superintelligence, which knows every detail of the state of the universe, we could calculate the outcome of every single drawing from initial conditions. Probability theory, for Laplace, reflects our shortcomings, not something fundamental about the world.[5]

But there is a curious feature of Laplace's conception. We could predict, or explain, the long-term outcome of the drawing procedure using probabilistic reasoning based on a prior knowledge of the equal numbers of balls—we could predict, in other words, what is most likely to happen in a long sequence of drawings, but not what must *necessarily* happen. By contrast, Laplace's imaginary superintelligence could predict the outcome with absolute precision through enormously complex calculations of the fully determined causal sequences bringing about each individual drawing. And the superintelligence's entirely causal prediction, or explanation, of the final outcome would make *no reference at all* to the structural fact of the numerical ratio of balls in the urn.

Laplace's probabilistic view of the reason why long-term regularities appear in apparently random events therefore involves a different type of explanation from the rigorously causal one that he thought it could approximate. It was the difference between likelihood and necessity.

Statistics as a Window on the Inaccessible: Quetelet

Probability is not the same as statistics. The concerted application of probability to aggregate statistical data first occurred on a large scale in the 1820s and '30s; until then, the term statistics designated a completely different field of study from probability even for work done by Laplace relating to large bodies of numerical data.

The English word "statistics" derives from the German *Statistik*, first used in 1749, which found its way into English and French and was fairly common by the beginning of the nineteenth century. But at that point it had nothing to do with specifically numerical data. The word simply meant the study of states, in the political sense, or more generally the description of social, political, and economic issues in the government and functioning of a state. Not until the 1820s was it associated specifically with numerical data rather than any other kind of information or description—a matter of looking at large bodies of numerical data, still relating to human social affairs.[6]

The serious collection of numerical data on human populations had started with the introduction of national censuses in Britain (1801) and France (1791); the first census for the United States of America, mandated by its new constitution, was made in 1790. These were motivated by governmental interest in determining the manpower available for the unprecedently large armies being raised in Europe; there was also the more general interest of increasingly bureaucratic and centralized states in information for planning and administration. As a consequence, by the early nineteenth century diverse quantitative information was being gathered systematically for the first time.[7]

Early attempts at taking censuses were often wanting, but by the 1820s especially detailed and scrupulous numerical surveys were being produced in France for the region of Paris (the Île de France). These studies were widely noticed because of their level of detail and the implications for rational administration.

This was the context into which a Belgian astronomer, Adolphe Quetelet, arrived when he visited Paris in 1823. He was the head of a new astronomical observatory in Brussels and had come to Paris to learn from gurus like Laplace and Fourier about the most advanced techniques for performing observational work. By this time the error curve was an established computational tool for reducing observations into properly refined positional data, and Quetelet was particularly impressed by it as a way of bringing order to a large collection of apparently messy, inconsistent numbers all purporting to measure the same value. This was also a period of increasing interest in numerical social statistical data, and Quetelet got the idea of putting the two together. In 1825 he gave his first paper on statistics to the Brussels Academy of Sciences, laying out some of the basic attitudes to the subject that ran through his important and widely read publications of, especially, the 1830s and '40s.[8]

Quetelet's memoir of 1825 was purportedly on applying probability to insurance statistics, but Quetelet used it as an opportunity to talk about the possibility of a true science of society, analogous to astronomy. Like other natural phenomena, human social phenomena showed periodic cycles and regularities (the 1825 paper looks at how numbers of deaths and births vary over the course of a year). The regularities and constancies of that kind of statistical data led him to formulate a general idea of "social physics" by 1835, dedicated to examining the statistical "laws" governing social phenomena.

Quetelet used two main mathematical techniques. One was Bernoulli's law of large numbers, as interpreted by Laplace: if statistics for social phenomena based on a large amount of data remained pretty constant year after year, that must be the effect of the law of large numbers revealing underlying causal regularities, or laws. Quetelet had in mind things like the constancy in numbers of suicides from year to year in a given city, or one of Laplace's often-cited examples: the constancy in the number of letters ending up at the dead letter office in Paris every year. Each instance of such an event would be utterly unpredictable, governed by uncontrollable individual causes and circumstances, but the overall aggregate numbers remained constant and apparently lawlike. (Another of Quetelet's favorite examples was the number of soldiers in the Belgian army killed each year from being kicked by a mule.)

Quetelet's other favored technique was the error curve, which governed the spread of deviations from the mean value. He used it to stress the stability of the mean value despite the variation in such things as the heights of individual men, or the number of murders in a year, scattered around that mean. He claimed to be able to calculate the true mean from the slew of variable data, just as astronomers calculated true stellar positions from varying observational measurements, by finding the curve behind the numbers.

Quetelet's view of statistical laws, then, focused on the stability of phenomena, or the stability of mean values, rather than on the variations about the mean. The distribution given by the bell curve is still an error curve, used just as astronomers did. Variations from the mean value in social statistical data represented for Quetelet the effect of accidental causes, whereas the mean value represented the constant, regular cause or causes (whatever they might be) that brought about the phenomenon. In the case of human behaviors, such as murders, those unknown causes would be essentially social.

Quetelet invented the idea that phenomena such as crime were the fault of society rather than of individual criminals. He said that a murderer is just someone through whom the constant statistical laws of the society, which depend on the structure of that society, are expressed: "It is society that prepares the crime.... The culprit is nothing but the instrument of its execution.... The wretched man who carries his head to the scaffold is ... an expiatory victim of society."[9] Quetelet's political views concerning social reform were of a piece with such an argument.

Quetelet, like Laplace, saw statistical regularities as manifestations of underlying determinate causal processes; large numbers of events filtered out the adventitious and accidental causes to leave exposed the underlying forces that caused the regularity. Human free will counted as an accidental cause, the supposedly free choices of individuals tending to cancel each other out when dealing with large numbers, leaving behind impersonal social phenomena.

Quetelet's thinking on social statistics became well-known and influential in the 1840s and '50s, especially in Britain. A long review essay in 1850 by John Herschel (astronomer-son of William Herschel and well-regarded legislator of science) promoted Quetelet's ideas and emphasized their wide applicability, including the capacity of the error curve to wash out random accidents in large volumes of data.[10] An important adopter of Herschel's views was an English historian and social thinker (as well as strong chess player), Henry Thomas Buckle, in his two-volume *History of Civilization in England* (1857–61). Buckle wanted to create a truly scientific approach to history, and he thought that the way to do it was to formulate laws of history. (Another seeker of such laws around the same time was a German philosopher called Karl Marx; such grand speculations were, like evolution, in vogue by the 1850s.)[11] Buckle's were laws of social development, the existence of which was evidenced by the sorts of well-known statistical regularities in which Quetelet—as well as, by this time, insurance companies—specialized.[12]

But Buckle continued to assume, like Laplace and Quetelet, that a true science must be causal and deterministic. Statistical probability was a way of revealing determinism, not replacing it. The fundamental difference between deterministic and probabilistic arguments, between necessity and likelihood, remained something that was largely unaddressed, with the notable exception of a French mathematician, Antoine Augustin Cournot, in the 1840s.[13] The model of science remained one that appealed to dynamic causation.

Maxwell and Statistical Physics

The development of probabilistic explanation as a legitimate form of scientific demonstration in its own right, rather than as an indirect way of approaching causal explanations, really occurred with Maxwell's work and his introduction of statistical methods to physics. One of the main reasons for Maxwell's interest in playing up probabilistic rather than causal mechanical explanation was to oppose determinism. People like Buckle had stressed determinism even in human actions that appear to be absolutely free.

Buckle's work attracted Maxwell's interest because of its challenge to free will, which was of great theological importance to Maxwell.[14] Buckle, like Huxley and other X Club members such as Darwin's cousin Francis Galton, enjoyed attacking religion as an obstacle to progress. Buckle drew explicit inferences about the nonexistence of free will that Quetelet had carefully avoided. A deterministic worldview had been a potential theological problem for a long time, but Buckle stressed it in explicit connection with human behavior based on the evidence of statistical science. Maxwell's reaction to this, on the other hand, was that order did not necessarily imply absence of free will.

Maxwell's invention of statistical physics was directly connected with the development of thermodynamics that took place during the 1840s and '50s. In an 1867 paper Maxwell argued for the applicability of the error law to gases, assuming the kinetic model, in which there are frequent collisions and regular mechanical exchanges of kinetic energy that continually alter the velocity of any given molecule. Although the complex interactions and paths of countless individual molecules in a volume of gas could not be tracked in detail, something could be said about the overall velocity distribution. That distribution would remain constant under conditions of unchanging temperature and pressure even though the velocities of individual molecules are constantly changing and exchanging energy. And that distribution, Maxwell suggested, is the binomial, or normal, distribution.[15]

Maxwell wanted to establish the point that thermodynamics, or energetics, is a fundamentally probabilistic science, and therefore not deterministic. He wanted to avoid strict determinism so as to leave room for free will, in accordance with his theological convictions. One line that he promoted in the late 1860s and early '70s was that his gas work showed that human understanding of real physical bodies is necessarily always probabilistic and contingent. All real bodies, which we can actually measure, are aggregates of particles, and we cannot know the details about every individual particle; we can address only the average properties

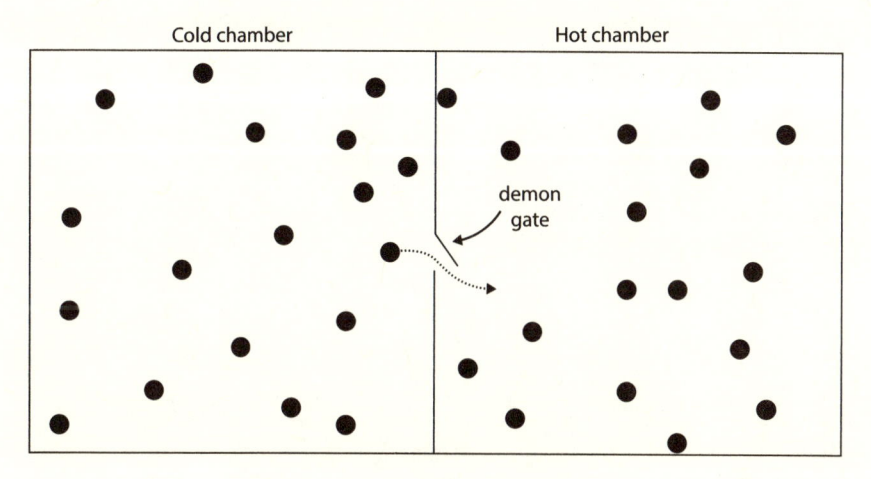

FIGURE 12.2. Maxwell's sorting "demon."

of the whole collection. There are even mechanical behaviors that refuse reduction to unequivocal prediction.[16] We cannot, as a matter of practice, ever come to know things through strict determinism. This was a fundamental human limitation that ran against the assumptions of scientific naturalism.[17] In 1867, in a letter to a fellow Scottish physicist, Peter Guthrie Tait, Maxwell invented his famous demon to illustrate the point further.[18]

One way of thinking of the second law of thermodynamics is to take it as a formalization of Carnot's fundamental principle that heat never spontaneously flows from a lower to a higher temperature. Heat will flow from a higher to a lower temperature, however, and work can be produced thereby. Remember that the law of conservation of energy would not be violated by heat spontaneously flowing uphill; the only violation would be of the entropy law. In Maxwell's kinetic model, gas temperature is a function of the average speed of the molecules (actually the mean square velocity): the faster they move, the higher the temperature.

Maxwell imagined a container divided by a wall into two compartments. On one side of the wall is gas at a high temperature; on the other, the same gas at a lower temperature. There is a little door in the wall, and a tiny demon next to it who can see the motion of individual molecules as they approach the door. Because of the wide spread of velocities in the gases, modeled by the normal distribution, there are some slower

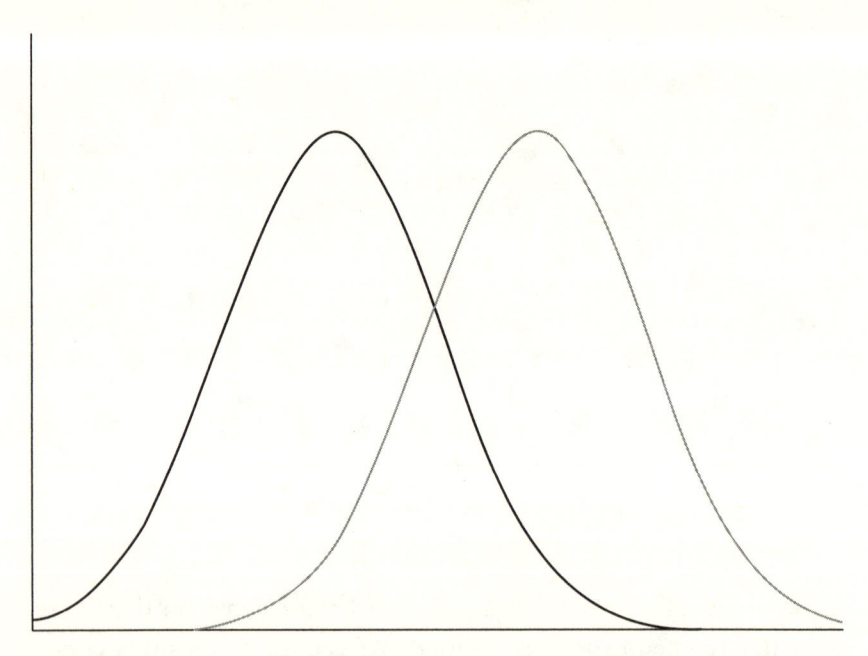

FIGURE 12.3. Two velocity distributions for two bodies of the same gas at different temperatures, variation as exploited by the sorting demon.

molecules in the hotter gas and some faster ones in the colder gas. When the demon sees a fast molecule in the colder gas approach the door, he opens the door and lets it through; when he sees a slow molecule in the hotter gas, he opens the door and lets it through in the other direction. (The energy involved in operating the door is assumed to be negligible, or as small as you wish.)

Gradually the faster molecules in the colder body of gas are exchanged for the slower molecules from the hotter body of gas; the hotter gas gets hotter and the colder gas gets colder. Heat is therefore flowing from the colder to the hotter gas, without the violation of any known mechanical laws and in direct violation of the second law of thermodynamics. The second law is thus not reducible to the laws of mechanics.

Maxwell used this thought experiment to argue that the second law of thermodynamics was only a statistical law, not a dynamical causal one, and that it has the status of a law only relative to human cognition: If we were the tiny demons, we could easily get work from heat by making heat flow uphill and then letting it flow down again through a heat engine, but our actual place in the universe precludes it. What Maxwell proposed was a conception of fundamentally statistical laws of nature, in place of the older assumption that statistical laws were ultimately reducible to underlying causal laws. This, he thought, left room for consciousness and free will.[19]

In Austria Ludwig Boltzmann attempted unsuccessfully to argue that the laws of thermodynamics could be derived from a version of Newton's laws applied to the kinetic theory of gases. He finally admitted defeat in the face of William Thomson's reversibility paradox of 1874, which pointed to the fact that the second law of thermodynamics would be violated by a movement of molecules proceeding from a very probable state of random motion and collisions to a highly improbable state in which the fastest molecules (hotter gas) were crowded at one end of a container while the slower molecules (cooler gas) occupied the other. Thermodynamics in effect denied that this separation of hotter from cooler gas could ever happen, understood in the kinetic theory as the consequence of the innumerable collisions and exchanges of energy among the gas molecules. Thomson saw that each of those collisions was reversible in time: Newton's laws would be followed even if all the molecules suddenly retraced their paths, as if time started running backward. We can imagine making a video of the molecules, represented perhaps by rubber balls bouncing around, and then playing the video backward to see that the collisions always follow Newton's laws in either temporal direction. The directionality of heat flow could therefore not be attributed to the mechanical behavior of the molecules. Newton's laws were not violated by the violation of the second law of thermodynamics. The second law was, in this view, fundamentally statistical in nature.

Galton, Heredity, and Eugenics

Physics was not the only scientific arena profoundly affected by statistics. In the life sciences, too, the error curve was a tool for understanding processes occurring in large aggregates. Instead of molecules, however, these were aggregates of biological organisms. Ironically, while Maxwell had used this approach to counter views of the determinism of the universe, the person centrally involved in applying these techniques in the living world was Darwin's cousin Francis Galton, a member of the X Club and a devotee of scientific naturalism. Galton was interested in human populations and eugenics, meaning human selective breeding aimed at improving human stock in much the same way that breeders attempted to improve farm animals—in fact, he invented the word, and first made eugenics intellectually respectable in that period.

In 1869 Galton published a book titled *Hereditary Genius*. He argued that, first of all, geniuses were people with intelligence a step above that of everybody else; they were not simply the top end of a continuous distribution. Second, he tried to show that genius was a hereditary trait. He looked at people generally regarded as being of superior mental ability and argued that they were much more frequently related to one another than one would expect by chance. One of his best examples was the Darwin family, to which he was related. Galton suggested, based on this data, that the stock of especially gifted people in the population could be increased fairly readily by judicious intermarriage. Darwin did not disagree, as *Descent of Man* makes clear.[20]

Thus, in considering eminent judges, Galton analyzed historical data to provide "another test of the existence of hereditary ability. It is a comparison of the number of entries in the columns of Table I."

> Supposing that natural gifts were due to mere accident, unconnected with parentage, then the entries would be distributed in accordance with the law that governs the distribution of accidents. If it be a hundred to one against some member of any family, within given limits of kinship, drawing a lottery prize, it would be a million to one against three members of the same family doing so (nearly, but not exactly, because the size of the family is limited), and a million millions to one against six members doing so. Therefore, if natural gifts were due to mere accident, the first column of Table I would have been enormously longer than the second column, and the second column enormously longer than the third; but they are not so. There are nearly as many cases of two or three eminent relations as of one

eminent relation; and as a set off against the thirty-nine cases that appear in the first column, there are no less than fifteen cases in the third.

It is therefore clear that ability is not distributed at haphazard, but that it clings to certain families.[21]

Galton's view of evolution is evident: he adopted the idea of discrete jumps—Huxley's saltations—as the material of evolution, rather than Darwin's slight variations; geniuses were a distinct level above everyone else.[22] Also clear is his stress on heredity as determining a person's characteristics—nature rather than nurture.[23] This attempt to find determinate, naturalistic explanations for what people were like went along with Galton's strong agnosticism. In fact, he suggested that eugenics might be used to replace Christianity; inheritance would replace immortality.[24]

But Galton's real applications and development of statistical methods came in his examination of the distribution of characters within human populations and the statistical characteristics of biological inheritance. Like almost everyone else, Galton assumed that blending inheritance was the right model. He applied the mathematics of error theory to the distribution of traits—height, hair color, and almost anything else that he could measure—throughout a population. Error theory was by then the obvious tool to use. Galton followed Maxwell's lead by reinterpreting the bell curve so that it represented not errors deviating from a correct (mean) value, but statistical variation of a character from the mean value for the population as a whole. Part of this adaptation of the curve involved histograms, ranks of distribution representing the relative numbers of individuals with particular degrees of any given character, so his project implied a very large empirical program. For a human population, that character could be anything Galton could think of—weight, height, intelligence, even artistic ability. Galton had enormous faith in the proposition that anything can be measured.

Given this distribution curve of a particular character in a population, Galton asked what would happen to it over successive generations. He initially examined this by doing experiments with sweet peas, looking specifically at the size of the seeds, but he did it explicitly as a way of gaining information to apply to *human* characters. In 1877 he published a paper, "Typical Laws of Heredity," based on the sweet pea results, reporting that he found that the overall distribution remained unchanged from generation to generation.[25] Individuals at the extreme ends of the curve were more likely to produce offspring closer to the mean than they were to produce offspring even further away from it. There was a constant tendency for these extremes to revert toward the mean in the following generation, which kept the overall distribution stable over the course of generations.

Galton set about testing this result for human characters. He gathered data by having, among other things, a so-called anthropometrics laboratory set up in the Science Museum in London during the 1884 International Health Exhibition, where visitors could voluntarily take part in the tests. Galton wanted to collect data from families on height, reflexes, color perception, and so forth, and he examined the results for the relationship of characters in offspring to those in their parents.[26]

Galton found in this human data the same statistical tendency as among sweet peas for reversion toward the mean of the entire population, with a consequent overall stability of the curve. That enabled him to calculate a figure called the reversion coefficient, which expressed for any character the statistically expected amount of reversion toward the mean by offspring depending on the extent of deviation from the mean of their parents.

Galton soon realized that these statistical laws applied to comparisons in data beyond those relating to parents and offspring, such as comparing sets of data relating to brothers, and he changed the name from *reversion coefficient* to *regression coefficient*.[27] Finally, Galton found that he could generalize his coefficient even more, to express relationships between two variables generally, as opposed to between different individuals in the same population; for example, he could show a similar statistical relationship between people's height and arm length. He finally came up with the general term for all such cases: *correlation coefficient*.[28]

The techniques that Galton developed became standard parts of statistical analysis, used routinely by scientists such as population biologists, experimental psychologists, and sociologists. But Galton developed them for the purpose of providing a statistical treatment of heredity as a basis for eugenics. However, his own results seemed to cause problems for his eugenic ambitions. Galton's results stressed the stability across generations of the population distribution, especially the tendency of the offspring of individuals, including those at the extremes of the distribution, to revert toward the mean. This gave no comfort to eugenists, because it appeared that such stability would prevent the perpetuation of exceptional qualities by breeding from individuals at the upper end of the curve. Galton wrote:

> The law of Regression tells heavily against the full hereditary transmission of any gift. Only a few out of many children would be likely to differ from mediocrity so widely as their Mid-Parent, and still fewer would differ as widely as the more exceptional of the two Parents. The more bountifully the Parent is gifted by nature, the more rare will be his good fortune if he begets a son who is as richly endowed as himself, and still more so if he has a son who is endowed yet more largely. But the law is even-handed; it levies an equal succession-tax on the transmission of badness as of goodness. If it discourages the extravagant hopes of a gifted parent that his children will inherit all his powers; it no less discountenances extravagant fears that they will inherit all his weakness and disease.[29]

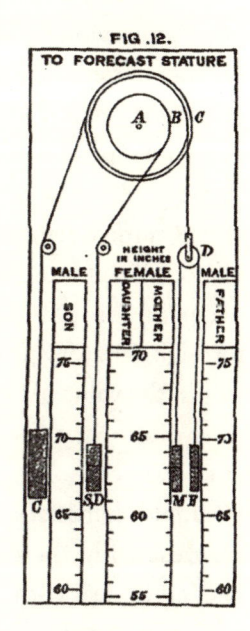

FIGURE 12.4. Francis Galton's "stature" instrument for comparing heights of parents with those of their children, from *Natural Inheritance* (1889), p. 107.

Galton regarded this stability of the population mean as consistent with his belief that small individual variations cannot fuel natural, or even artificial, selection, because species and their characters are quite stable due to reversion. Only saltatory jumps, such as Galton claimed existed between ordinary people and geniuses, could be the basis of a shift, and such jumps (sports) did not form part of Galton's statistical population work. Consequently, they were not constrained by statistical laws of heredity, and were, Galton thought, still worth spreading by selective breeding.

In the 1890s Galton's most important follower, Karl Pearson at University College London, pursued a research program called biometrics, aimed at measuring the same sorts of characters in populations. Unlike Galton, Pearson was an orthodox Darwinian, believing that evolution occurred through slight, individual variations among organisms, not through saltatory jumps. But Pearson was also an ardent eugenist, and he saw the value of his work as being a way to assist humanity in deliberate evolutionary development. In effect, this required selective breeding at the top end of the distribution curve (representing a high level of desirable characteristics) and suppression of breeding at the lower end (representing the undesirable) to shift the mean upward. Pearson needed to deal with Galton's phenomenon of reversion and show that it did not prevent gradual change over generations. He managed it simply: he said that reversion meant a tendency to return to the *existing* mean; there was no biological reason to suppose that the mean itself could not be shifted by selective breeding.[30]

Pearson's project in biometrics provided a powerful push to eugenic ambitions in Britain and elsewhere from the 1890s onward, using the newly established statistical techniques as its scientific foundation. With Maxwell, thermodynamics defied reductionist determinism and elevated probabilistic physics, while biometrics defied fixity of the human species and created ambitions for controlled human improvement through eugenical programs. Both projects characterized the trajectory of science in the twentieth century toward the descriptive rather than the explanatory.[31]

13

Electromagnetism, Action at a Distance, and Aether

THE FELT need for a concept of energy and its conservation arose from consideration of processes that could be seen as tracing some force from one form to another. Crucial among these processes were newly discovered electromagnetic phenomena: the production of magnetism from electricity, electricity from magnetism and motion, and motion from electricity and magnetism, which had been found in the 1820s and '30s. They all tied in with the discovery, in 1800, of electric currents. The discovery of electromagnetic phenomena followed a path of inquiry beginning with the invention of the electric battery.

Frogs' Legs and Electric Current

The electrical research of the eighteenth century had concentrated on what are now called *static* electrical effects; the idea that there might be more to electricity than that simply didn't exist. The story of how that changed starts with work by an Italian professor of anatomy in Bologna, Luigi Galvani, in the 1780s, published in a book of 1791. While examining muscular responses to electricity in the legs of dead frogs, Galvani noticed that keeping the frogs impaled on brass hooks hung on iron railings sometimes resulted in apparently spontaneous muscular convulsions. He soon found that he could produce the effect intentionally by touching a frog's leg with a piece of each metal simultaneously.

Galvani interpreted this effect as being a manifestation of what he called "animal electricity."[1] As a physiologist and anatomist, he took it to be a property of organic tissue, not an effect of ordinary electricity. His discovery attracted a great deal of interest in the 1790s, especially in France, and the phenomenon was dubbed Galvanism. At first it continued to be seen as a physiological rather than a physical phenomenon. That finally changed with the work of Alessandro Volta, an Italian natural philosopher at the University of Pavia, who was not part of the medical faculty and therefore uncommitted to a physiological interpretation. Disliking the idea of animal electricity, Volta wanted to show that the effect was not essentially physiological at all.

However, any demonstration that animal tissue was not an essential part of Galvanism faced a paradox: the means of detecting the effect, producing muscular spasms in animal tissue, *was* the phenomenon, and therefore by definition not separable. So Volta shifted the problem sideways by inventing a surrogate phenomenon that would closely resemble the original but dispense with twitching frogs' legs. He devised an artificial detector, soon called the Voltaic pile, in an attempt to argue that what was involved was nothing but the passage of ordinary electrical fluid, the same fluid involved in regular electrical effects. The Voltaic pile consisted of disks of two different metals, such as zinc and copper (or, early on, zinc and silver coins), placed together in pairs, with each pair separated by a piece of moistened cardboard—a sort of surrogate frog's leg. In discussing the device, Volta made explicit comparisons with electric eels and torpedo fish to reduce the apparent distance between living and nonliving bodies.[2]

Volta believed that the contact of the two different metallic surfaces somehow released electrical fluid from the metals. This involved things like showing that the contact of the two metals could produce a small charge (Volta published the result in 1794). He proceeded to suggest that, once released, the fluid was impelled along the length of the pile. The damp cardboard allowed the fluid to pass through it, so the contributions from each pair of disks added together. That the electric fluid was actually being propelled through the pile was shown by the strong, continuous "current" that could be felt when the experimenter's hands touched the two terminals. Volta thought that Galvani's effect was simply a physiological response to a contact-action electrical phenomenon, which could be produced and intensified in the pile.[3]

That was in 1800, and Volta's arguments were not immediately convincing. The trouble was that, at first, the only way of detecting the effect remained the production of muscle spasms, albeit in the hands of experimenters, which meant that animal tissue was still centrally involved. It took a few years for the identification of the new effect with regular (static) electricity to become generally accepted.

The crucial development that encouraged that identification was the rapid development of electrochemistry. French investigators, including Biot (a confirmed Laplacian), and people in England like Humphry Davy at the Royal Institution, soon noticed chemical effects being produced in the pile. When the pile's ends were connected (by wire, not by the experimenter's body), the damp cardboard generated small bubbles, and the surfaces of the zinc disks displayed oxidation after the pile had been running for a while.[4] Around 1807 Davy turned to investigating the effects of electric currents through salt solutions, with the attendant decomposition of salts and evolution of products at the electrodes (the decomposition of water had already been observed by British chemists in 1800). Electrochemistry remained an important research interest for Davy's eventual successor as head of the Royal Institution in London, Michael Faraday.

Faraday, Oersted, Ampère

The Royal Institution was something of an anomaly in this period. Founded in 1798, it served as a center for the public dissemination of the sciences, but was also a research institution in its own right; it had laboratory space in the basement, where Davy and Faraday both worked. Faraday first came to the Royal Institution in 1812 as a laboratory assistant, straight from having been a bookbinder's apprentice. Davy gave him the job after Faraday kept pestering him, and soon Faraday turned into an excellent experimental researcher himself, even though he lacked any formal scientific education. Faraday never learned any mathematics, and all his work remained purely qualitative, but he nonetheless became an important theoretician as well as experimentalist.[5]

Like Davy, Faraday focused initially on electrochemistry, but since that necessarily involved electricity he was just as concerned with electrical phenomena generally as with properly chemical ones. Electromagnetism caught Faraday's attention, as it did everyone else's, in 1820. A Danish chemist who had been trained in Germany and France, Hans Christian Oersted, published an account of his discovery, made the previous year, of a relationship between electric current, generated chemically by a Voltaic pile, and magnetism. He found that a compass needle would align itself transversely to a current-carrying wire, like a tangent to a circle around the wire. If the needle was moved around the wire it would be drawn in a circular or spiral path.

Oersted had been looking for some relationship between magnetism and electricity, prompted by a late eighteenth- and early nineteenth-century movement in German thought called *Naturphilosophie*, or nature-philosophy. It emphasized a kind of spontaneity and sympathy in the natural world with which the properly attuned philosopher could engage—a kind of romantic conception of nature.[6] In particular, it encouraged a search for the unity and interconvertibility of forces in the world. Oersted decided to seek such a connection between electrical and magnetic forces. The relationship that he found was nonetheless anomalous; he had assumed a direct distance force similar to those associated with gravity or electrical charges.[7]

But if Oersted might have had some idea of what he was looking for, it took others completely by surprise. The French were the most taken aback, because their approach to electricity and magnetism assumed distinct electrical and magnetic fluids. When François Arago, Biot's collaborator in work on refraction in gases, presented the new phenomenon to his colleagues, they were incredulous.[8] Everyone knew that magnetic and electrical phenomena were similar, but for the French there was no reason to suppose that they interacted. The French would never have looked for Oersted's phenomenon, or remarked on it had they seen it.[9]

In his work with magnets and current-carrying wires in 1821, in the wake of Oersted's discovery, Michael Faraday made the first of his major

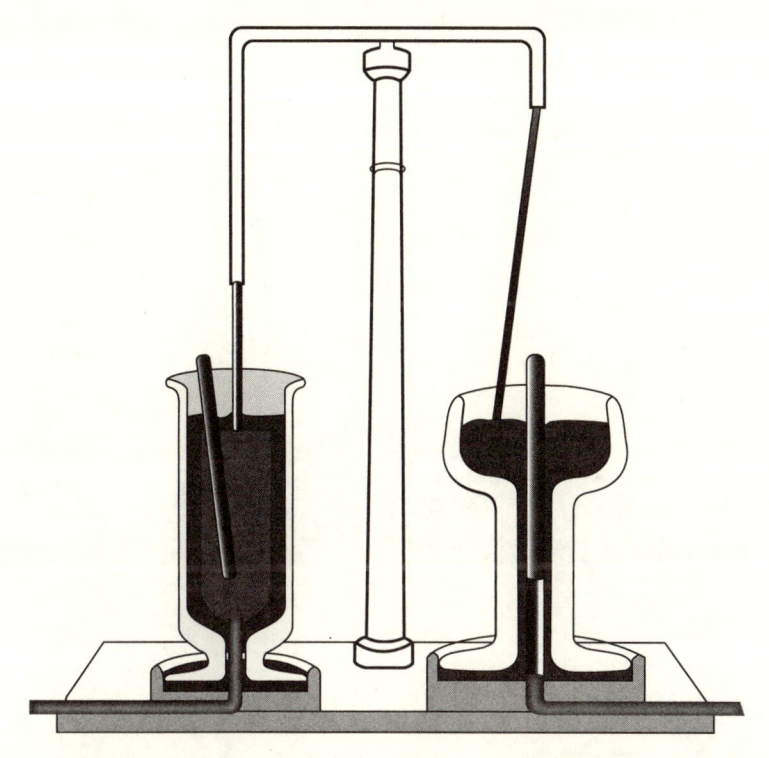

FIGURE 13.1. Faraday's electric motor. The vessels are filled with mercury.

discoveries in electromagnetism. Oersted had interpreted his phenomenon in terms of some kind of vortex spinning around the wire to explain the appropriate magnetic directional effects. Working on that idea, of motion being somehow involved, Faraday found that a magnet brought near a current-carrying wire would push the wire from side to side in such a way that the wire tended to circle around the magnet. It was then quite easy for Faraday to rig up a device whereby a wire could rotate around a fixed magnet to yield a simple motor driven by electrical current.[10]

Faraday's approach was based on the idea of circular magnetic action *around* the wire. The year before, the Frenchman André-Marie Ampère had produced an inverse square law mathematical formalism to describe the magnetic forces around a current-carrying wire, and he disliked Faraday's interpretation. Ampère saw Oersted's effect as having its physical origins *inside* the wire, propagated outward as action at a distance.

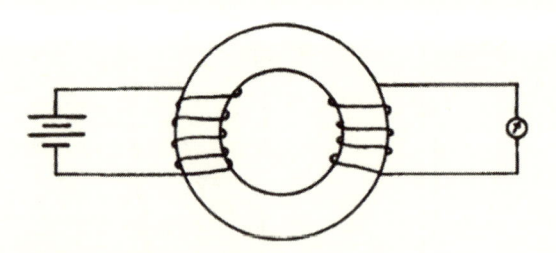

FIGURE 13.2. Faraday's induction apparatus.

FIGURE 13.3. Photograph of Faraday's induction coil (1831). This is an iron ring around which are wound two insulated coils of copper wire, one on each side. This coil is now on display at the Royal Institution, London. Photograph courtesy of the Royal Institution of Great Britain/Science Photo Library.

One of the problems with Faraday giving primacy to circular forces outside the wire was that they were not easily mathematizable. That bothered Ampère much more than it did Faraday, who always thought in concrete three-dimensional terms and never touched mathematical formalisms.[11]

FIGURE 13.4. Modern model of Faraday's dynamo for producing an electric current mechanically. Photograph by Daderot (September 4, 2013). Available under Creative Commons CC0 1.0 Universal Public Domain Dedication.

No sooner had Oersted found that magnetism could be produced from electricity than people began to look for the opposite effect—the production of electricity from magnetism—but with little success. In 1831 Faraday attacked the problem by considering the magnetic effect of a current-carrying wire on another wire—in other words, treating magnetism as basically an intermediary between two electrical phenomena. (See Figure 13.2.)

In this setup Faraday found, after some false starts, that if the secondary circuit was already completed, then when the primary circuit was closed there was a momentary current in the secondary. When the primary circuit was broken again, there was another surge of current, this time in the opposite direction. Faraday soon found that he could get an induced current of this kind just by moving the circuits relative to

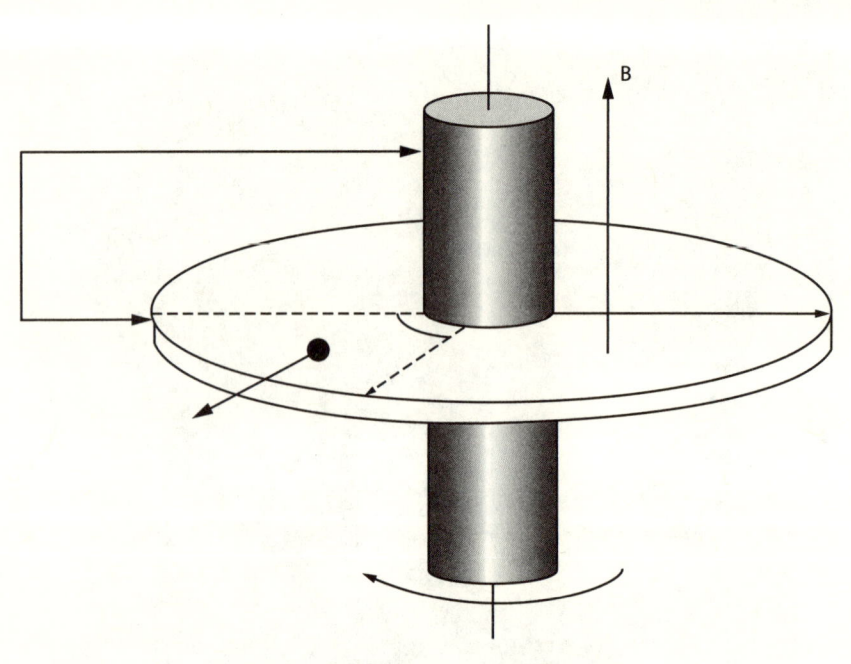

FIGURE 13.5. Faraday's spinning disc.

one another; there was no need to turn the primary circuit on and off. The motion of an ordinary magnet near a test circuit also turned out to produce a current, which indicated that it was a change in magnetic intensities, not simply the presence of magnetic forces, that produced current. Faraday used this idea to invent the dynamo, a mechanical means of generating electric current.[12]

From Action at a Distance to Field Lines

Faraday almost single-handedly invented the idea of magnetic fields. They first arose in his work in connection with a phenomenon related to the discoveries of 1831: now, the magnet and the test circuit both move, but they remain stationary relative to one another other (Faraday did this through early 1832). The entire apparatus is rotated about the axis of the magnet at the disc's center. Faraday again finds an induced electric current, even though no part of the apparatus is moving between regions of greater and lesser magnetic intensity. (If the magnet alone is rotated about the axis, no current appears at all.)[13]

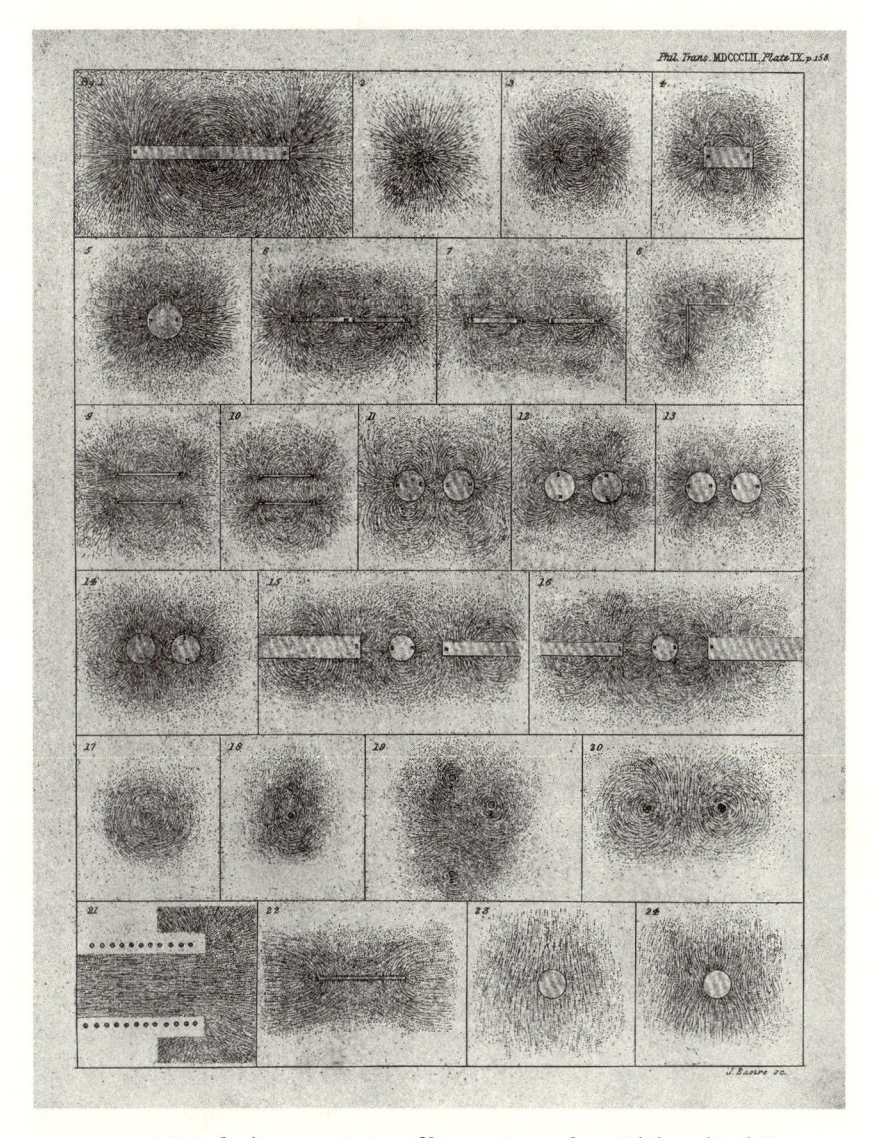

FIGURE 13.6. Faraday's magnetic iron filing patterns, from *Philosophical Transactions of the Royal Society* (1852). Photograph courtesy of the Royal Institution of Great Britain/Science Photo Library.

Faraday explained these curious results using an idea of what he called "field lines." These field lines could be represented by the patterns he knew iron filings would make when sprinkled on a card in the vicinity of a magnet. Faraday codified the production of an induced current in these setups by saying that current is induced when field lines associated with the magnet cut through the circuit. In figure 13.6, this happens because the cluster of field lines associated with the magnet remains stationary even when the magnet itself rotates around its axis. Faraday identified the magnet with a stationary bunch of field lines so that the rotating disc is still cutting through field lines; the lines themselves do not move with the rotation of the physical magnet. In the other setups current is induced in a similar way by field lines briefly expanding or contracting to cut the circuit when the primary circuit is switched on or off; between those events there is no induced current because the lines are sitting stationary in their new positions. At this stage, however, Faraday seems not to have thought of such field lines as having a physical existence in space; instead, they were a way of thinking about these phenomena in terms of the manipulation of the apparatus. Faraday's field lines were real for him in the sense that he manipulated them as he did wires and bar magnets.[14] Their reality as things with independent existence in the world was a question he held in abeyance.

Faraday's shift to a more concrete sense of the physical reality of field lines emerged from a different area of his research over the following few years. During the 1830s Faraday devoted much time to electrochemistry, as had Davy before him, and by its very nature this work focused attention on the medium between two electrodes. Prior to Faraday's work on the subject, most ideas about electrochemical effects had tended to conflate two different sorts of action in the medium. One was action at a distance, with the electrodes pulling on all regions in the medium; the other was contiguous action between the particles of the electrolytic solution itself: transport phenomena, rearrangement of particles, and so forth. Faraday became convinced that *no* action at a distance was involved; the electrodes did not pull particles directly toward them.[15]

His main argument against action at a distance in electrochemical decomposition referred to the apparent fact that the shape of the electrodes made no difference to the phenomena: the products of decomposition appeared on all sides of the electrodes, not just on the side facing the other electrode. So, thinking in terms of lines again, it looked as though the lines running between the electrodes were curved, looping around the backs of the electrodes, like magnetic field lines. The fact that the lines were curved was evidence for Faraday that no action at a distance was involved, because he thought that such a force could only exert itself in straight lines. In this electrochemical case, the lines appeared to Faraday to be most likely chains of particles in the solution.[16] By 1833 Faraday's view of electrolytic decomposition had adopted the following sequence: first, the electrodes, connected to the terminals of a Voltaic pile, polarize the particles in the solution. These particles then arrange themselves in chains between the electrodes. The strain on these chains is such that the particles constituting them continually break apart and recombine into new molecules. Loose positive and negative fragments at the electrodes themselves come out of solution rather than recombining, due to the physical interposition of those electrodes.[17] The poles are actually barriers at which the processes of decomposition and recomposition occurring in the medium are interrupted, obliging the fragments of decomposition to come out of solution.

The significance of this electrochemical picture is that, on its basis, Faraday began to consider whether ordinary electrostatic effects might work in a similar way. Perhaps a strain is set up in the medium between capacitor plates, for example, rather than there being direct action at a distance between the plates (the usual interpretation). So he did experiments to find the effects of different media put between two charged plates on the capacitance of those plates. To use somewhat later terminology, he looked at the effects of different dielectrics. He interpreted the differences he found as being due to the differing capacities of the medium to take on the strain between the plates. In both cases there are lines of tensed particles linking the plates, and these particles are polarized, just as in the case of electrolytes.

Notice that all the action occurs in the medium. The difference between insulators and conductors in this model is that insulators have a high capacity for holding strain, whereas conductors have a low breakdown threshold, analogous to electrolytes. Also by analogy with electrolytes, the lines of strain would tend to be curved and to spread apart from each other. The fact that the medium made any difference to electrostatic phenomena was in itself evidence for Faraday that action at a distance was the wrong model. The classic case of action at a distance was gravity, and gravitational attraction between two bodies was not reduced or affected in any way by the presence of a third body between them. Gravity couldn't be *screened*, whereas electrical forces could be; consequently, Faraday argued, they do not resemble gravity.[18]

That was Faraday's theory of electrostatics by 1838. There were still problems with it, and Faraday remained undecided about how to understand magnetic lines of force. He spent the 1840s in coming up with solutions to these difficulties.

Faraday, Matter, and Forces

All of this research was originally published in the Royal Society's journal, the *Philosophical Transactions*, which meant that it was widely known even far away from England. A fundamental objection to Faraday's electrostatic ideas was raised by an American, Robert Hare, of the University of Pennsylvania, in correspondence with Faraday in 1841. The air in the gap between two capacitor plates can be compressed and expanded, and therefore, said Hare, must have considerable empty space between its particles. So what does it mean to talk about chains of contiguous particles linking the plates? Surely action at a distance would still need to occur between successive, but necessarily separated, particles? Faraday's immediate response was to say that such action might indeed occur between particles separated by as much as, perhaps, half an inch. Clearly this was an unsatisfactory and ad hoc answer that appeared to undermine Faraday's entire approach to explaining electrostatic effects, because it allowed action at a distance in through the back door. Faraday seems to have realized this, and in 1844 he came up with a much less compromised solution. It took an exceedingly radical direction, however, involving a very different conception of the nature of matter.[19]

Faraday argued that atoms (ultimate particles of matter) should not be understood as being small solid lumps, as most people assumed. Had they been of that nature, the empty space between them would have to be at the same time an electrical insulator and a conductor. The atoms in materials, he observed, with shades of Hare in evidence, would always have to be separated by some empty space, because any material body can be compressed if it is subjected to enough pressure. Given that (questionable) conclusion, there should be no difference between insulating and conducting materials because either empty space conducts or it does not; it cannot do both at once.

So Faraday suggested, as an alternative to the usual view, that atoms should be identified with the forces associated with them: the idea of solid matter is rejected in favor of matter as an empirical concept inferred from a resistance to external forces. The resisting forces exerted by such atoms are centered on dimensionless points. Regarding Hare's objection, we see that in this view atoms in the air are not separated by empty space at all; they are in contact insofar as they are nothing more than the forces that enable their identification. Matter in space is reduced to force patterns interacting with one another. Such ideas, representing the apparent solidity of material particles purely by reference to the forces by which they are known, were not of Faraday's invention; versions had been propounded in the eighteenth century by Joseph Priestley and the Dalmatian Jesuit Ruggiero Boscovich, among others.[20] For Faraday, electrostatic chains constituted electrical "lines of force."

Faraday's initial parallel between magnetic field lines and these electrostatic lines of force, and his concern to understand the interrelationships between electrical and magnetic forces, came together in his subsequent attempts to account for magnetic field lines: he now wanted to show that they, too, had independent *physical* existence. In effect, this idea of point atoms, that is, atoms seen as dimensionless points identified with the forces traced from them, led Faraday to give more precedence to the lines of force themselves, treating them as the primary reality rather than the particles that the forces were supposed to constitute. Furthermore, force lines needed to be demonstrated to exist independent of ordinary matter and its configurations if they were to be more than ways of representing the consequences of material physical processes. This was no small challenge.

Faraday believed that he had finally managed to establish the real existence of magnetic lines in 1845. He reckoned that he could show that light interacted with magnetic lines of force, independent of ordinary matter, and thus that the lines were not merely ways of thinking about magnetic forces that were attributable to something else. His central

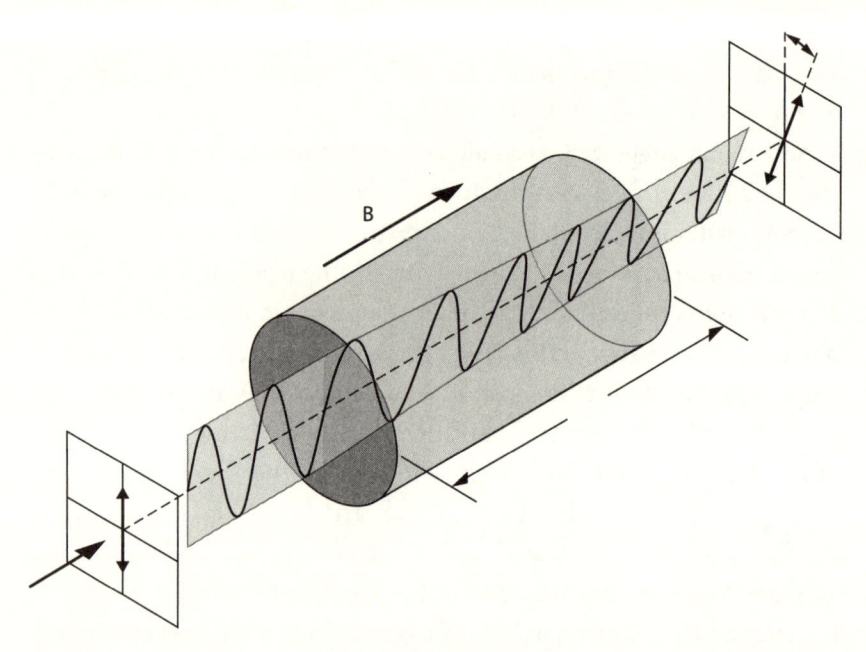

FIGURE 13.7. Magneto-optic rotation. Modified from image by DrBob, Wikimedia Commons (April 14, 2007). Available under Creative Commons Attribution-Share Alike 3.0 Unported license.

experiment involved passing polarized light parallel to magnetic lines through glass.

Faraday found that the plane of polarization would twist to the left when passing the light in one direction and to the right when passing it in the opposite direction. In other words, he got an absolute rotation depending on the mode of interaction between the light and the magnetic lines.[21] Crucially, the rotation appeared not to be a function of the arrangement of the particles of the glass; in fact, the effect was the same even if the glass was spinning.

These experiments with light convinced Faraday that magnetic lines of force had a real, independent, physical existence. That provided his field concept. Field lines were real things, and magnets were just clusters of field lines; the emphasis was taken away from matter itself. Electric currents created magnetic lines of force, which could affect matter, but were not themselves material. In 1852 Faraday brought all his arguments together in a paper called "On the Physical Character of the

Lines of Force." In it he described the propagation of field lines, mysteriously, as being by a "condition of space": "With regard to the great point under consideration, it is simply, whether the lines of magnetic force have a *physical existence* or not? Such a point may be investigated, perhaps even satisfactorily, without our being able to go into the further questions of how they account for magnetic attraction and repulsion, or even by what condition of space, aether or matter, these lines consist."[22] Magnetic lines were supported either by matter filling the space that they occupied, or else by the always mysterious aether that was otherwise known for its role in transmitting light and any other action that might require it. "Space" was never a void, in Faraday's view, and the substrate occupying it was either aether or matter, depending on the properties called for. Either way, there were physically real lines in space, and Faraday's experimental research brought him into constant interaction with them.

The question, for Faraday's scientific colleagues and for historians ever since, has therefore remained: What exactly was Faraday's notion of fields, and how did it relate to subsequent ones? The idiosyncrasies of Faraday's ideas are made clear by the very different approach to fields soon adopted by James Clerk Maxwell. Maxwell tried to interpret Faraday's field lines in terms of a mechanical model of the aether.[23] Obviously, describing field lines as the effect of motions in a material medium is different from talking about them in relation to a "condition of space." To that extent, the development of Faraday's ideas by his fellow British physicists was rather different from the way Faraday thought of them. But the crucial distinguishing feature of fields was the focus on what was going on in space, whether through a material aether or not. That sharply distinguished the British approach from that of Continental physicists, especially Germans. The latter remained committed to an action-at-a-distance view of electromagnetism for a good while longer than the British.

Action at a Distance and Aether

Faraday thought about physics in very concrete terms, probably in part because of the way he did practical science. In a sense, Faraday thought in the laboratory, with experimental apparatus, rather than at a desk, with abstract mathematical concepts. Physical lines of force were just as real for him as current-carrying wires, because from an operational point of view it seemed to him that he was manipulating both in his laboratory. Faraday had a hands-on, tangible feel for these lines of force.[24]

His contemporaries, by contrast, wanted a more satisfactory explanation of what field lines, assuming they existed, actually *were*. It was all very well for Faraday to claim that they had a physical existence, but talking about field lines did little to explain the phenomena of electromagnetism, at least from the point of view of most British physicists at the time: electromagnetism could only be explained via field lines if field lines themselves could be explained. They might exist, but what were they?

Faraday provided few answers. The idea of point atoms, or talk about field lines being propagated through a "condition of space," met with little enthusiasm. Furthermore, Faraday could give no indication even of what an electric charge was, or what it would mean for a particle to be electrically polarized, as his view of electrostatics required.

One of Faraday's ideas that Maxwell approved of was his dislike of action at a distance. Both Faraday and Maxwell tried to show that Newton had agreed with them on this point, at various times quoting a passage from a then-recently-published letter of Newton's from the 1690s in which Newton had talked about how he did not regard gravity as an essential property of matter (it was imposed by God; see chapter 1), and that it was ridiculous to think that matter could interact with matter in any way other than by contact.[25]

In the 1850s Maxwell approached these issues by imagining some kind of aetherial medium to carry action between or around ordinary matter in electromagnetic phenomena. He knew it would be difficult to imagine one that would have all the necessary properties, but he was certain that something of the kind was needed. He explained his reasons in a lecture at the Royal Institution: he suggested that it was "more philosophical to admit the existence of a medium which we cannot at present perceive, than to assert that a body can act at a place where it is not."[26]

The idea of a pervasive aether was by no means invented to deal specifically with electromagnetism. Fresnel's development of his wave theory of light in the late 1810s called for a mathematically characterizable medium through which light waves might travel, and acceptance of Fresnel's theory seemed to most people to include the implication that an appropriate medium, an aether, really did exist. This medium had to have rather inconvenient properties, however. Unlike the longitudinal pressure waves of sound, Fresnel's light waves had to be transverse (wiggling back and forth) to explain the phenomenon of polarization in terms of the plane in which the waves oscillated. This implied that a simple fluid aether would be inadequate. Fluids could support longitudinal pressure waves, but they could not carry transverse waves; some kind of rigidity in the medium was needed to sustain shearing forces. In fact, Fresnel's wave theory appeared to call for an aether that would be an elastic solid—a medium that would wobble.

By the time that Maxwell came to consider this issue of the aether, there was already a well-established conception of an aether as a light-bearing medium. But Maxwell and other British physicists differed in their approach from earlier mathematicians who had worked on light-bearing aethers in the 1830s.[27] Those mathematicians had treated aethers as hypothetical—as mathematical fictions. But in the 1840s, physical scientists began to worry about the general properties of an aether as something physically real.

The conviction that action at a distance was not acceptable was shared by almost all British physicists in the second half of the nineteenth century, though action at a distance was still alive and well on the Continent with major theorists such as Helmholtz. Unlike Faraday, most of them, including Maxwell, thought that in order to make electromagnetic phenomena truly intelligible, some kind of mechanical aetherial medium was needed to sustain them. One of those who shared Maxwell's conception of the problem was William Thomson. Like Maxwell, he thought that physical explanations ought to be grounded in Newton's three laws of motion, plus the new concept of energy. In other words, to use the appropriate Victorian term, he thought that physical theories, and the physical picture of the world generally, should be *dynamical*.[28]

The desideratum was to develop explanatory theories of nature, not just mathematical descriptions of phenomena. The point was to show that the phenomena of electromagnetism were explicable in mechanical, dynamical terms, even if it might not be possible to show that any particular mechanical model was the only possible one. Thomson was the earliest British physicist to argue this position, in the 1840s; the important thing was always to be able to show the possibility of representing physical phenomena by a mechanical model compatible with dynamical principles, however contrived and artificial it might appear.

Faraday's field line ideas were first mathematically expressed in Maxwell's work of the late 1850s and early '60s. In themselves they represented descriptive, empirical generalizations that Maxwell expressed in formal mathematical terms. Maxwell nonetheless said that Faraday's concept of lines of force had defined "with mathematical precision the whole theory of electromagnetism, in language free from mathematical technicalities."[29] Maxwell tried to supply the technicalities. But to make physical sense of Faraday's discoveries, he sought a way of representing Faraday's field lines in terms of mechanical action in a medium.[30] One of the most striking early examples of this approach was a model devised by Thomson, published in 1856, to account for Faraday's discovery in 1845 concerning the rotation of the plane of polarized light in a magnetic field, which Faraday had used as a strong argument for the physical reality of field lines. Thomson's "dynamical illustration," as he called it, of this property of a magnetic field involved arguing that light, taken as waves in a fluid aetherial medium, could be twisted in its plane of polarization by the mechanical rotation of vortices in this medium. But the precise mechanical nature of the aether remained an open question; Thomson had simply shown the possibility of mechanical explanation.

Maxwell's Mechanical Aether

In the 1850s, with Faraday's ideas about lines of force well-known and a dynamical ideal of explanation that regarded action at a distance with disfavor, Maxwell set himself the task of mathematizing Faraday's lines of force and developing a concept of an electromagnetic aether that would make his mathematical formulations mechanically intelligible.

Maxwell published his first paper on this subject, "On Faraday's Lines of Force," in 1856. The paper sets up his approach in terms of physical analogies, in this case an analogy with the motion of an incompressible fluid flowing in tubes. Maxwell didn't think that lines of force really were a matter of fluid flow; the analogy was just a way of deriving a mathematical treatment to be explained later in physical terms. But although Maxwell was unable to say what, in mechanical terms, lines of force really were, he, like Faraday, regarded them as having an independent physical existence.

He tried to carry through what he considered to be a proper physical treatment of electromagnetism in a later paper, "On Physical Lines of Force," published in four parts in 1861 and '62. Maxwell saw the equations that he had derived in the earlier paper as having been deduced from the phenomena; their validity did not rely on the correctness of any particular model. He described the status of the equations as being like Kepler's laws and therefore still requiring explanation. Kepler's laws of planetary motion from the early seventeenth century had described the motions of the planets around the sun in mathematical terms—with elliptical orbits, laws governing speed of planetary motion, and so forth. Later in the seventeenth century, Isaac Newton had, in Maxwell's view, provided an explanation of those descriptive laws in terms of his fundamental laws of motion and the inverse square law of gravity. In comparing his own earlier work on lines of force with Kepler's laws, Maxwell meant that the motions in the aether had merely been stated, with no analysis of the forces that would have to be acting to produce them. The purpose of the new article was to do the entire job—to be both Kepler and Newton.

Part 1 abandoned the fluid-flow analogy for a model incorporating what Maxwell considered to be the general mechanical properties of the medium necessary for producing the phenomena of a magnetic field. Like Faraday, Maxwell thought that magnetic lines of force had to represent some state of tension, and he followed Thomson in thinking that the phenomenon of the rotation of the plane of polarized light required mechanical rotation in the aether itself. Those two features were not simply representational models to illustrate the phenomenon or to show how it might be; at this point, Maxwell thought that they were required by the experimental evidence taken together with the presupposition that the phenomena were ultimately mechanical. This picture, as Maxwell presented it, was not a fully articulated mechanical (dynamical) model of Faraday's lines of force; it just stated the minimum necessary properties of any acceptable model.

Then Maxwell presented a detailed model of an electromagnetic aether that would show how a mechanical medium might serve to produce all the phenomena of electromagnetism, and thereby show at the least that there was no contradiction between electromagnetic behaviors and the received laws of dynamical explanation. This model took the magnetic field as being the consequence of rotating tubes, or tubular vortices, rather like Thomson's idea from 1856. These tubes were the reality behind magnetic lines of force, a concept that Maxwell took as seriously as had Faraday.[31] Filling space, they were closely packed together, like a bunch of uncooked spaghetti, and there were many of them in the volume of a molecule of ordinary matter (the conjectured size of which Maxwell did not care to specify). For mathematical purposes, Maxwell treated these long tubes in segments, as small cylinders. For any one of these cylinders the direction of the axis of rotation would correspond to the direction of the line of force at that point in space, and the intensity of the field at that point was a function of the cylinder's rate of rotation. The important thing for Maxwell was to render the phenomena physically intelligible by inventing mechanical models that would produce the same behaviors as those found by the experimenter.

Maxwell's full mechanical model of the aether presented in the 1861 article was pretty comprehensive. Having decided that he needed these tubular vortices for magnetic field lines, he now needed to explain exactly how they worked mechanically as the moving parts of a mechanical device. He saw a serious difficulty: since the spinning tubes needed to be tightly packed together, Maxwell had to explain how they could be rotating in the same direction in any particular region of space. If they were mechanically linked, interacting and rolling against each other, each successive contiguous vortex would be rolling its neighbors

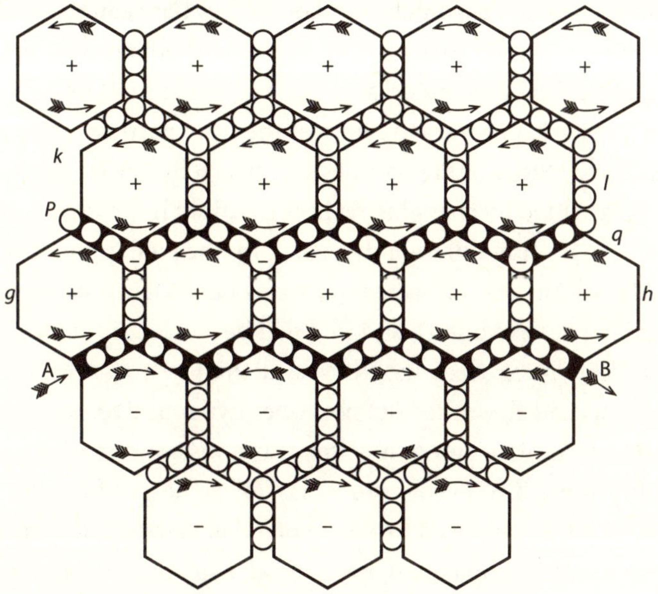

FIGURE 13.8. Maxwell's mechanical aether model. Maxwell's diagram from the *Philosophical Magazine* of 1861 showing the rotating vortices represented by hexagons and the idle wheels between them. (Arrow direction errors in original.)

in the *opposite* direction. Maxwell took this mechanical necessity as seriously as it was possible to do: he introduced small "idle particles," making up layers between successive vortices, which acted like little gear wheels to ensure that the vortices in any particular region all rotated in the same direction.

Having established their mechanical necessity, Maxwell then identified these unexpected ball-bearing particles with particles that, when in translational motion, constitute electric current. These particles, as they move along between vortices, correspond to the flow of an electric current, and the tangential action of the particles against the vortices affects the rate of spin of the vortices themselves. (Notice that these are not charged particles, like electrons, but purely mechanical devices.)

Imagine, then, the vortex portions of the aether seen in cross-section as cells, separated by layers of idle-wheel particles. The cells, when not spinning as vortices, represent the absence of a magnetic field. If the idle-wheel particles are somehow made to flow between the cells, however, they set the cells rotating (by dragging them around, in effect), which corresponds to the creation of magnetic field lines. In other words, an electric current (the flow of the little particles) has set up a magnetic field (the rotating vortices). Conversely, a changing magnetic field is represented by changing rates of rotation. If there are successive vortices with slightly different rates of spin, they will provide a net directional push on the idle-wheel particles sandwiched between them (the particles are extruded), and that will be an electric current.[32]

Remarkably, Maxwell managed to get all this to work so that it satisfied his equations and all worked out in proper dynamical terms, abiding by the three Newtonian laws of motion plus conservation of mechanical energy. But something even more remarkable emerged from the model. By looking at the measured forces involved in electrostatics and correlating them with the imagined mechanical distortion of the aether (the dielectric constant for a vacuum), Maxwell derived a figure for the aether's coefficient of rigidity. Using existing mathematical analyses, he then obtained a figure for the velocity of transverse waves in the aetherial medium, determined solely in terms of empirically determinable electrostatic and electromagnetic constants. It turned out to be close to the measured speed of light in a vacuum.[33]

The identification of the speed of light with the speed of waves in Maxwell's electromagnetic aether was not accepted immediately. Among other things, it required a high degree of confidence in the empirically measured speed of light in a vacuum, as well as acceptably accurate determinations of the relevant electromagnetic quantities. Only if those two sets of figures could be brought into agreement could Maxwell claim to have demonstrated that light was probably just this disturbance in the aether. This was an empirical issue, a matter of measurement and the techniques involved in making measurements to a satisfactory degree of precision, as well as a theoretical one. The question was how close the quantitative agreement needed to be in order to justify a claim that the numbers revealed an identity between light and vibrations in Maxwell's aether. After all, such an agreement would never be perfect, and even if it were it would not logically entail that one thing was in its nature the same as the other.

Maxwell and his British colleagues did not regard absolute precision on this point as crucial. In the first edition of his famous book of 1873, *A Treatise on Electricity and Magnetism*, Maxwell sidestepped the question by simply saying there was no evidence that the relevant figures were significantly different. In effect this was trying to shift the burden of proof away from himself and put it on the shoulders of anyone who might want to deny that light was a disturbance in Maxwell's electromagnetic aether. And most British physicists were quite happy to go along with Maxwell on this point. At that stage, quantitative precision in the relevant measurements was not, for them, an issue. German physicists were less readily satisfied.[34]

Measurement

A related concern in British science at this time concerned standards of measurement in general. How accurate did "close enough" have to be? There were fierce debates, for example, over whether the metric system would be better than the British Imperial system of feet and yards and miles and pints. In 1863 John Herschel gave a talk on the subject that complained that the French metric system was not any more scientific, or rooted in nature, than the Imperial system of weights and measures. There was, he said, nothing intrinsic to the meter that made it preferable in any absolute sense to the yard. And after all, the Imperial system was *British*: "England is beyond all question the nation whose commercial relations, both internal and external, are the greatest in the world.... Taking commerce, population and area of soil then into account, there would seem to be far better reason for our continental neighbours to conform to *our* linear unit."[35] Such nationalistic concerns were consciously injected during this period into debates over the determination of standards and the required accuracy of measurements of natural constants.[36]

Maxwell's mechanical model was never intended to represent the true nature of the aether. Its purpose was to show that the phenomena of electromagnetism were consistent with dynamical assumptions. There was an aether, certainly, but there was no way of telling whether its precise mechanical structure resembled the model. Because of this, Maxwell was rather embarrassed by the fact that his derivation of the rigidity of the aether was hard to justify *without* the model, which involved an ad hoc conception of molecules of matter that included the introduction of the so-called displacement current when electric potential was applied across a region of space. He tried to reformulate his ideas without using a specific model, but the derivation of the velocity of light looked fairly persuasive anyway.[37]

Maxwell certainly liked the general notion of a mechanical aether. This is what he said at a Royal Institution lecture in 1873:

> The vast interplanetary and interstellar regions will no longer be regarded as waste places in the Universe which the Creator has not seen fit to fill with the symbols of the manifold order of His Kingdom. We shall find them to be already full of this wonderful medium; so full, that no human power can remove it from the smallest portion of space, or produce the slightest flaw in its infinite continuity. It extends unbroken from star to star, and when a molecule of hydrogen vibrates in the Dog Star, the medium receives the impulses of these vibrations and after carrying them in its immense bosom for three years, delivers them in due course, regular order and full tale into the spectroscope of Mr. Huggins at Tulse Hill.[38]

(This number for the distance of Sirius in light-years is sufficiently different from the modern figure of about 8.6 light-years as to show again the problems that could be involved in making precise empirical measurements.)

Maxwell never seems to have considered that electromagnetic waves could be generated by variable electric currents, although that was later found to be implicit in his mathematical theory. He seems to have thought of light waves as being generated by physical vibrations of molecules in the aether. Maxwell died in 1879, and it was the German physicist Heinrich Hertz who showed in the 1880s that variable currents should generate electromagnetic waves (that is, radio waves), and who detected their production experimentally in 1886 and '87. His work rendered the earlier measurement issue moot, because after its publication no one doubted that light was an electromagnetic disturbance—although not necessarily one in a mechanical aether.

In England, nineteenth-century physicists after Maxwell continued devising mechanical aethers, always in the same spirit as Maxwell: to show that electromagnetism could in principle be reduced to mechanical, dynamical terms. They had to explain how ordinary bodies could pass unhindered through what seemed to be a necessarily solid aether, and they wanted to include gravitational attraction as well. This kind of model building continued through the 1890s.[39]

The acceptance of Maxwell's electromagnetic theory on the Continent did not typically carry with it the mechanical interpretation common in Britain, but Continental interpretations varied quite a bit. Helmholtz, who had been central in the creation of the principle of conservation of energy in the late 1840s, took Maxwell's theory as an essentially descriptive mathematical account of phenomena. He wanted to separate the system of equations from any particular associated physical interpretation, like Maxwell's aether, and continued to use action-at-a-distance conceptualizations, focused on the interactions between charged bodies, rather than field ideas that placed action in the space around bodies. Hertz had started out in his own work from the perspective of Helmholtz's approach, and it was on that basis that he performed his experimental production and detection of electromagnetic radiation. But the result of that work for Hertz was that when he found that electromagnetic radiation was an empirical reality, he became convinced that there really were fields in an aetherial medium; they were not simply conventional contour lines drawn on a map for the purposes of calculation, but things that could be manipulated. Hertz adopted the more authentic Maxwellian approach in place of Helmholtz's, in other words, and his discovery convinced most other Continental physicists of the existence of an aether of some kind—although not necessarily a *mechanical* aether—to support electromagnetic waves.[40]

By the end of the nineteenth century, then, there was a general tendency among Continental as well as British physicists to jettison action at a distance. The field itself was now taken to be a real thing in space—a primary physical reality.

14

The Chemical Use of Atoms

VIEWS OF the underlying structure of matter varied widely during the eighteenth century, and continued to do so throughout the nineteenth century. The most common view was Newton's particulate conception, with distance forces operating between the particles. But, as we have seen, not until the work of French natural philosophers in the late eighteenth century did such views become closely integrated with predictive theories. Even so, the work of such as Fresnel on light showed that the particulate view of nature need not always be adhered to, and that some kind of continuum might on occasion do better.

Nonetheless, in all the physical sciences, including chemistry, reduction to particles remained the default choice of most practitioners. Lavoisier's list of the "simple substances" involved in chemical composition failed to characterize them in terms of unique particles, preferring to restrict their discussion to phenomenological properties experienced in the laboratory. But as we have seen with Berthollet, and with Lavoisier himself in his earlier consideration of combustion, particulate conceptions of material bodies were always in play among chemists themselves. Accordingly, new ideas about chemical particles associated with weighing procedures that had become increasingly common among eighteenth-century chemists would attract considerable interest when they were put forward by John Dalton.

John Dalton

John Dalton (1766–1844) was an English Quaker who spent most of his scientifically productive life, from 1793 onward, in Manchester, in the industrial north of England. He earned his living as a teacher of mathematics and natural philosophy and was not, at least to begin, primarily a chemist.[1] What led him into chemical questions and prompted his famous atomic ideas was meteorology. He was interested, in the years around the turn of the century, in problems to do with the behavior of the atmosphere, the expansion of gases and its relation to heat, and the amount of water vapor that gases at different temperatures and pressures could hold. He did experimental work on these things, especially concerning the water-holding capacity of gases—although not at a very refined level quantitatively—but he was continually trying to develop a theoretical conception of what was happening physically.[2]

That involved thinking about the gases in the atmosphere in terms of their particles. Dalton took it for granted that the atmosphere is made up of a variety of distinct gases, which was a newly accepted idea thanks to Lavoisier and Priestley. There was also nothing remarkable in conceptualizing them as made up of particles: that was the long-standing, standard Newtonian view. In the "Queries" Newton had spoken of aerial fluids made up of mutually repelling particles. Lavoisier's fixed air escaped from its fixed state by virtue of the particles becoming surrounded by matter of fire, which made them push one another apart; in taking analogous views Dalton was being entirely conventional. In the opening years of the nineteenth century, Dalton extended a particulate conception of gases into questions about the solubility of various gases in water. He thought of this in terms of the different kinds of gas particles being kept in solution by the pressure of the same gas on the surface of the water. The particles of each kind of gas were mutually repelling, as with Newton's model of gases, but exerted no forces on the particles of other kinds, hence the independence of their pressures. But that way of thinking posed a problem, because Dalton knew by experiment that some gases were intrinsically more soluble than others. Carbonic oxide was particularly soluble, for example.[3]

Given Dalton's generally Newtonian, forces-and-particles approach to such matters, he thought of the differences in solubilities as being related in some way to the differences in the sizes, associated with weights, of the particles constituting the various gases. Bearing in mind that, after Lavoisier, gases were chemical substances just like solid and liquid substances, Dalton should be described as seeing chemical substances as being made up of particles of different, characteristic weights; they might or might not be gases.

That was in 1803, and these ideas pointed Dalton in the direction that led finally to his 1807 lectures in Edinburgh and Glasgow—soon published as a book—on what he called his "doctrine of atoms."[4] These were supposed to be (again, his term) the "primary elements" of bodies.

Chemically, in post-Lavoisierian chemistry, Black's fixed air was called carbonic oxide in the form of a gas, which meant that somehow those two substances, carbon and oxygen, were chemically combined in the gas. Dalton's conception of the particulate nature of gases led him to think of the fundamental units of carbonic oxide as being compound particles made up of, in the simplest interpretation, one atom of carbon joined to one atom of oxygen; the exact numbers of each had to be guesswork on his part.

This atomic interpretation of chemical combination, and the turning of Lavoisier's "simple substances," his principles, into full-fledged, fundamental elements, extended quite easily beyond gases by simply pointing to work that had already been done by some Continental chemists on the constant proportions of weight that seemed to be involved in the formation of compounds. Of course, not all of the "simple substances" could be transformed in this way: light and caloric headed Lavoisier's list, but, having no weight, did not lend themselves to the same treatment: "The most probable opinion concerning the nature of caloric, is, that of its being an elastic fluid of great subtilty, the particles of which repel one another, but are attracted by all other bodies."[5]

But Dalton argued that the constant proportional weights in the formation of compounds (although this was by no means uncontested data) were due to the combination of constant small-number ratios of atoms, and that the actual weights reflected the relative weights of the

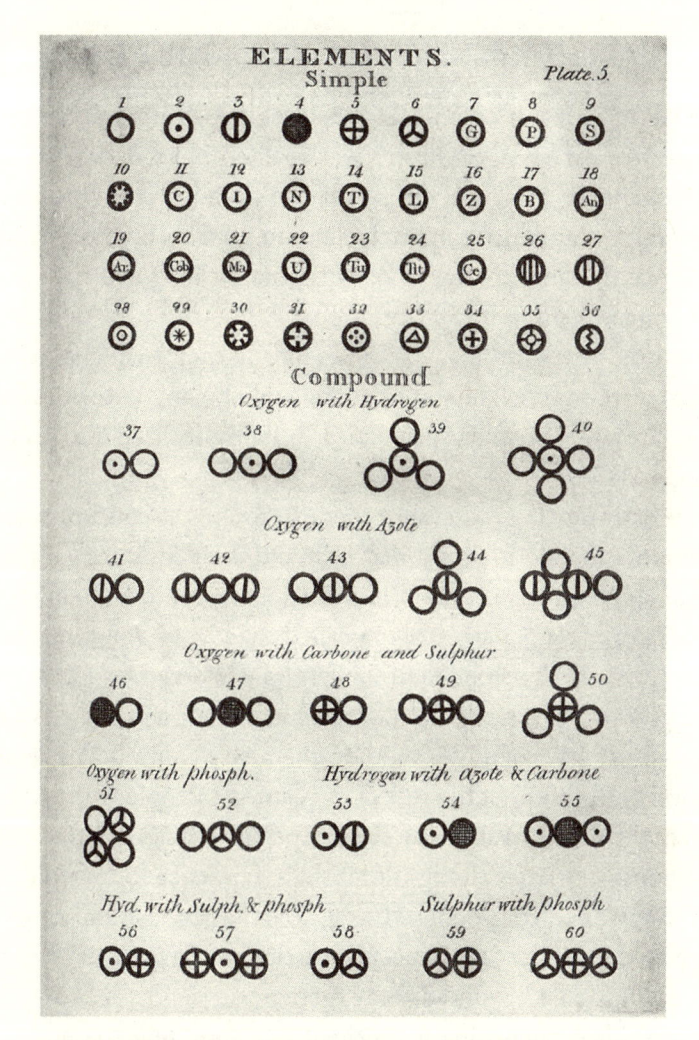

FIGURE 14.1. John Dalton's "atomic" particles in *A New System of Chemical Philosophy* (1810). Courtesy of Science History Institute.

atoms themselves. The regular, constant weight ratios found when two elements combined chemically would then be a function of the simple numerical combination of the elementary atoms to form "compound atoms." Dalton assumed that these would usually be the simplest possible combination, of one atom of each constituent element. Water, for example, would be one atom of oxygen and one of hydrogen.

The Early Career of Daltonian Atomism

John Dalton's idea of chemical atoms, set out definitively in 1808 in his book *New System of Chemistry*, was an attempt to make sense of the relative quantities of substances involved in chemical combination. Dalton thought that identifying each element with its own fundamental particle was the simplest way to explain what each of Lavoisier's numerous chemical "simple substances" actually was and how they formed compounds. These "atoms," as well as having their own characteristic weights, were indivisible and unchangeable by any known means—rather like Stahl's "chemical principles," but physicalized into property-bearing particles.[6]

In the early nineteenth century chemists found the concept of chemical elements and atomic combinations (and their constant weight proportions) increasingly useful. While nineteenth-century chemistry was certainly a product of Lavoisier's so-called chemical revolution, Dalton's work at the same time tended to subvert it by asking the sorts of fundamental physical questions that Lavoisier had made a point of avoiding. These distinct ways of understanding chemistry continued to generate controversy throughout the rest of the century. People who were convinced of the reality of atoms as discrete particles that were the ultimate constituents of matter thought that their existence followed directly from the evidence. But people who rejected the proposition argued that atomism was no more than a speculation that, while perhaps sometimes useful, went too far beyond direct experience.

Dalton's ideas, then, did not meet with a resounding acceptance. His work certainly made a big impact—Dalton changed chemistry, and modern chemical formulas and symbolic expressions for chemical reactions were developed as a result of his approach—but most chemists were reluctant to accept his theory at face value. Part of the problem was having so many different types of elementary particle, each with its own characteristic weight and chemical properties; this was not economical matter theory. Practically all natural philosophers, not just those in Britain, assumed that matter had some kind of particulate structure, presumably with forces between the particles, in accordance with Newton's model in the *Opticks*. But the point about that view was that all matter would be fundamentally the same. One might have particles of matter in different combinations and arrangements, but not fundamentally different types of particles, as Dalton was suggesting. Daltonian atomism, taken literally, was counterintuitive, or inelegant. But at a methodological level, chemists such as Humphry Davy said that there was simply no direct evidence for Daltonian atomism.

This was a particularly appropriate thing for chemists to be saying in this period, given how chemistry had developed during the preceding century. Following Stahl, chemists had focused increasingly on real chemical operations and less on speculation about the true nature of matter and the physical character of chemical combination. That had culminated in Lavoisier's so-called chemical revolution, which was more of a culmination of those tendencies in eighteenth-century chemistry than a true revolution. Lavoisier's "simple substances" were explicitly presented as purely pragmatic chemical principles rather than true elements. They represented the practical limits of current chemical analysis; Lavoisier did not claim that they were fundamental constituents of the universe or give any suggestion of what they were at the particulate level.

In the wake of Dalton's work, most chemists held that chemical experience still provided insufficient grounds for the inferences Dalton tried to make about fundamental particles. Dalton had drawn attention to a new regularity in chemical behavior, namely the constancy of weight ratios in chemical combination; however, he also wanted to interpret those weight ratios as reflecting underlying *atomic* weight ratios. Most chemists saw no justification for taking that step, however tempting it might be. In practice, the standard way, at first, of taking Dalton's ideas seriously was to talk not about "atoms" but about "equivalents"— the term was originally introduced by a leading British chemist, William Wollaston, in a paper in the Royal Society's *Philosophical Transactions* in 1814.[7] Where Dalton would speak of one atom of oxygen combining with one atom of hydrogen to produce water, Wollaston talked instead about one oxygen equivalent combining with one hydrogen equivalent, defined by the relative weights of the combining reactants.[8] So the weight of an equivalent, as measured in the laboratory, actually meant operationally the same as what Dalton called "atomic weight," without needing to speak of atoms.

Oxygen was most often used as the standard for such measurements because it was involved in so many reactions, although Dalton had used hydrogen, the lightest element, as his basic unit. Chemistry textbooks from this period onward often talked about atoms and atomic weights when they were discussing these aspects of chemical combination. But they always made it clear that the terms should not be taken literally and that the real meaning of the Daltonian atomic idea lay in chemical weight measurements—the things that equivalents highlighted.

There were, however, some people who took Dalton's ideas literally. In 1815 William Prout, who later wrote one of the Bridgewater treatises on evidence of the divine in nature, came up with what came to be called "Prout's Hypothesis."[9] Prout suggested that what Dalton called atoms were not literally indivisible elementary particles, but were themselves aggregate particles made up of hydrogen atoms, which were the only truly simple atoms. The implication was that all atomic weights should be exact multiples of the atomic weight of hydrogen. Unfortunately, many elements did not in practice yield integral atomic weights, although some came close. But this failure of Prout's empirical research program to corroborate his hypothesis, while providing no support for a literal interpretation of chemical atomism, did not hurt it either. It just meant that his version seemed inadequate.

Volumes, Equivalents, and Avogadro's Conjecture

Besides Prout's, another hypothesis from the second decade of the nineteenth century had in the long run a great deal more significance for chemical atomism. Although "Avogadro's hypothesis" was by no means ignored, most chemists found it unpersuasive, for various reasons, for about half a century.[10] Avogadro's hypothesis emerged from work published in 1809 by the French chemist Joseph Gay-Lussac, a member of the Arcueil circle headed by Laplace and Berthollet. He had been looking at reactions between gases and measuring not weights but the relative volumes of the reactants, and of the product as well, if gaseous. Gay-Lussac found that the volumes were almost exactly either equal or in simple ratios.[11]

For the reactions that he studied, the ratio of volumes approximated either one to one or two to one. The closest was the reaction to produce water, for which Gay-Lussac got a hydrogen-to-oxygen volume ratio of 1.97:1. The nitrogen and oxygen reactions gave the two forms of oxide, also with outcomes close to small whole-number ratios. He also investigated reactions between ammonia and acidic gases like marine acid gas (later called hydrogen chloride) to form salts.

Gay-Lussac suggested that these results were too close to exact whole-number ratios to be simply accidental or fortuitous, and that they seemed to be good evidence for the theory recently proposed by John Dalton. Rather remarkably, Dalton responded by being quite offended; he was critical of Gay-Lussac's results, and tried to suggest that they were due to experimental error.[12] Dalton's complaint was that although the volumes were simple ratios, they weren't the *right* simple ratios. Reacting nitrogen with oxygen to saturation to get nitrous gas left the same volume of product as there had been of initial reactants: one hundred volumes of nitrogen reacted with one hundred volumes of oxygen to yield two hundred volumes of nitrous gas. The modern representation of that, in post-1860 chemistry, after the general acceptance of Avogadro's hypothesis, is $N_2 + O_2 = 2NO$. But Dalton thought that elementary gases like oxygen and nitrogen had to be *monatomic*; it made no sense according to his conception of gases that identical gas atoms could join together into compound atoms. Dalton's commitment to a Newtonian static-gas model, where atoms of the same gas repelled each other, made it implausible that they would attach in that way. And if monatomic oxygen combined with monatomic nitrogen, then a volume of nitrous gas, with one oxygen and one nitrogen stuck together, would occupy *half* the combined volumes of the reactants, if Gay-Lussac and Dalton were both right. Since that was not the case, Dalton decided that Gay-Lussac was wrong; after a bit of reflection, Gay-Lussac decided that Dalton was wrong.

Gay-Lussac's conclusion was not, however, simple retaliation, although questioning an experimentalist's skill was, and is, no slight matter. Instead, it relied on an acceptance of Dalton's own inferences about the implications of his findings. Gay-Lussac just had greater faith in the reliability of his own experimental work. He had to reject a Daltonian atomic interpretation because it looked as though nitrous gas particles would have to consist of a *half* atom of oxygen and a *half* atom of nitrogen, and half-atoms would be contradictions in terms. So Gay-Lussac adopted the position that most chemists in England, including Wollaston, were in the process of choosing: he decided to adhere to the experimentally accessible idea of "volumes" and leave atomic inferences aside.

Although it really applied only to chemical species in the gaseous state, this approach was the same in practice as the idea of equivalents for weights, and both terms became common. When the Swedish chemist J. J. Berzelius introduced what amounts to the modern notation for chemical reactions, in 1813, he cited the half-atom problem as his grounds for rejecting a Daltonian interpretation. So he stuck to talking about "volumes," or, strictly speaking, the weight of a unit volume, as the actual referent of a chemical symbol, such as an O for oxygen: the O would stand not for an atom of oxygen but for the weight of oxygen in a standard unit volume. Instead of half-atoms, he could then simply talk about half volumes, which presented no conceptual problems.

Meanwhile, Amedeo Avogadro, an Italian teacher of physical sciences in Turin, had published a paper in 1811 criticizing Dalton's criticisms of Gay-Lussac. It was clear to all who paid attention that a Daltonian atomic interpretation of Gay-Lussac's law of combining volumes assumed that equal volumes of the relevant gases, under the same conditions of temperature and pressure, would have to contain the same number of atomic particles or molecules (Avogadro preferred the term *molecule* over Dalton's word *atom*, perhaps because of its emphasis on weight). Dalton saw no reason why that should be the case (and, indeed, none was offered, by Avogadro or by Gay-Lussac himself), and used the point as another stick with which to beat Gay-Lussac.[13]

But Avogadro took a different tack: he accepted the rule of equal numbers of particles as an empirical implication of Gay-Lussac's discovery, whether or not it could be explained, and criticized Dalton for trying to deny it or to explain it away. Avogadro simply stated, as a general principle applying to all gases (not just the particular sets of reactants involved in Gay-Lussac's and others' experiments), that a given volume of any gas, at a given temperature and pressure, contains the same number of molecules as any other gas under the same conditions. Crucially, he avoided the half-atom problem simply by saying that gas molecules could be split; unlike Dalton, he had no objection to "diatomic" molecules of the same kind, as with oxygen or nitrogen molecules. Such axiomatic treatment of allegedly empirical facts allowed Avogadro to skirt theoretical difficulties by ignoring them.[14]

An additional advantage of Avogadro's idea was that it enabled the determination of molecular formulae in a fairly direct way, rather than assuming the simplest possible numerical ratios, as was the norm. The usual assumption was that water was HO, but based on combining volumes and Avogadro's version of the atomic hypothesis, it would be H_2O, because twice the volume of hydrogen was used up as of oxygen (this inference also required assuming that both oxygen and hydrogen gases are composed of diatomic molecules). By the same principles, ammonia was NH_3.

Avogadro's idea quickly became known. William Prout used it, for example, and so did the French student of electromagnetism André-Marie Ampère, although each claimed to have invented it himself—which is possible, since it is almost presupposed by Gay-Lussac's work. But it failed to win general acceptance until after 1860. This may seem surprising, because it looks as though Avogadro's hypothesis, including diatomic elementary gases, makes perfect sense of Gay-Lussac's results in terms of atoms.

But it was an appealing idea only for someone determined to use a chemical atomic theory. Assuming that there are Daltonian chemical atoms, Avogadro could show how they must behave if Gay-Lussac's law were to hold; but without that assumption Avogadro's hypothesis is a mere curiosity. Avogadro's hypothesis provided no additional grounds for believing in chemical atomism than existed already. It was a completely peripheral idea for decades, and thinking in terms of "volumes" or "equivalents" seemed to take care of practical matters for most chemists anyway.[15]

In fact, there were serious objections to a literal acceptance of a doctrine of chemical atomism. The French chemist J. B. Dumas, who invented the modern terminological distinction between *atom* and *molecule*, supported atomism and Avogadro's hypothesis in the 1820s, in connection with his work on vapor densities. But by 1837 experimental anomalies found in the cases of mercury, sulfur, and phosphorus vapors made simple acceptance more complicated, related specifically to uncertainties related to the distinction between atom and molecule.[16] Furthermore, while the Swede Berzelius had become converted to the idea of atoms of some kind in the 1820s, he nonetheless rejected Avogadro's idea because, like Dalton, he had his own theoretical reasons, to do with an electrical theory of chemical combination, for rejecting diatomic elementary molecules.

By 1860 these sorts of objections and problems had largely vanished. But because Avogadro's hypothesis had never really been in the forefront of most chemists' minds, somebody had to draw attention to that fact before the hypothesis could be seriously considered again. Chemists had become used to disregarding it or thinking of it as having various problems to which they seldom devoted much attention.

One of the problems that had quietly disappeared was to do with the vapor anomalies investigated by Dumas. Sulfur and phosphorus, the two chief culprits, gave strange results that seemed to imply variable atomic weights depending on temperature. But it had since been determined that this was probably because they had different *forms* in the vapor phase at different temperatures—molecules with different numbers of atoms. That had become a sensible interpretation after work in the 1850s showing a general phenomenon of thermal dissociation of other kinds of chemical substances, not simply elements, in the vapor phase.[17]

Similarly, experimental evidence had been developed in favor of the crucial idea that a gas like oxygen really was diatomic, whether or not that idea fitted one's favored theory of bonding. In the 1840s two French physical scientists, Pierre Favre and Johann Silbermann, found that carbon burning in nitric acid gave off more heat than when it burned in oxygen. They interpreted the difference in terms of the greater heat required to dissociate the oxygen molecule. It was the first experimental evidence that such an entity might exist—that gaseous oxygen should indeed be represented as O_2.

In 1860 a major international chemical conference took place at Karlsruhe in Germany. It saw the reemergence of Avogadro's hypothesis as something worth taking seriously, now that most of the theoretical and experimental difficulties were dwindling away. An Italian chemist, Stanislao Cannizzaro, circulated a paper he had printed beforehand detailing how the old objections to Avogadro's hypothesis were no longer valid and pointing out its advantages. He had clearly timed his intervention well; very quickly the consensus of most chemists—although certainly not all—was in favor of it.[18]

One of the consequences of accepting Avogadro's hypothesis was that one could now talk sensibly about relative atomic weights (ultimately determined from measuring vapor densities), exactly as Avogadro had suggested. Atomic weights no longer seemed arbitrary once there was in effect a way of counting, at least comparatively, the number of molecules in a given volume of gas and the number of atoms making up those molecules. (There was still no theoretical physical explanation of why there should be equal numbers of molecules in equal volumes of gas at the same temperature and pressure, until the British physicist James Clerk Maxwell showed it to be an implication of the new kinetic theory of gases.[19]) In any case, by the 1860s the majority of chemists were prepared to think in terms of atoms of some kind. This was mostly a result of various kinds of structural chemistry, centering on organic chemistry and isomers.

Thinking about Atoms: Visible Dyes and Invisible Molecules

By the 1820s a number of chemists (including Michael Faraday) had found that conventional analysis of a variety of organic compounds turned up some groups of substances possessing the same proportions of the same elements: things like acetic acid and cellulose, or starch, gum Arabic, and sugar—so-called isomerism. Before then it had been generally assumed that the properties of a compound were determined by the proportions of its constituents. Isomerism obviously violated that assumption; acetic acid and cellulose are clearly different things. Berzelius, Gay-Lussac, and many other chemists decided that the most plausible explanation of isomerism was that, although the same proportions of elements were present, they should be regarded in terms of atoms, arranged and joined together into molecules in structurally different ways. As organic chemistry took off as a research field in the 1820s, new isomers continued to be found.

But these considerations, and the general acceptance of Avogadro's hypothesis, failed to end the controversy over the reality of atoms. During the second half of the nineteenth century most chemists talked easily about molecules, atoms, and atomic weights; now, with a definitive means of determining atomic weights, there followed the construction of the periodic table by Dmitri Mendeleev in the late 1860s.[20] Nonetheless, there were still some chemists who held the position that atoms were just a hypothesis because there was no direct evidence for them: nobody had ever seen an atom, and molecules and molecular structures, they thought, were just convenient fictions for organizing real chemical experience. Chemical formulae could always be interpreted formalistically, without talking about what they "really meant" at the level of matter theory. That was what a number of British chemists, especially, argued in the 1860s.[21] Even the German inventor of the famous benzene ring idea, August Kekulé, asserted firmly in 1867 that the idea of chemical atoms should not be taken literally "in a philosophical sense."[22]

The first major exponent of organic chemistry in the nineteenth century was Kekulé's teacher, Justus Liebig. He set up the first dedicated university chemistry laboratory for turning out not just results but also students who, once having received their doctorates, left to populate other universities and take those interests and techniques with them. Liebig's importance lies in his establishment of research chemistry, specifically organic chemistry, in German universities.[23] He set the pattern at the generally rather undistinguished University of Giessen, one of the smallest of the independent principalities before German unification in 1871.

Liebig came to Giessen in 1824 after having spent some time studying in Paris, especially with Joseph Gay-Lussac. Over the course of the next few years he used his increasing personal reputation, especially through the medium of offers of professorships at other universities, to get ever better research facilities for both himself and his doctoral students. He developed a kind of organic chemistry factory, he and his students publishing their results (chiefly analyses of organic compounds), and the university therefore also turning out a lot of doctorates in organic chemistry. This organizational structure for scientific research was to become a model for research science in the nineteenth century.[24]

Liebig's success as an academic entrepreneur emphasized two principal areas of utility for chemical studies. One was agricultural chemistry. In 1840 Liebig published a book titled *Chemistry as Applied to Agriculture and Physiology*. This was the first proper examination of the elementary substances that plants needed other than those from water and carbon dioxide (then known as carbonic oxide). Liebig argued that they needed other things from the soil as well, especially nitrogen compounds. The other main area where practical uses of academic chemistry seemed promising was in producing a major—indeed, prototypical—chemical industry: the aniline dye industry.

Aniline was derived from coal tar, which had long been an object of interest because it was a plentiful industrial by-product. Coal gas started to be used as a source of lighting around 1810.[25] Its production involved the destructive distillation of coal to produce a mixture of inflammable gases such as ethane and ethylene. The process left behind a sludge called coal tar, for which profitable uses became an immediate focus of interest. By 1820 the more volatile ingredients were being separated out by distillation and used as a turpentine substitute or a solvent for rubber. This was the origin of the raincoat known as the mackintosh, made by treating fabric with rubber dissolved in a coal tar derivative. Creosote was made from it, and other fractions were used for waterproofing paper. By the 1840s organic chemists, the first being the Liebig-trained and then London-based August Wilhelm Hofmann, were taking an interest in coal tar derivatives, and their commercial importance was a significant factor. Hofmann first isolated and named aniline (for its bluish appearance, ironically seen only when impure).

A number of chemical substances were isolated from coal tar during the 1840s and '50s that were already known from other organic sources. Benzene, for instance, had been discovered by Michael Faraday as a product of the gas evolved by decomposing fish oil. Benzene was one of the most abundant coal tar constituents, and Hofmann's aniline, present to some degree in coal tar, was easily produced from benzene. This plentiful raw material was a principal reason for the subsequent enormous growth of the aniline dye industry.[26]

The emergence of this new science-based industrial chemistry coincided with major developments in theoretical organic chemistry. These centrally involved the elaboration of structural chemistry with the idea of valency. One of the major inventors of these theoretical approaches in the 1850s and '60s, which became, and remains, absolutely fundamental to organic chemistry, was August Kekulé, himself a German chemist and a former Liebig student.

One of the most important developments in organic chemistry, giving chemists firm theoretical control over what was happening in their (hitherto rather hit-or-miss) dye syntheses, was Kekulé's invention in 1865 of the benzene ring theory, which gave the first effective structural account of the components of these synthetic dyes. Their molecules could be understood as being made of a number of benzene ring structures linked together, with various auxiliary groups like amino groups attached to them—hence the importance of aniline, which in Kekulé's terms comprised a benzene ring with an NH_2 group taking the place of one of the hydrogens.

The new model-building techniques of structural chemistry inevitably made the atoms and molecular structures tangible and immediate. Isomers, for example, as molecules made up of the same atoms arranged differently with respect to one another, surely made the idea of chemical atoms more credible even before Stanislao Cannizzaro's reintroduction of Avogadro's hypothesis. Nonetheless, even Kekulé expressed doubts about the literal reality of chemical atoms.[27]

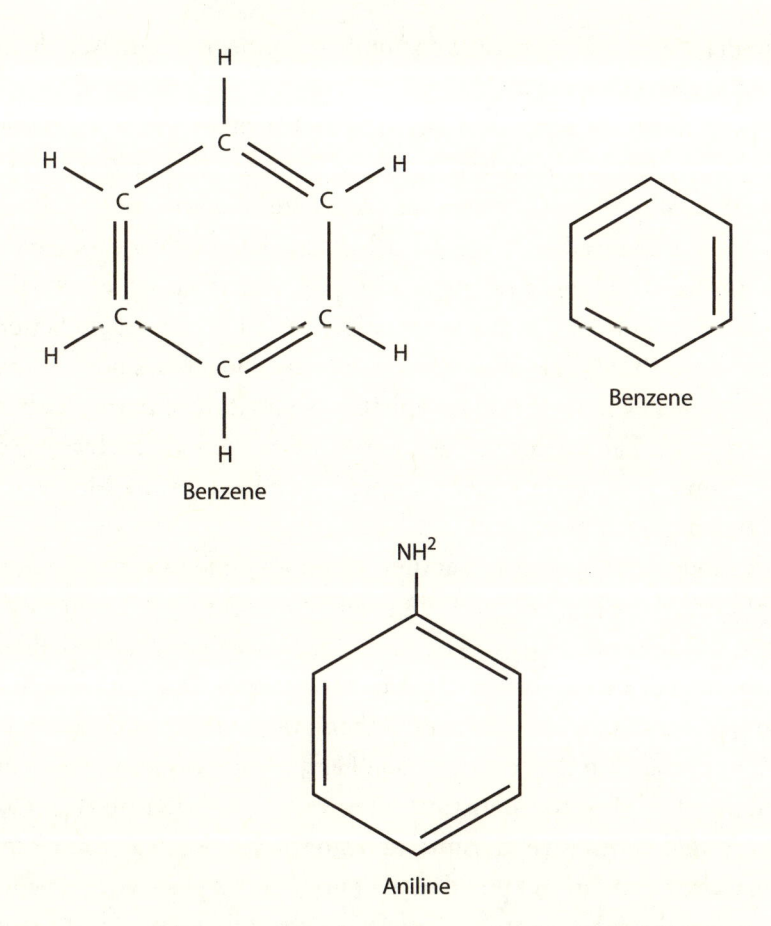

FIGURE 14.2. A benzene ring and an aniline molecule, after Kekulé's work.

One might suppose that, even if there were chemists who rejected Dalton- or Avogadro-style chemical atomism, no one by this time would reject some kind of physical atomism, at least in the sense of material substances being ultimately made up of particles. The development of the kinetic theory of gases by Maxwell and Ludwig Boltzmann in the '60s and '70s provided powerful new resources for the physical sciences. But there were a number of significant dissenters into the first decade of the twentieth century. They included major chemists, such as the German-Latvian Friedrich Wilhelm Ostwald, in many ways the founder of physical chemistry, and quite a number of physicists, chiefly in France and Germany.

The philosophical position shared by such people was known as positivism, and it was particularly strong in France in this period but had adherents all over. Broadly speaking (and there have been many different versions), it involved making a sharp distinction between phenomena on the one hand and theories on the other. Unobservable—theoretical—explanatory entities like atoms are either treated as convenient fictions that help with thinking about the phenomena but should not be taken as real, or they are rejected completely as illegitimate parts of a true, "positive," science. This second, hard-line version characterizes the antiatomic positivism widely adopted at the end of the nineteenth century.[28]

Although it was, even at that time, a minority scientific position (except perhaps in France), part of the attraction of that position by the 1890s was the availability of a new area of science that could be held up as a model of a true, positivistic science of phenomena. Classical thermodynamics eschewed conjectural theoretical entities such as atoms, being restricted to the coordination of empirical measurements. From that point of view the statistical interpretation of thermodynamics, which *did* hypothesize invisible atoms or molecules, was completely unnecessary, and in fact polluted the purity of this positivistic science. The pinnacle of this form of thermodynamics at that time was the work in the 1870s of the American J. Willard Gibbs, who applied thermodynamics to phase changes, and Ostwald's own work on chemical equilibria.[29] This was experimentally founded science that made no reference to anything other than observable, measurable phenomena such as quantity of heat, and temperature; no conjectures were needed about the structure of matter. (We shall see later that Einstein was to bring similar attitudes to his work on relativity.)

Hence the rejection of the doctrine of atoms. People like Ostwald said that it was unnecessary to treat matter as particulate, and that there was no direct evidence for the proposition anyway. The interesting thing about the hard-line positivist position in this period is not that it was dominant, but that it was viable. It was only in the first decade of the twentieth century that the kinetic theory of heat and the idea of atoms, including Einstein's work on Brownian motion and the studies of the French experimentalist Jean Perrin, became so firmly integrated with direct evidence that even Ostwald backed down and finally allowed the existence of atoms in 1908.[30]

By then the focus had already shifted from questions about the atomic constitution of matter to the internal structure of atoms themselves, along with the development of quantum theory stemming from Max Planck's statistical work on heat radiation.[31] Atoms started being taken for granted as real things in the world only when the question of what they were like became more interesting than whether they existed.

15

Laboratories of the Heavens

PHYSICS IN THE OBSERVATORY

THE STUDY of the universe inside a laboratory was a central part of science in the nineteenth century, just as it was in the eighteenth and twentieth. But not all sciences were laboratory sciences: not natural history, and not even one of the physical sciences, astronomy. Astronomy in the nineteenth century was the most prestigious of all the sciences in that it represented the ideal of a true science. The great statesman of British science around midcentury, William Whewell, called it the "pattern science" for that reason: it provided the methodological model for the sciences.[1] Astronomy in that context meant positional astronomy, concerned with the positions of celestial objects in the sky and how those positions change over time—primarily, therefore, the astronomy of the solar system. This was a system governed by Newtonian mechanics and universal gravitation, with sophisticated mathematical techniques to deal with the motions of planets and comets—often called celestial mechanics—as well as techniques to deal with sources of error. A science, in this view, was a deductive structure based on principles drawn from careful observation of nature, which yielded precise, accurate predictions.

Positional Astronomy and the Stars

The first half of the nineteenth century saw some notable achievements in positional astronomy that served to confirm it as the model any true science ought to emulate. The most impressive was the prediction of the existence of a new planet, with specification of its position and its orbit, on the basis of observed irregularities in the orbit of the newly discovered planet Uranus. The computational work that achieved this result was carried out independently by the mathematicians John Couch Adams in England and Urbain Jean Joseph Le Verrier in France in the mid-1840s, followed by German observational confirmation of the new planet, soon named Neptune, in 1847. Adams was credited only in retrospect, his predictions not having been pursued until after Le Verrier's.[2]

The other major achievement in positional astronomy in the first half of the century was especially notable because it went beyond the solar system to deal with the stars, for the first time providing a direct measure of stellar distances. Unlike the prediction of Neptune, this was an achievement that relied on a high level of instrumental and observational refinement rather than on purely mathematical skill.

Measuring the distances of the stars held tremendous potential significance for understanding the large-scale structure of the heavens.[3] But by the time that observational measurements appeared, no one was in the least bit surprised. This was crucial: the acceptance of the measurements depended on the fact that no one found them implausible. The difficulty of the observations and the scope for error required an enormous amount of trust in the skill of the observer.

The technique involved the measurement of annual stellar parallax. This is a slight shift in the apparent position of a star in the sky as the earth, carrying the observer, moves from one side of the sun to the other. Because over this period the star is observed from different positions, it will appear to be in slightly different directions in the sky—the parallactic shift. A number of people had tried in the later seventeenth century and the first three decades or so of the eighteenth century to measure stellar parallax. The main incentive in the seventeenth century was not so much to measure stellar distances as to provide direct evidence that the earth orbits the sun, but by the beginning of the eighteenth century that had ceased to be an issue; natural philosophers no longer needed convincing. Astronomers stopped bothering after the late 1720s, when the British astronomer James Bradley decided on the basis of his own

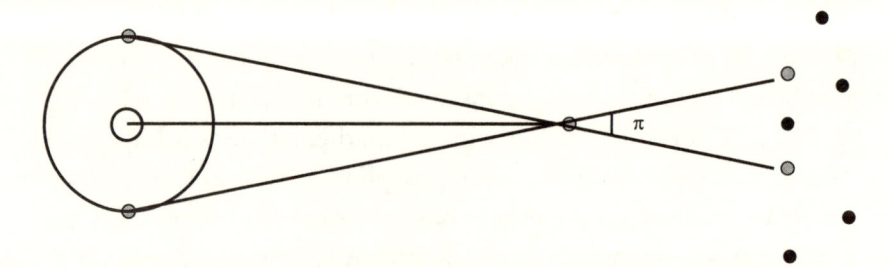

FIGURE 15.1. Annual stellar parallax. π is the parallactic shift.

observational studies that stellar parallax would have to be less than one second of arc, because otherwise he would have been able to detect it with his instruments, and he could not.[4]

But the question of the stars' distances were not by this time regarded as especially interesting. Simply demonstrating that the earth orbits the sun was no longer of concern, and natural philosophers and astronomers believed that they already knew the kinds of distances that separated the stars, including the sun, from one another. This conviction was based on nothing more than the assumption—well-entrenched by 1700—that the stars were other suns. On the assumption that all stars, including the sun, are of about the same intrinsic brightness, variation in the apparent brightnesses of stars as seen in the sky would be due to their differing distances. So a star's distance from the earth could be estimated in comparison to the distance of the sun by comparing their respective brightnesses. One person who attempted this in the later seventeenth century was Isaac Newton.

It was his estimate, published posthumously in 1728, that made the biggest impact in the eighteenth century—both because it was Newton's and because it was consistent with Bradley's parallax estimate at almost exactly the same time. Newton calculated the nearest, which is to say brightest, stars as being about one million earth-sun distances (what are now called astronomical units) away. Bradley said that if measurable stellar parallax was less than one second of arc, then the nearest stars had to be at least four hundred thousand astronomical units away. Such broad agreement appeared to settle the question to much better than an order of magnitude.[5]

Interest had been revived by the beginning of the nineteenth century, when various astronomers began thinking that measurement of stellar parallax might now be feasible. This was because of increasing refinement in instrumental capabilities and especially in new techniques for processing the data. It was a matter of making all the appropriate adjustments for sources of error like atmospheric refraction, instrumental variations at different temperatures, and observer biases—the so-called personal equation to describe the consistent, systematic measurement errors of individuals, They could also make use of the new error curve. The task was certainly not just a matter of jotting down numbers.[6]

Quite a few failed attempts were made during the first couple of decades of the new century, but they were serious attempts, and techniques were steadily refined. One of the main questions concerned the choice of stars in looking for parallax. The obvious ones were bright stars, which seemed most likely to be closest on the assumption that brightness indicates nearness (which had also been central to William Herschel's stellar-distribution work). But by 1800 quite a few faint stars had been found with large proper motions (motions across the sky with respect to other stars); it was reasonable to assume that if a star, even a faint one, showed a large proper motion, then it was probably quite close. So a limited number of candidate stars for parallax work started to emerge: ones with large proper motion or with strong apparent brightness combined with at least detectable proper motion.

At the end of the 1830s, three different observers had announced figures for stellar parallax (all of around a third of a second of arc): the Germans F. W. Bessel (61 Cygni, 1838) and Otto Struve (Vega, announced 1840), and the English astronomer Thomas Henderson (working at the Cape of Good Hope on Alpha Centauri, 1839). Henderson announced his result just after Bessel; he might well not have been credited if his results had not generally agreed with the others. He had not been working at it for as long, with such good equipment, or produced the same volume of data, but he had the British scientific community to promote his claims. All three astronomers had reported results in keeping with what was generally expected (and therefore with little grounds for doubting their competence); no one had had serious doubts about stellar distances for over a century. The successful measurement of annual stellar parallax was primarily an achievement in observational and instrumental techniques rather than something with real cosmological significance. But at the time it was the pinnacle of classical positional astronomy. Motions and distances of stars were becoming accessible in a way similar to the case of planets in earlier astronomy. For all the effort and difficulty, however, it only amounted to a modest extension.[7]

Nebulae and Reflecting Telescopes

Astronomy in the broader sense—investigations of the variety and nature of objects in the sky, as well as cosmological questions concerning the large-scale structure of the universe—remained much more speculative and tentative. For a long time these issues focused on a specific problem area that descended from work done in the late eighteenth and early nineteenth centuries by William Herschel, to do with the nature of nebulae.

Herschel had been interested in the evolution of the heavens as well as its overall static structure, and he pursued both of those interests by paying particular attention to nebulae. His central question was whether or not nebulosity—patches of apparent cloudiness in the sky—was due to some kind of glowing gas or vapor in space distinct from stellar systems (true nebulosity) or was just an optical effect of enormous agglomerations of stars at distances too great to allow for optical resolution into separate component stars. The kind of nebulae called globular clusters appeared clearly to be agglomerations of stars, so from a practical point of view the issue for Herschel was to determine whether *all* apparent nebulae were ultimately resolvable into individual stars.

This was not easy to determine. Using telescopes with ever-larger apertures capable of revealing greater detail and fainter objects might finally result in the resolution of some particular nebula into constituent stars. However, that would not show that there was no such thing as true nebulosity, if there were still plenty of other nebulae that had not been resolved. On the other hand, if subjecting some nebula to examination using ever-larger telescopes did *not* succeed in resolving it into stars, that could not show that there *is* such a thing as true nebulosity, because the possibility always remained that a still larger telescope would be able to resolve it.

William Herschel's opinions on the matter changed during the course of his career: in the 1780s, he thought that there was no true nebulosity, but starting in the 1790s he decided that there probably was. He built that view into his ideas about stellar evolution, with stars condensing out of nebulous gas and associating in groups due to gravity (see chapter 2). Herschel's ideas on stellar development were paid relatively little heed by most astronomers during his lifetime (he died in 1822). This was chiefly because they did not count as astronomy—the precise, certain, quantitative Newtonian science that served as the model for all others—by the standards of the time. Real astronomy, for most astronomers in the earlier nineteenth century (and to some extent throughout the century) was the sort of thing that Adams and Le Verrier, or Struve and Bessel, did, concerning accurate and precise measurement of celestial motions. Describing and classifying nebulae, what Herschel called the "natural history" of the heavens, scarcely counted as astronomy at all for the university-trained mathematical astronomer.[8]

It is only fitting, then, that it should have been a nonmathematical amateur enthusiast who turned the spotlight back on nebulae in the 1840s. William Parsons, who in 1841 became the third Earl of Rosse, was an Irish nobleman and landowner (the local village in Ireland was called Parsonstown). He had more money than he knew what to do with, so he built telescopes. But these telescopes were not designed for precision positional measurement. They were reflectors, as William Herschel's had been, designed to detect very faint objects and to have high resolving power—to see new things, or new aspects of things, in the sky rather than to measure accurately their positions. For precise positional work, like producing star catalogues or measuring parallax, astronomers normally used precision refractors.

But there were severe practical restrictions on how big objective lenses for refractors could be without getting bad distortion from the sheer weight of the glass making the shape of the lens sag. So although a big reflector had no advantage over a refractor for positional work, it was much easier to make one with a very large aperture, which means a high degree of light-collecting power. The bigger the aperture, the fainter the objects that can be seen; big mirrors could be made and mounted more easily than big lenses. Parsons was interested in the sorts of qualitative aspects of the heavens that made big reflectors worthwhile. Although William Herschel had kept to himself details on how he made his telescopes, Parsons tried and got very good at it.

Telescope mirrors at this time were usually made of an alloy of copper and tin called speculum metal (*speculum* being Latin for mirror); among other advantages, speculum metal tarnished less quickly than silvered glass. Parsons's first big telescope, in 1839, had a mirror three feet in diameter—much bigger than Herschel's eighteen-inch-aperture reflecting telescope (although Herschel had also made one with a mirror four feet in diameter that he had found too unwieldy for regular use). But Parsons produced his real claim to fame in 1845: a mirror six feet in diameter, the telescope for which was so large and unwieldy that he had to have it slung between two thick brick walls because wooden frames were not strong enough to support it. This monster was dubbed the Leviathan of Parsonstown: the mirror alone weighed four tons, and the tube was fifty-six feet long. The setup meant that it could be properly moved only up or down, with little room to swing it from side to side. The observer had to wait for the object of interest to come up, as the heavens revolved, toward the meridian, the telescope hung between walls aligned north-south.

Parsons's project began paying off immediately. Observing as many theretofore unresolved nebulae as he could (he had not picked the best location for the weather), Parsons claimed to be resolving large numbers of them into stellar systems. Unlike Herschel, Parsons made the telescope available to other observers, so there were plenty of experienced astronomers to confirm the veracity of his nebular drawings (astronomical photography still lay in the future). The most sensational claim of all was made in 1846 regarding the great nebula in Orion. The Orion nebula is the largest nebula visible in the sky and one of the earliest identified as such in the seventeenth century, when telescopes had first come into use.[9] Parsons claimed to have resolved it into stars, and

FIGURE 15.2. Lord Rosse's six-foot reflecting telescope. *Source*: De Luan/Alamy Stock Photo.

that claim was soon confirmed by other observers at Harvard, using a large refractor. That established, as the conventional wisdom by around 1850, that there was no such thing as true nebulosity. If even the great nebula in Orion was a massive star cloud (many branches of which, however, still remained hard to resolve), then there was every reason to suppose that all nebulae were.

Some astronomers, including John Herschel, still thought there might be such a thing as true nebulosity, but this became an unsettled question in the face of the new claims about resolvability. Instead, the standard picture of the stellar universe reverted to the one that Kant had first advocated in the eighteenth century, of countless island universes like our own Milky Way. If all nebulae were really, or probably, collections of stars, then the obvious explanation for those still unresolved was that they were very far away and therefore very large. Parsons had seen some of them as having a spiral structure, and spirals and disk shapes fitted the model of the Milky Way as we see it in our own island universe very well.[10]

FIGURE 15.3. A Rosse drawing of the Orion nebula as seen through his big telescope (1867).

But that picture lasted only into the 1860s. From then until the 1920s, the island universe idea was abandoned again in favor of the assumption that all nebulae are formations of some kind at similar distances to those of ordinary stars. The rejection of the island universe occurred as a direct result of an innovation in astronomy that allowed entirely new techniques to be applied to learning about the nature of stars and nebulae—techniques that formed no part of traditional astronomy. This innovation was the invention of astrophysics, which itself rested on the development of spectroscopy. Spectroscopy provided convincing new

FIGURE 15.4. A modern photograph of the Orion nebula taken by NASA/ESA's Hubble Space Telescope (2006). Notice how modern photographs do not show "how things really are"; they are just the products of different ways of creating an image.

arguments that true nebulosity existed, thereby removing the grounds for the island universe conception entirely.[11]

Spectral Images

Spectroscopy as part of astronomy developed in the nineteenth century in two principal stages. The first was the discovery of what came to be called the Fraunhofer lines in the solar spectrum. Joseph von Fraunhofer was an optical instrument maker specializing in precision lenses in

Munich in the first half of the nineteenth century; Struve used Fraunhofer's instruments for his parallax investigations. In 1814 Fraunhofer was testing different types of glass to determine their differential refractive properties, using the sun as one of his light sources. He noticed that the solar spectrum (produced by a prism) was striped with a large number of fine dark lines.[12]

Fraunhofer mapped out the lines, giving letters to the most prominent ones; he counted a total of 576. The lines drew a lot of attention (and the English chemist William Wollaston had already noticed a few in 1802, in fact). But despite this, and even though Fraunhofer also looked for lines in bright stars like Sirius, he had no special interest in them. Nobody else knew what to make of them, either. As a consequence, there was little further study of the lines for several decades.

In the wake of Fraunhofer's discovery, the idea developed that there was a link between the dark lines on the solar spectrum and spectra that could be made in the laboratory. The laboratory spectra were made by passing the light from incandescent material through a prism. They showed discrete bright lines against a black background—the inverse of solar spectra. It was generally assumed that the spectrum of a chemical element was somehow characteristic of it, akin to a fingerprint, but there seemed to be too many anomalies. Depending on the physical conditions under which a chemical species was made to yield a spectrum, especially whether an electrical spark or a flame generated the incandescence, the spectrum of that element would look different. Also, different elements seemed to have lines in common, thereby weakening the idea that each one could be characterized by its unique array of lines. The bright yellow line corresponding to Fraunhofer's solar D line, for instance, appeared everywhere. Not until well into the 1850s was a consensus established that this line was really just a sodium line that kept popping up because of impurities—the presence of common salt—in the materials under study.

By the end of the 1850s chemists generally agreed that spectra did give unique fingerprints to chemical elements, but there was still no theoretical basis for understanding it. The difference between these discrete bright lines and the dark lines cutting across the otherwise

FIGURE 15.5. Chromolithograph of a table of spectra by Heinrich Schellen in *Spectrum Analysis in Its Application to Terrestrial Substances, and the Physical Constitution of the Heavenly Bodies* (1872). Courtesy of Science History Institute.

continuous solar spectrum was a mystery, too, although it looked as though at least some of the lines coincided in wavelength. Only when sense was made of those correspondences could astronomical spectroscopy begin in earnest.

The door was opened in 1859 when the chemist Robert Bunsen (of Bunsen burner fame) and the physicist Gustav Kirchhoff, both at the University of Heidelberg, compared solar and laboratory spectra. They concluded that the dark lines in the solar spectrum were *absorption* lines, representing wavelengths in the spectrum that were absorbed by the gaseous medium through which the light of the sun passed after its initial production on the sun's surface. These dark lines could be intensified by passing the light through different substances in the gaseous state in the laboratory, or even entirely created by passing an artificial continuous spectrum from an ordinary flame through gas.[13]

Over the next couple of years Kirchhoff established the following picture: incandescent solids or liquids generally give off light that displays a continuous spectrum (this is in the visible region), whereas incandescent gas yields a spectrum showing discrete bright lines on a black background. That latter case occurs with laboratory flame spectra, for instance, where the substance is being vaporized in the flame. But when a cooler, nonincandescent element in the gaseous state has a continuous spectrum passed through it, it selectively absorbs the same wavelengths of light that it emits when incandescent.

Kirchhoff's interpretation of the solar spectrum followed directly: the sun has an incandescent liquid surface surrounded by a cooler gaseous atmosphere; that gaseous envelope produces the absorption lines in the continuous spectrum of the light passing through it from the glowing surface. The characteristic patterns of those lines enabled the identification, by comparison with the arrays of bright emission lines found in spectra produced in the laboratory, of the chemical elements present in the solar atmosphere.

Not surprisingly, application of these results both to investigation of the sun and to other celestial objects followed quickly. In 1863 Kirchhoff published a diagram of the solar spectrum identifying absorption lines with the chemical elements (mostly metals) corresponding to them.

Others took up the job, too, attempting to produce better wavelength scales for spectra. Diffraction gratings soon came to be used in place of prisms, since the former provide spectra with linear dispersion; thus the lines in different investigators' spectra could easily be compared with each other.[14]

Amateur Astrophysics

Investigation of the sun was quickly supplemented by attempts to look at stellar spectra to see if other stars were made of the same stuff. Fraunhofer had noted fairly casually that the spectra were not identical, but now it became possible to look at the problem in a serious way, with some idea of what the spectral lines meant. The English amateur astronomer William Huggins was the first to move into this area, beginning in 1863, by trying to apply Kirchhoff's work to stars. He set up a laboratory in his garden at Tulse Hill in south London to produce comparison spectra as well as having his astronomical instruments and spectroscopes. In a sense, his was the first purpose-built astrophysical laboratory, and it was only fitting that a new disciplinary label was soon to be attached to this new kind of astronomy, one with its own claims for the first time to being a laboratory science.[15]

The work Huggins set for himself was challenging. As he said in a paper of 1864 in the Royal Society's *Philosophical Transactions*, the faintness of stars made it very difficult to see anything but the strongest lines in their spectra; not until the 1870s were they successfully photographed. Huggins's most spectacular discovery at this stage therefore took the form of a broadly qualitative observation made possible by the spectroscope.

Huggins decided to use his equipment to look not just at individual stars but also at some nebulae. He looked at a "planetary nebula; very bright; pretty small; suddenly brighter in the middle, very small nucleus. In Draco."[16] (Planetary nebulae are so called because they appear as disks with measurable diameters, like planets and unlike stars.) Huggins expected to see a continuous stellar spectrum with some lines in it, because everyone knew, thanks to Lord Rosse, William Parsons, that nebulae were generally collections of stars.

On August 29, 1864, I directed the telescope armed with the spectrum apparatus to this nebula. At first I suspected some derangement of the instrument had taken place; for no spectrum was seen, but only a short line of light perpendicular to the direction of dispersion. . . . I then found that the light of this nebula, unlike any other ex-terrestrial light which had yet been subjected by me to prismatic analysis, was not composed of light of different refrangibilities, and therefore could not form a spectrum. A great part of the light from this nebula is monochromatic, and after passing through the prisms remains concentrated in a bright line occupying in the instrument the position of that part of the spectrum to which its light corresponds in refrangibility.[17]

Emphasizing the role of laboratory work in this novel science of the heavens, Huggins explained, "The positions of these lines in the spectrum were determined by a simultaneous comparison of them in the instrument with the spectrum of the induction spark taken between electrodes of magnesium. . . . The faintest of the lines of the nebula agrees in position with the line of hydrogen corresponding to Fraunhofer's F [line]."[18]

Kirchhoff's interpretation of such phenomena informed Huggins's understanding of what he had seen: "Subsequent observations on other nebulae induce me to regard this faint spectrum as due to the solid or liquid matter of the nucleus, and as quite distinct from the bright lines into which nearly the whole of the light from the nebula is concentrated. . . . The colour of this nebula is greenish blue."[19]

Huggins's spectroscopic initiatives soon defeated the island universe view of the stellar universe. Most unresolved nebulae, like the spirals, showed some emission-type characteristics superimposed on continuous spectra, suggesting at least a mixture of stars and true nebulosity. Since these nebulae therefore no longer needed to be seen as huge, distant star systems, they ceased to be seen as exceptionally far away or particularly enormous, whereas the Orion nebula, which had large apparent size, turned out to be chiefly composed of true nebulosity.[20]

Astronomical Photography

The explosion of activity in astronomy that followed the advent of stellar spectroscopy involved the establishment, especially in the United States, of new observatories equipped with ever-larger telescopes. It was a matter of the virtual invention of a new branch of science: spectroscopic analysis of starlight could be used to investigate an almost unprecedented field: the experimental physics of the heavens, or what quickly came to be called astrophysics.

In the twentieth century the universe that was conceptually accessible to astronomers grew enormously larger. At the same time, helping to make that possible, the instruments, apparatus, and size of research groups also grew enormously larger. The money represented by such expansions also expanded. The emergence of what has been called "big science" first occurred in its starkest form with the advent of big telescopes in the late nineteenth and early twentieth centuries. They were expensive, and they were American.[21]

Observational astronomy, including the new field of astrophysics, relied above all on technical instrumental capabilities: the ability to make very accurate measurements, whether of precise stellar positions or, now, of precise spectral line wavelengths, and the ability to detect very faint objects. The basic technical requirement for this new area of science was a means of intensifying faint optical images. In the specific cases of stars and nebulae, not only were the original images as seen through the telescope often very faint in themselves, but if that light was spread out into a spectrum it became much fainter still. Being able to identify absorption or emission lines in the spectrum depended on somehow getting more light to get a brighter image.

The fundamental first step was to use a telescope with a large aperture so as to grab and focus as much light from the observed object as possible. After that, the only way to proceed further was through long-exposure photography. The telescope's mount would be driven by a motor (clockwork or electric) so that it could track the object's apparent motion through the sky due to the earth's rotation. Then, by letting the light from the object fall on a photographic plate for minutes or even hours at a time, an image of the object, or of particular features of it, could be obtained that might be invisible to an observer looking directly through the telescope; so similarly for spectra generated from that light.

The rapid changes in stellar astronomy and astrophysics in the last two decades or so of the nineteenth century perhaps owed even more to new developments in photography than to the building of larger telescopes. When William Parsons had been making his claims about the resolvability of nebulae in the 1840s, photography was a brand-new invention, and the early techniques simply were not sensitive enough to create a decent photograph of a telescopic image—the light was just too faint to affect the plate. So Rosse's claims, and those of others using large telescopes, had to be taken on the word of the observer and the evidence of the observer's drawings.[22]

In 1851 a new photographic process was invented, wet plate photography, that was more light-sensitive and could be used for some astronomical purposes (such as photographing the moon). But it was still of limited use for long exposures, because wet plates dry out. Exposures were limited to no more than ten or fifteen minutes, and the limitations in plate sensitivity were still such that the resultant image was no more revealing than what could be seen by the eye directly through the telescope. Astronomical photography had become possible, but it represented no advance, no qualitative jump, in what astronomers were capable of seeing.

The innovation that opened up a new era in astronomical and astrophysical research was so-called dry plate photography, developed by 1874. It permitted exposures of practically unlimited duration, so that the effective light-collecting power of the apparatus—telescope plus camera—was vastly increased. Faint objects like nebulae, in particular, could now be photographed to show new levels of detail. And spectra of faint objects could be produced that allowed improved identification of absorption and emission lines.[23]

The difference this made to the kinds of things astronomers could investigate was potentially enormous. In 1884, just as these new techniques were being fully recognized and exploited, an American astronomer, Samuel P. Langley, at the U.S. Naval Observatory, remarked that the traditional job of the astronomer, "until very lately indeed, has still been to say *where* any heavenly body is, and not *what* it is."[24] But things were changing; the nature of celestial objects was now a direct object of investigation. The popularizing book in which Langley said this was *The New Astronomy*, a title also used by Huggins (and, canonically, in 1609 by Johannes Kepler).

Strikingly, however, in 1888 another American, Simon Newcomb, also a specialist in precision positional astronomy, claimed that it "would be too much to say with confidence that the age of great discoveries in any branch of science had passed by," yet "so far as astronomy is concerned, it must be confessed that we do appear to be fast reaching the limits of our knowledge."[25] Newcomb regarded the progress of astronomy as a matter of cataloging things and categories that were already established, and this included the study of stellar spectra (part of what he called "physical astronomy"): a great observational description and classification of the universe.

Astrophysics and Large Telescopes

Huggins established the existence of genuine nebulosity once and for all with his production of emission spectra from some nebulae. But he soon showed the uses of spectroscopy for other kinds of investigations, too. Sirius is the brightest star in the sky, and thus the best one for revealing absorption lines clearly. Huggins claimed in 1868 that the light from Sirius indicated that it was moving away from the earth. He inferred this by measuring a slight displacement of the array of absorption lines toward the red, or lower-frequency, end of the spectrum, compared with their terrestrial or even solar counterparts.

Huggins interpreted the shift as being due to the so-called Doppler effect. This had been first proposed by Christian Doppler, at the German university in Prague, in 1842: the observed frequency of light waves emitted by an object should vary depending on the object's motion. The usual analogy for this is the same effect in the case of sound waves: a stationary observer will hear a train whistle as a higher note as the train approaches and as a lower note after it passes and recedes even though the speed of the waves through the medium carrying them is unaltered. Similarly, the light from a star will appear to be shifted to a higher frequency if the star is approaching the earth and to a lower frequency if it is receding. That constitutes a so-called red shift, marked by the move of familiar arrays of spectral lines toward the red, or lower frequency end of the spectrum. Huggins reported that he had found such a shift in the light emitted by Sirius. With advice and assistance from Maxwell, Huggins calculated that the shift corresponded to a radial velocity (that is, a velocity directly toward or away from the observer) of about thirty miles per second.[26]

The possibilities opened up by the new astrophysics, actively explored in the closing decades of the century, were enormous. Spectroscopy provided a basis for determining stellar chemical composition, stellar radial velocities, and details of physical conditions on the surface of stars (inferred from the effects of temperature and pressure on the absorption lines produced in the laboratory). Closer to home, the nearest available star, the sun, could be studied for its surface details, including things like magnetic field characteristics in the vicinity of sunspots, also found through examining the absorption lines in those regions.

In this spectacular new field, success was determined by access to elaborate equipment and the money to pay for it. At this period the United States, as well as having an established tradition of observational and experimental science, had the best resources of any country to move into a new area rapidly. The United States had started to take the lead in the building and use of large telescopes around the middle of the nineteenth century. In the 1850s a portrait painter in Massachusetts, Alvan Clark, who had taught himself lens grinding as a hobby, started to develop a reputation as one of the best lens makers in the world (Huggins used one of Clark's eight-inch lenses for his spectroscopic studies in the 1860s).[27]

In 1873 Clark made a telescope with a twenty-six-inch lens for the U.S. Naval Observatory, beating another recent English telescope as the world's largest by one inch. Soon, after the appearance of a couple of even bigger telescopes in Europe, the Clark firm's production of a thirty-six-inch refractor was set in motion.

The significance of this new refractor is not just in its size but also in the way it shows the direction of American science at the time. The thirty-six-inch telescope was a big, expensive project, and it was paid for with private—not governmental or institutional—money. A San Francisco businessman, James Lick, decided in the 1870s that he wanted something big and impressive with his name on it, and that the largest telescope in the world would be just the thing. He died in 1876, and soon afterward, following the terms of his will, the new Lick Observatory was set up on Mount Hamilton, near San Francisco. The new telescope (the lens again made by the Clark firm, now run by Alvan's son) started operations in 1888.[28]

The Lick Observatory not only illustrates the continued growth of an American astronomical tradition focused on powerful instruments but also represents the role of philanthropy as a major force in funding this new kind of big science. The United States was famous, or notorious, in the rest of the world in the later nineteenth century (as it remained in the twentieth century) as the land of millionaires. That reflected the enormous economic expansion of the period and the money that American businessmen were able to make from railroads, telegraphs, or any number of other enterprises where there was the opportunity to corner the market. In the days before antitrust legislation in the so-called Progressive Era of the early twentieth century, immense private wealth was a cultural fixture. Many extremely wealthy Americans chose to become benefactors of enterprises or institutions that might contribute to the public good, often in such a fashion as to immortalize the philanthropist's name, rather like aristocratic patrons of the arts in earlier periods in Europe. Ezra Cornell, to take an example from the 1860s, founded and paid for a university, which is a not unusual example for the postbellum period, and Andrew Carnegie founded public libraries. Big, expensive telescopes became, in the wake of Lick's bequest, another route that some wealthy businessmen chose for donating their money. (Lick was entombed in the concrete support on which the thirty-six-inch telescope was erected.)

The use of the new spectrographic and photographic tools in astronomy, as we have seen, had started by this time to be called "the new

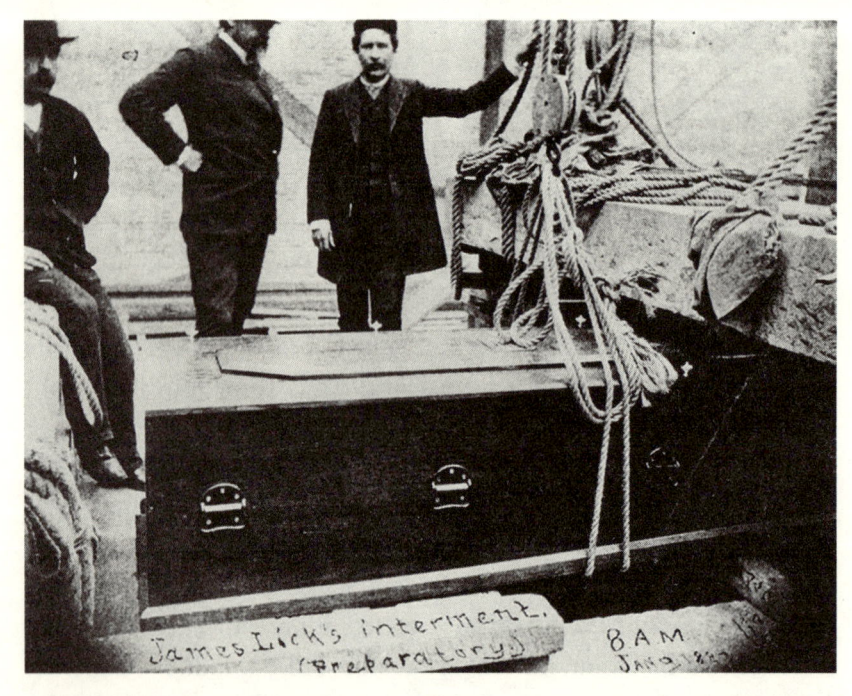

FIGURE 15.6. James Lick's telescopic tomb: a modern monument. Courtesy of Special Collections, University Library, University of California Santa Cruz, Lick Observatory Records.

astronomy." The word *astrophysics* also dates from this period; the earliest use of the word mentioned in the *Oxford English Dictionary* dates from 1870, with usage becoming more common in the 1890s. By the time Lick's telescope began operations, the explosion of astrophysics, and American dominance in world astrophysics, was underway.

George Ellery Hale

The history of early American astrophysics is dominated by George Ellery Hale. Hale specialized in the study of the sun—"our local star," as he routinely explained in popular science articles. He first got interested in it when he was a student at MIT. While still an undergraduate he invented the first version of what quickly became the fundamental instrument for solar astrophysics: the spectroheliograph. This device allowed photography of the sun's surface in any desired wavelength. Hale graduated from MIT in 1890, and in 1892 became the first director of the astronomical observatory at the new University of Chicago.[29]

But Hale knew where he could get real money. He managed to persuade one of Chicago's wealthiest businessmen, C. T. Yerkes, to found a major new observatory for the university. Yerkes had made his money in streetcars, and he now spent some of it on the Yerkes Observatory. The observatory was built out in the Wisconsin countryside and equipped with yet another large telescope. This one had a Clark objective lens, again, this time forty inches in diameter—still the largest refracting telescope in the world, even today. It commenced operations in 1897.

The chief reason that it remains unsurpassed in size is its sheer bulk. The two components of the Yerkes compound lens weigh about 225 kilograms. The unwieldiness of having that at the far end of a telescope tube, and the risk of the optical shape distorting under its own weight, varying with position, looked to be getting excessive (although this telescope worked perfectly well). Reflecting telescopes, the kind that William Herschel and William Parsons had used, now began to look to Hale and others as a better bet for very large telescopes because the mirror, unlike a lens, could be supported from below.

Reflectors had further advantages: glass lenses absorb light at the blue/violet end of the spectrum, which is the kind of light to which photographic plates were most sensitive. Hale showed the clear advantage of reflectors for photography, finding that he could see stars with forty-minute exposures using a twenty-four-inch reflector that were totally invisible to the Yerkes forty-inch refractor no matter how long the exposure. So reflectors, he concluded, were clearly the way to go.

Hale soon decided that his ambitions were being thwarted by inadequate funds for staff and expenses; C. T. Yerkes had paid for the observatory initially, but the University of Chicago was responsible for the running costs. Hale applied to the Carnegie Institution of Washington, a research foundation funded on a grand scale by the industrialist Andrew Carnegie, and persuaded them to fund a new observatory to be built in southern California on Mount Wilson, overlooking Pasadena.[30]

The Mount Wilson Observatory began operation in 1904. It was equipped with a number of smaller telescopes specifically for solar work, and later a sixty-inch reflector, the glass disk for which Hale brought from the Yerkes Observatory, where he had lacked the money to have it figured and set up as a telescope. He also brought a number of astronomical colleagues with him from Yerkes.

By 1910 the sixty-inch was in operation. It meant that a full range of stellar as well as solar astronomy could now be done, and Mount Wilson almost immediately became the leading astronomical center in the country, perhaps the world. Leading astronomers from abroad as well as the United States made extended visits there to use the new reflector because of the clear seeing conditions on the mountain as well as the telescope's quality. The permanent members of the observatory were themselves doing cutting-edge work in stellar astronomy because of the big reflector.

One of the consequences of William Huggins's determination that there was such a thing as genuine nebulosity had been to weaken severely the idea of island universes. Since Huggins found that there were genuine nebulae rather than just unresolved star systems, the idea of distant galaxies lost its justification. Meanwhile, observational work with large telescopes in the late nineteenth century had continued to record new spiral nebulae, which had first been identified as such by Lord Rosse. By the turn of the century it was common to suggest that these spirals, most spectacularly the famous Andromeda nebula, might actually be solar systems in formation. The attractiveness of this idea ultimately flowed from a suggestion made by Laplace in the early nineteenth century known as the nebular hypothesis.[31] It explained the condensation of nebular stuff into the solar system that we know today in fully naturalistic terms, and the spiral arms of such nebulae could be seen as the formation of planets in process.

But the relatively large apparent size of the spirals, suggesting that they were fairly close to us, gradually came to persuade astronomers that they probably were not single solar systems in formation. The usual view by around 1910 was that they were perhaps the birthplace of a whole group of stars (particularly because stellar-type spectral characteristics

FIGURE 15.7. Modern photograph of the Andromeda nebula. Together with the Orion nebula, this object has long served as a touchstone for cosmological interpretation. Photograph used with permission from Willem Jan Drijfhout, AstroWorld Creations (September 20, 2020).

could be detected in them as well as emission lines). But either way, they were still part of our galaxy; they were not island universes.

Cepheids and Nebulae

In 1908 a member of the Harvard College Observatory, Henrietta Leavitt, had published a catalogue of 1777 variable stars, that is, stars that varied in brightness, often cyclically, over time. As a woman, Leavitt represented a departure from the norm for most official scientific activity in this as well as earlier periods. But in astronomy in the United States in the late nineteenth and earlier twentieth centuries, there were quite a few women researchers (although they were still very much in the minority). The main place where this was the case was the Harvard Observatory—there were no women on the Mount Wilson astronomical staff, for example. A notable aspect of this situation, however, is the kind of work that women astronomers undertook: like William Herschel's sister Caroline Herschel in the late eighteenth century, women astronomers at Harvard did almost exclusively routine observational work—mapping and cataloguing, like Henrietta Leavitt's work on variable stars.[32] Men did that kind of thing as well, but some of them also developed higher-status theoretical studies, in which women astronomers played little part. This was the manual labor aspect of astronomy, as contrasted with the prestigious intellectual side, so it was apparently seen as acceptable to allow women to do it. They certainly were not allowed to be astronomers on a level with men, a point revealed starkly by the patronizing label jokingly applied to the women astronomers at the Harvard Observatory, whose director was Edward Pickering: they were known as "Pickering's harem."[33]

Henrietta Leavitt's variable star catalogue concerned stars located in the Small Magellanic Cloud, an offshoot of the Milky Way visible in the Southern Hemisphere. This catalogue summed up a vast amount of work done on the basis of comparing photographic plates of the cloud and giving the light curves (brightness plotted against time) of variable stars detected in them. In subsequent work published in 1912, Leavitt noticed that the shape of the light curves for some of these stars showed a remarkable logarithmic relationship between their periods of brightness variation and their apparent brightness: the brighter the star, the longer the variation cycle. That the period was in turn a function of the intrinsic, not merely apparent, brightness (or absolute magnitude) of these stars followed from the fact of their similar distances out in the Small Magellanic Cloud: if one such star appeared fainter than another, this could only be because it truly was. What remained to be determined was a means of calibrating this relationship between period or brightness and absolute distance, since the actual distance from us of the Small Magellanic Cloud was unknown, as was the size of the Milky Way.[34]

In 1913 the Danish astronomer Ejnar Hertzsprung reported that the light curves of Leavitt's variables in her 1912 paper were similar to those of stars called Cepheid variables, named after a prototypical variable star in the constellation Cepheus that showed this particular signature light curve. Hertzsprung now attempted to turn Leavitt's discovery into a basis for determining actual stellar distances from Cepheid observations.[35] His ingenious and elaborate technique involved inferring the absolute distances of relatively nearby Cepheid variables (ones showing measurable proper motion) via parallax measurements for nearby ordinary stars. Taking the common components of the proper motions of nearby stars to be reflections of the earth's own motion through space (a technique used by William Herschel), Hertzsprung derived a baseline of the earth's motion (statistical parallax) that could be used to determine some stellar distances inaccessible to annual parallax and that could be correlated with annual parallax determinations for stars near enough to show it. Then, for the few Cepheids that happened to display proper motions, an absolute distance measurement could be determined, albeit one with a large degree of uncertainty.[36] This could, finally, be used to infer the star's absolute magnitude.

One astronomer greatly interested by this new work was Harlow Shapley, who had joined the Mount Wilson staff in 1914 after finishing his PhD at Princeton. His particular concern was the study of globular clusters—globe-shaped clusters of faint stars that had been noted since William Herschel's time. Shapley wanted to measure the magnitudes and colors of the stars in these clusters to learn more about their nature, and part of that project required some way of estimating their distances. One of the techniques he used for making distance estimates involved Cepheid variables.

From about 1916 onward, then, Shapley applied the new Cepheid technique to globular clusters, which tended to contain quite a number of Cepheid variables. That way he could determine the distances of globular clusters; some of them were very small and faint and, it turned out, a very long way away. These great distances made the galaxy much larger than many other astronomers were prepared to concede, and Shapley's reliance on Cepheid variables was controversial. But if the Milky Way galaxy was really that big, it made it even less likely that the nebular spirals and disks were outside and of similar size to it. Shapley therefore used his work as an argument against the island universe idea: he believed that there was no evidence of other systems comparable to our galaxy, and that our own galactic system was therefore unique as far as observational evidence could tell. For the next decade and more the debate continued, including an intermediate position whereby the spirals were taken to be outside the Milky Way but of smaller size and closely associated with it.[37]

Hubble and a New Universe

The argument over the nebulae was finally resolved through the acceptance of work done by another Mount Wilson astronomer, Edwin Hubble, who joined the observatory in 1919 (Shapley left to join the Harvard Observatory in 1921). Hubble specialized in faint nebulae, which he had studied for his PhD at Yerkes, Hale's old observatory. In 1924 Hubble found something remarkable in the most prominent of the spiral nebulae, the one in Andromeda: he detected a Cepheid variable.[38]

The instrument Hubble used had only come into operation at Mount Wilson in 1918. Hale had set to work on its acquisition in 1906 through channels he knew well: soliciting private money. John D. Hooker was a wealthy businessman in Los Angeles, and Hale got him to part with forty-five thousand dollars to get the new telescope underway. The project was long, with much difficulty in casting a flawless mirror. This one was one hundred inches in diameter, by far the largest in the world, and the financing was subsequently supplemented with Carnegie money. The final cost was around half a million dollars—a lot at the time—owing in part to special custom-made grinding and polishing equipment, the cost of the glass disk (from France), and a purpose-built air-filtered shop for polishing and testing. That was apart from turning the mirror into a telescope. In 1918 it was done, and it was called the Hooker telescope, entirely funded from philanthropic sources.[39]

By 1924 novas (what we would now call supernovas) were known to appear frequently in the Andromeda spiral. But a regular variable star was something different. Hubble found it through comparing photographic plates of the nebula, using special apparatus that alternated two plates so that any change in the brightness of a particular star would appear to the eye as the star flickering between the plates. He first thought that he had found another nova, but further comparisons showed it to be a very bright variable with the characteristic curve of a Cepheid. By the end of the year Hubble had found several more in other spirals as well. That gave him a basis for calculating their distances in a way that could convince even Shapley, the king of the Cepheid method, that these were external to the galaxy and of at least comparable size (although Shapley held out on the last point for a while longer). The Cepheid method allowed distance measurements to be made on six closer "island universes," while for others—ones for which it was still possible to make out individual especially bright stars—Hubble made estimates on the basis of the stars' apparent brightness, assuming that the very brightest stars were all of about the same magnitude. Hubble's findings were publicized at the beginning of 1925. The result was a full-fledged reacceptance of the island universe idea, with our galaxy being just one of many spiral-shaped galaxies (as well as many that were not spirals).

These new ways of determining distance also enabled Hubble to make correlations between distance and speed. He measured the radial velocity of a galaxy by looking at the Doppler shift in its spectrum, much as Huggins had done for the star Sirius in 1868. For that Hubble needed a big telescope that could enable spectral photographs that showed enough lines, which he had in the hundred-inch. He found that the newly identified "galaxies" were all (except for members of our own local group, which included the Milky Way) receding from us, all displaying a red shift in their spectra. In 1929 he announced his famous Hubble constant, which related the speed of recession of a galaxy to its distance: the farther away a galaxy, the faster it was receding. That fitted into recent theoretical ideas based on Einstein's general relativity about an expanding universe, and it made Hubble famous. It also gave us more or less the universe that we live in today, populated by countless galaxies often equal to or larger than our own island universe.

In 1948, after twenty-five years of planning and building and testing, the celebrated two-hundred-inch telescope at Mount Palomar came into operation. The disk for its mirror was made by Corning Glass, an American glass manufacturer that had benefited from the cancellation of German industrial patents during the First World War.[40] It was called the Hale telescope, after the person primarily responsible for bringing it about, and was paid for by money from the Rockefeller Foundation. But that was the point at which big science was already becoming too big even for very wealthy private sources to fund.

These achievements were especially American and especially dependent on large sums of money for unprecedentedly expensive equipment. It was really the beginning of big science in the form we know it today, not only with space telescopes and particle accelerators (such as the one at CERN in Switzerland), but also geographically distributed collaborative projects like the human genome project, which in some cases are not simply funded by one national government but are often internationally supported.[41] The development of astrophysics signaled a massive change in the scale of many scientific enterprises in the twentieth century, in which the work pursued often could not be done at all without massive funding. Our world is made by modern achievements of scale that are entirely unprecedented.

16

New Modes of Natural Philosophy

POPULAR IMAGES of science and its cultural character in the twentieth century were remarkably unaffected by the radical changes in physics represented by the advent of relativity theory and quantum mechanics. The image of science in the twenty-first century still owes its basic character to the nineteenth century. One of these nineteenth-century views sees science as making unrelenting progress toward Truth, moving closer and closer to knowledge of absolute reality. This was an integral part of Victorian views among so-called scientific naturalists such as John Tyndall or Herbert Spencer, and in Helmholtz's visions of the interconnected cosmos, and continues to appear in journalistic accounts of cosmological theories or discoveries in particle physics as seeking to uncover "ultimate reality." The confirmation of the Higgs boson accompanied talk of it as the "God particle."[1] Such hype serves the professional interests of scientists in its insistence on the cultural preeminence of their work in cases where it is difficult to point out practical technological uses, the most common justification of public support for science. The irony in this popular notion of the relationship of science to reality is that the physics that defined the natural philosophy of the twentieth century—relativity and quantum mechanics—implicated at its birth conceptions of scientific truth and scientific inquiry that ran directly counter to the nineteenth-century conceptions that are still typically applied to physics and to science more generally.

Prologue to a Revolution

There is a standard story about the origins of Einstein's special theory of relativity that relates directly to Maxwell's concerns regarding electromagnetism and the aether. Heinrich Hertz had become converted to the idea of an aether as the medium carrying electromagnetic action as a result of his prediction and detection of radio waves in 1886–87. In 1890, in order to simplify his theoretical models, Hertz made the assumption that the aether shared the motion of bodies moving through it. That assumption meant that he could disregard complications in treating the electrodynamics of *moving* bodies; ordinary, ponderable matter and the aether would always be treated as if they were at rest relative to one another.

In 1892 the Dutch physicist H. A. Lorentz wrote an article, "Maxwell's Electromagnetic Theory and Its Application to Moving Bodies," in which he tried to deal with a particular problem resulting from Hertz's simplifying assumption. A phenomenon known as the aberration of light, first explored by James Bradley, the British Astronomer Royal in the first half of the eighteenth century, suggested that the relative motion of the earth and the light rays coming from a celestial object shifts the apparent position of that object in the sky to an extent readily measurable by an eighteenth-century astronomer.[2] However, Hertz's simplifying assumption could not incorporate that well-known phenomenon because it ignored the relative motion between the earth and the direction of external light rays. Accordingly, Lorentz took the opposite tack, using the idea of a perfectly stationary aether rather than having, as in Hertz's picture, the aether traveling like an atmosphere surrounding moving bodies (to which light rays would accommodate themselves before entering the astronomer's instruments). But Lorentz's approach immediately encountered difficulties with a particular piece of experimental evidence: the famous Michelson-Morley result of 1887.

In 1881 the American physicist Albert Michelson (who would receive the Nobel Prize for Physics in 1907) had tried to detect directly for the first time the earth's motion relative to the lumeniferous (light-bearing) aether. Two beams of light are sent out from a common source, one

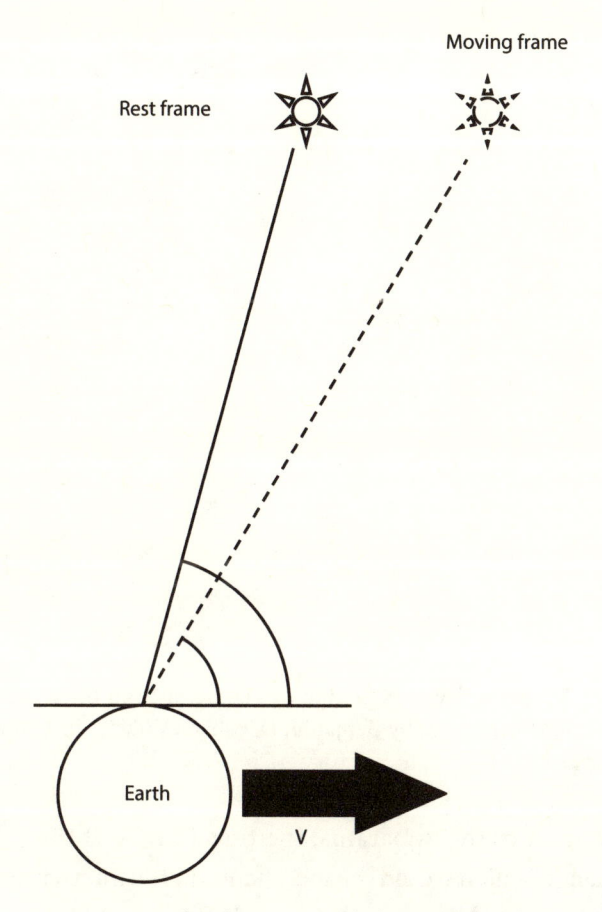

FIGURE 16.1. The aberration of light. Modified from image by inkscape, BlankAxolotl, available under Creative Commons Attribution-Share Alike 3.0 Unported License.

traveling along a path perpendicular to the earth's direction of motion through space (and, presumably, through the aether), and the other on a path of equal length parallel to the earth's direction. The beams are then reflected back along their respective paths and recombined to show an interference pattern.

Michelson expected that the light would have a different time of travel depending on the orientation of its path with respect to the earth's motion through the aether. If that orientation was altered by changing

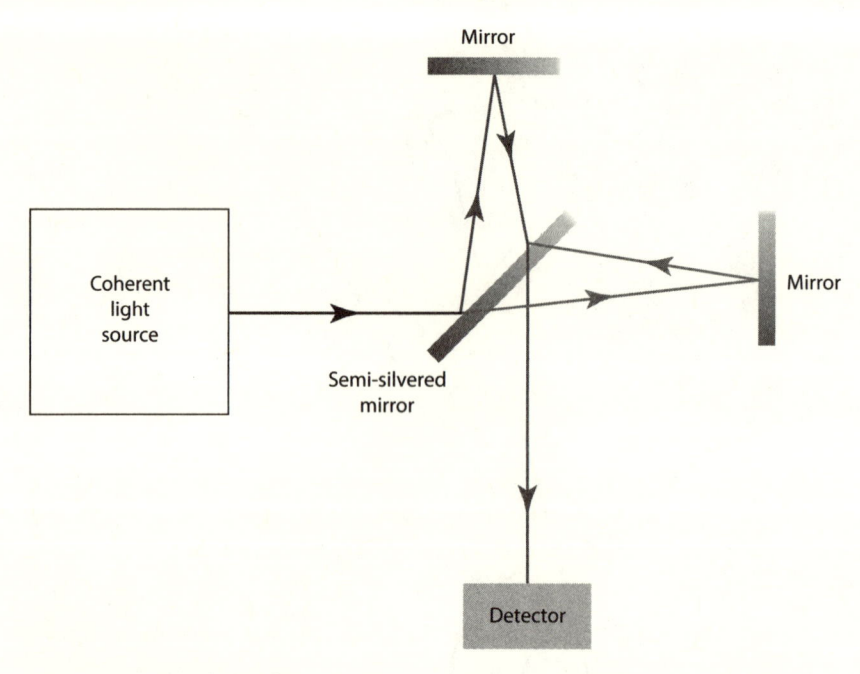

FIGURE 16.2. Michelson's experiment to detect the earth's motion through the aether. Modified from image by Benjamin D. Esham, Wikimedia Commons (September 22, 2020).

the orientation of the apparatus, the travel times of the two beams should change, which would mean a change in the interference pattern displayed when the beams were reunited. They would have gotten out of sync, so to speak.[3]

But in fact, against expectations, Michelson found no difference at all. His intention in devising this experiment was less to provide proof of the earth's motion through the aether (or even to confirm the existence of the aether itself) and more to display his technical skills as an experimentalist. The experiment was, from that point of view, a failure; the assumption had been that there would be a measurable motion through the aether, so failure to measure it was taken to be the mark of an incompetent experiment. Accordingly, Michelson tried again in 1887: he repeated the experiment with a collaborator, Edward Morley, this time at a much higher elevation, on the assumption that the original experiment had been compromised by a possible aether drag near to the earth's surface that might not occur further out. Improved inferometric techniques also contributed to this attempt at repeating and improving upon the 1881 experiment.[4] But Michelson and Morley got the same null result as before. The conclusion seemed to be that the aether was strongly dragged along by the earth, so that the aether around the apparatus remained immobile even at a considerable height.

Einstein's Relativity: Principles and Experiments

In 1905 Albert Einstein published a now-celebrated article, one of three landmarks that year, titled "On the Electrodynamics of Moving Bodies."[5] This article introduced what became known as the special theory of relativity. Einstein attempted to solve some of the problems with Lorentz's theory of matter and the aether, which seemed to be incompatible with the Michelson-Morley result, by entirely recasting the approach to the issue.

Einstein regarded Lorentz's theory as what he later came to call a "constructive" theory, one built up on the basis of hypotheses: it was, he thought, similar to the kinetic theory of gases, which was developed from a hypothesis about the nature of gases, the implications of which were compared to observation and experiment. Its predictions proved to fit such results quite closely, which was generally taken to betoken its truth. But Einstein disliked this kind of theory, regardless of its empirical success. He wanted instead to adopt the approach that he thought was involved in classical thermodynamics, which he called a "theory of principle." That kind of theory would be developed from basic principles based on universal experience, such as the impossibility of perpetual motion, or the assertion that heat will not spontaneously flow from colder to hotter temperatures.

In effect, Einstein's natural-philosophical sensibilities could never be invalidated by empirical results. His well-known fondness for thought experiments as ways of testing theoretical models had a long lineage going back to the demonstrations of the ancient Greek mathematician Archimedes. The ideal was always to show how things *must* be in the world rather than simply showing how they *happen* to be. This involved establishing fundamental principles and showing what would follow from them once accepted. A famous, although trivial, example by Einstein concerns the predicted behavior of a candle flame in an elevator. Einstein asked what would happen to a candle flame in a closed elevator that suddenly began to fall freely: Would the flame continue to rise in the usual way? The answer was that the flame would cease to rise, instead taking on a spherical shape around the wick. This followed because the hot combustion products would no longer be displaced upward by descending cold air; in free fall there is no weight.[6] The question can and must be answered through a process of reasoning from a principled understanding of the fundamental processes involved; there is no suggestion of performing an experiment to investigate or test the proposed phenomenon. Solid principles would yield solid conclusions. For his new theory, Einstein chose two fundamental principles of this kind.[7]

The first was the principle of relativity. This stated that all electromagnetic laws remain valid in all reference frames for which the laws of mechanics are valid. This means that the electromagnetic laws (those derived from Maxwell's work, in practice) that govern the interactions of a set of bodies—imagine there are electric charges involved—will be valid regardless of whether the assemblage of bodies is at rest or moving along uniformly in a straight line. This latter case thus concerns motion in a nonaccelerated or inertial reference frame—one that could be associated with a body having no measurable forces acting on it.

Einstein's second principle was that the measured velocity of light in free space is independent of the motion of its source. So it will be measured as being the same regardless of whether the source is moving rapidly away from (or toward) the observer doing the measuring or is stationary with respect to the observer.

Notice that these two principles give no consideration at all to issues of motion in a light-bearing aether. One of the advantages of these principles, for Einstein, was that they were basically kinematic rather than electromagnetic. They had to do with relative motions, so they applied equally to laws of mechanics and laws of electromagnetism, and could therefore unify them under a common set of physical principles. This is in contrast to Lorentz's approach, which distinguished between mechanically defined ordinary matter and an aether characterized exclusively in terms of its electromagnetic properties. Einstein explicitly rejected the notion of an aether as redundant; it served no function in his theory, and could therefore be ignored: "The introduction of a 'light ether' will prove superfluous, inasmuch as in accordance with the concept to be developed here, no 'space at absolute rest' endowed with special properties will be introduced, nor will a velocity vector be assigned to a point of empty space at which electromagnetic processes are taking place."[8]

Einstein's theory came up with many of the same results as Lorentz's electron theory, such as the relationship between a body's length and mass to its velocity.[9] It was the length-contraction implication, whereby the length of a measuring rod will contract in the direction of its motion, that both Lorentz and Einstein used to make sense of the Michelson-Morley result. Lorentz's theory proposed ideas concerning the propagation of forces through the aether between the particles constituting a body, thereby relating motion through the aether to the forces that determined how closely those particles were bound together: stronger forces in the direction of motion would make the length in that direction contract. Lorentz considered that the reduction in length of the arm of Michelson's apparatus that points in the direction of the earth's motion through the aether would exactly counteract the anticipated longer travel time of the light back and forth along that arm as compared with the travel time for light along the perpendicular arm, the consequence being that no net difference should be detected.[10] Einstein's theory yielded essentially the same result, without having to talk about theoretical entities like the aether or the propagation of forces acting between molecules; measurement was all.

Einstein said in later years that when he wrote his 1905 relativity paper he was not directly aware of the Michelson-Morley experiment. But the experiment was not irrelevant to Einstein's innovations: Lorentz knew about it; it was part of the problem area that produced Lorentz's work, which was in turn well-known to Einstein in 1905. And Einstein gave Michelson-Morley an important role in his account of these issues in a review article of 1907.[11]

Einstein regarded his approach as better than Lorentz's because it relied only on measurements, the significance of which was inferred through their relation to general principles derived directly from firsthand experience of the world. Einstein's new theory eschewed speculation about the composition of the world at a more fundamental, ontological level; It just coordinated measurable phenomena and inferred resultant behaviors.

Einstein credited his methodological approach at this time to the ideas of Ernst Mach, a German philosopher-scientist who was one of the most influential positivists of the late nineteenth and early twentieth centuries.[12] But Einstein's position on scientific knowledge throughout most of his life, and especially the views that put him in conflict with Niels Bohr and Werner Heisenberg over quantum mechanics, went beyond a methodological injunction about how to make theories.

Perhaps ironically, the quasipositivistic Einstein wanted to discover the true nature of the universe; he typically expressed that ambition by reference to God. While not religious in any conventional sense, the nearest that Einstein could get to explaining what he thought science—specifically physics—should be about was by using rather mystical talk about grasping the "secrets of nature." He was just as likely to use the word *God* as *nature*; for him, they seem to have meant the same thing. And it reflects the sense in which the German physics community in this period was still centrally concerned with natural philosophy.

Responses to Relativity

In his "Autobiographical Notes" from 1946, toward the end of his life, Einstein wrote:

> It is quite clear to me that the religious paradise of youth . . . was a first attempt to free myself from the chains of the "merely-personal," from an existence which is dominated by wishes, hopes and primitive feelings. Out yonder there was this huge world, which exists independently of us human beings and which stands before us like a great, eternal riddle, at least partially accessible to our inspection and

thinking. The contemplation of this world beckoned like a liberation, and I soon noticed that many a man whom I had learned to esteem and to admire had found inner freedom and security in devoted occupation with it. The mental grasp of this extra-personal world within the frame of the given possibilities swam as highest aim half consciously and half unconsciously before my mind's eye.[13]

This might at first seem to be a trivial point, that a scientific theory, or a scientific worldview, should be about an independently existing reality separate from human knowers. But the debates that accompanied and followed the development of quantum mechanics focused on exactly this issue and set Einstein in opposition to what became, and remained, the dominant approach to the physics of the subatomic realm.

Nonetheless, popular perceptions of relativity theory came to see it as an example of the opposite view of scientific truth to the one that Einstein himself championed. Following the First World War, relativity theory quickly became well-known by nonscientists in the 1920s as a supposed justification for a quite different, although superficially similar, idea relating to all areas of human knowledge, known as relativism. People started referring to Einstein's relativity as a sanction for the claim that all knowledge is relative to one's point of view, and that therefore there can be no such thing as absolute knowledge or absolute truth; it all depends on your perspective.[14]

That superficial reading of Einstein came from noticing that, according to relativity theory, things like the length of an object, or its mass, or the simultaneity of two events taking place at different locations, are not themselves absolute or fixed. Their numerical values vary depending on the relative states of motion of the observer and the observed; there is no longer a reference frame of absolute rest that would be distinguishable from all the others, as there had been when there was an aether to act as the stationary backdrop against which could be defined the true values of things.

Relativism, by contrast, claimed that cultural and moral standards, involving judgments of the nature of human beings and human societies, of right and wrong, were ultimately relative to the particular culture in question rather than being timeless truths. The new trendy theory from physics called "relativity" looked from this standpoint like a good reference to bolster such claims. Physics had become in the nineteenth century the most prestigious touchstone of scientific truth, so if even *physical* truths were "relative," then no one ought to deny that truths about human society were also a matter of your point of view.[15]

Einstein rejected such arguments. In 1932 he remarked, "I believe that the present fashion of applying the axioms of physical science to human life is not only a mistake but has something reprehensible to it."[16] For Einstein there was indeed a sort of objective reality, a God's-eye view of the world, that he wanted to discover. Relativity theory was not a matter of believing whatever you wanted to believe or saying that one person's truth might be another person's falsehood. If there was no longer any privileged reference frame for determining a body's mass, for example, that did not mean that there was no objective thing called "mass" existing independent of human observers. It simply meant that the old Newtonian concept of "mass" was an incorrect representation of the world.

In effect, Einstein set up laws of nature as his absolutes, rather than properties of things in the world. So the absolutes in special relativity are the transformation equations that say precisely how measurements of mass or time or simultaneity will vary when shifting from one reference frame to another. Those values might be relative to the observer's reference frame, but the way they vary *between* reference frames is absolutely determinable. The transformation equations themselves were invariant, not dependent on a point of view, and they were what described how the world is. In a letter written in the early 1930s Einstein explained to a correspondent that "the theory says only that the general laws are such that their form does not depend on the choice of the system of coordinates."[17]

In fact, Einstein did not initially call it the theory of relativity. That term was coined by the eminent German physicist and founder of quantum theory Max Planck in 1906, and Einstein did not much like it. In correspondence around that time he tended to refer to it as his "invariant theory," a label that puts a very different gloss on it and points up the equivalence of different reference frames rather than any arbitrary aspect of choosing among them. For some years, before he became resigned to it, Einstein referred in print to the "*so-called* relativity theory." From his point of view the use of the word *relativity* itself was unfortunate.[18]

General Relativity

The major reason for the explosion of hype about Einstein's theory was the empirical confirmation in 1919 of the general relativity theory of 1915 (by which time Einstein had accepted the term *relativity*). The general theory was an attempt to extend relativity from cases of reference frames moving with uniform relative velocities to cases of accelerated reference frames. One of its predictions was that light rays ought to be bent by a gravitational field. The effect would be small, but Einstein predicted a detectable shift in the apparent positions of stars seen very close to the edge of the sun because of the sun's gravitational effect on the light: 1.7 seconds of arc. In order to see stars properly that close to the sun, it was necessary to wait for a total eclipse. In 1919 a British expedition went to West Africa, to an island in the Gulf of Guinea over which the path of totality would pass, and confirmed the effect.[19] The leader of the expedition, Arthur Eddington, sent Einstein a cable to inform him of the result. When Einstein showed it to one of his students at Berlin, she enthused over the news, but Einstein seemed rather indifferent, saying, "But I knew that the theory is correct." When she asked him what he would have done if it had not come out right, he said, "Then I would have been sorry for the dear Lord—the theory *is* correct."[20]

Einstein's remark raises the question of how he thought that the truth about what he regarded as an independent, objective universe could be acquired in the first place, if an experiment to test a theory about it could be regarded as beside the point. The answer returns to theories of principle. Einstein had a powerful belief in the ultimate simplicity of the universe. After establishing the right foundations of a theory, the rest should follow. No simple correlation of experimental and observational evidence to "prove" statements about the world would suffice, since experiments were fallible and the logic of experimental confirmation was fundamentally indecisive.[21] In the case of general relativity, as with special relativity, Einstein established his foundations on basic universal principles. Just as special relativity rested on the principle of relativity plus the invariance of the speed of light, so general relativity, to deal with acceleration and to extend the theory to gravitation as well as electromagnetism and mechanics, had in addition the principle of equivalence. Einstein had first come up with this idea in 1907 (although the name came a little later); the principle derived from a consideration of what, in Newtonian mechanics, looks like a remarkable and inexplicable coincidence: the equality of inertial and gravitational mass.

Newton conceptualized a kind of mass that was defined by his second law, usually expressed as force equals mass times acceleration; consequently, mass equals force divided by acceleration. This kind of mass is called *inertial* mass, a property of a body that resists applied forces; when the body is pushed, it pushes back. But Newton also recognized another property of bodies that involves forces: gravitational attraction, exerted by all massive bodies (bodies with mass) on all other massive bodies.[22] A responsiveness to gravitational force is simply added on to the forces involved in directly applied (contact-action) accelerations. Gravity is in no way implied in the concept of inertial mass, and responsiveness to gravitational force is what defines *gravitational* mass.

For Newton, it just so happens that the weight of a body near the earth's surface, due to the gravitational force acting on it by the earth's attraction, is quantitatively the same as its inertial mass. There is no principled explanation of why this should be so, but it is a property of universal experience that it *is* so. Galileo had argued that all bodies fall at the same (accelerating) rate if sources of frictional resistance and differential buoyancy can be neglected, as in a vacuum. So, for example, a two-pound weight falls at the same rate as a one-pound weight, even though it is accelerated by twice the gravitational force—it weighs twice as much. It also has twice the inertial mass, meaning that it resists acceleration twice as much. It is because the two effects cancel out that the two bodies exhibit the same rates of fall; a body's inertial mass equals its gravitational mass in all circumstances. But Newton's theory does not explain why that should be the case.

Einstein's idea here was typical of his approach to physical problems. In the case of the invariance of the speed of light, he had not asked why it should be invariant (as in the Michelson-Morley experiment, for instance). He just made its invariance a fundamental, universal principle, vindicated by all known experience without exception. Similarly, he made no attempt to discover why inertial and gravitational mass should be equal; instead, he set that equality as a principle. He treated the two as always and necessarily the same, because they are fundamentally the same thing; they are of necessity indistinguishable. Hence the principle of equivalence: any physical system undergoing uniform acceleration will behave in exactly the same way as if it were in a uniform gravitational field. Inertial and gravitational forces are just two manifestations of the same thing, and gravity becomes a property of space and time relative to a particular coordinate system.[23]

The elegance, simplicity, and economy of the foundations of a theory acted for Einstein as the hallmarks of its truth. That was how he could express such confidence in the truth of general relativity even without empirical confirmation: it had to be true, because it had all the conceptual hallmarks of truth. The connection with Einstein's well-known predilection for thought experiments as ways of testing the plausibility of a physical idea is clear.

But elegance and simplicity of a theory's foundations were not sufficient: Einstein also held that true knowledge should be in some sense causal. By that he meant that everything that happens in any physical system ought, if understood completely, to be perfectly predictable. In other words, Einstein thought that determinism was an essential feature of any scientific worldview. This was, of course, the view that had been common throughout most of the nineteenth century, and was especially associated with Laplace. Laplace had imagined a superintelligent being that could know the state of all particles in the universe and calculate their entire past and future motions by applying the invariable laws of nature that governed them. That was a way of dramatizing the idea that the physical world is entirely deterministic; nothing happens by chance; everything is rigidly, causally linked to what came before. With all appropriate adjustments and qualifications, Einstein's deterministic view of the universe retained just that idea. It was the basis of Einstein's opposition to the kind of quantum theory that was developed and championed by Niels Bohr and Werner Heisenberg.

Quantum

The idea of quanta of energy had first emerged from work done by Max Planck in 1900. Planck came up with a view of a spectrum of electromagnetic radiation as being made up of discrete units or steps of energy for each frequency, instead of having a continuous distribution of energy. Planck made no claim to know what brought about this quantization of energy; his result emerged from a statistical approach that happened to fit experimental data, and he arguably did not originally assert that quantization is literally true rather than a mathematical trick producing correct answers.[24]

However, Einstein used Planck's quantization in another famous paper of 1905 on the photoelectric effect, in which Einstein treated light as if it were made up of discrete quanta of energy, later to be called photons: $E = hv$, where E is energy, h is Planck's constant (derived empirically in Planck's work on black-body radiation), and v is the frequency of the light.[25] Subsequently, the Danish physicist Niels Bohr produced in 1913 his model of the atom, building on a picture suggested by Ernest Rutherford, a New Zealander working at the University of Manchester in England. Rutherford had envisaged the atom as resembling a tiny solar system, with a positively charged nucleus orbited by negatively charged electrons, the latter being a conceptual invention of the 1890s. (This solar system representation has endured in popular culture in various schematic forms.) Bohr developed this picture by proposing that the electrons orbit only at discrete, quantized energy levels derived from Planck's $E = hv$.

Bohr's model enabled for the first time an understanding of the discrete emission and absorption lines found in the spectra produced by chemical elements.[26] Some of these sets of lines had already been found in the 1880s to correspond to arithmetically expressible regularities, and Bohr managed to derive them as consequences of energy absorption

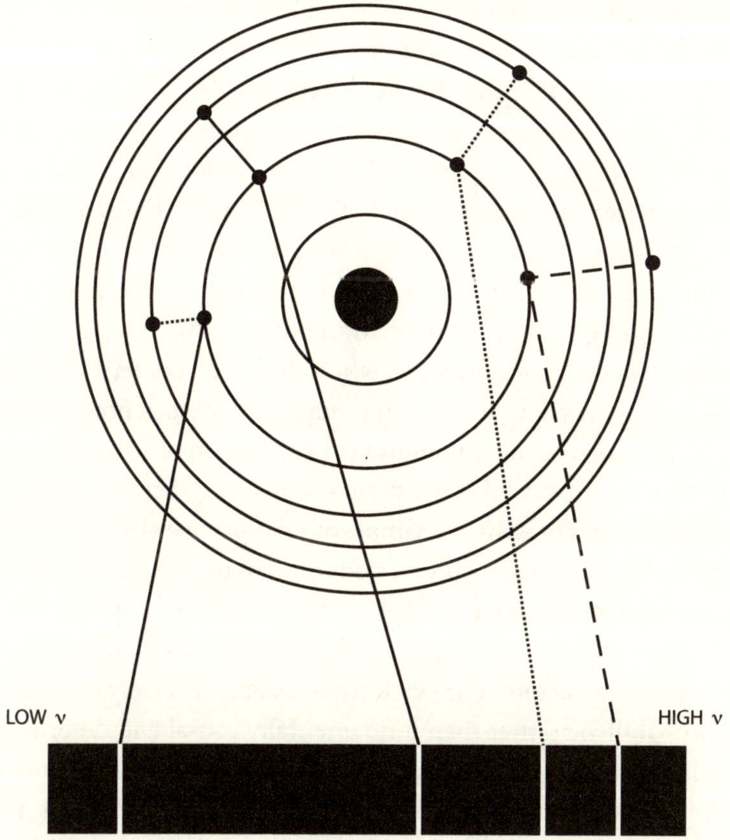

FIGURE 16.3. Discrete (noncontinuous) energy levels in the Bohr atom.

and emission as the orbiting electrons acquired or lost orbital energy by jumping up or dropping down between discrete energy levels around the atomic nucleus. (Why there should be such a restriction in allowable energies remained a mystery.)

The development of quantum theory during the following decade or so, especially as conducted by Arnold Sommerfeld and his students at Munich, led in 1926 to one of them, the German physicist Werner Heisenberg, producing a full-fledged "quantum mechanics" to replace Newtonian, or "classical," mechanics as a way of describing the atomic-scale behavior of matter relating to energy and forces. Almost simultaneously, the Austrian physicist Erwin Schrödinger produced an equivalent theory, known as wave mechanics, which quickly became the most accepted and widely used version of the new quantum mechanics, no doubt because of its more immediately picturable approach.[27]

From Einstein's perspective, the problem with the development of quantum mechanics was that Bohr and his allies elevated the kind of statistical treatments that had initially led Planck to the quantum idea to a fundamental status. Instead of treating atomic events as in principle determined by prior states, they were simply regarded as more or less likely, according to strict mathematical probabilistic rules. In the case of radioactive decay, for example, where an atom's nucleus sometimes releases an alpha particle (a helium nucleus), there is nothing in the prior state of that atom that allows the prediction of when decay will occur. When, as in actual experimental situations, there are enormous numbers of atoms present in a sample of a radioactive element, it is possible to make very accurate predictions of what proportion of them will have decayed after a given interval of time. But even a Laplacian superintelligence would be unable to say when any *particular* atom will decay. Quantum mechanics in Bohr's view treated such processes as fundamentally probabilistic rather than fundamentally causal and determinate, regardless of the completeness of available knowledge of the situation. For Einstein, those probabilistic features of quantum mechanics meant that it was an inadequate theory. The world wasn't like that, and being satisfied with probabilistic knowledge of the structure of reality was simply giving up on the real task of physics. He famously remarked—on more than one occasion—when talking about these matters that "God doesn't play dice with the universe."[28]

Einstein's principled opposition to these developments in physics ran him directly up against Bohr, the leading champion of the new views. Part of the impetus driving Bohr to develop a philosophically coherent justification of probabilistic quantum mechanics in the 1930s was Einstein's opposition to it; Bohr's approach amounted to a radical departure from Einstein's. Bohr's position came to be called the "Copenhagen interpretation" of quantum mechanics because of the location of his physics institute, although it was taken up by people like Heisenberg elsewhere as well. Bohr argued that scientific knowledge cannot, by its very nature, tell us about how the universe really is, independent of human knowers, which was Einstein's ambition. Instead, scientific knowledge is the product of human interaction with the world, and is as much determined by human characteristics and limitations as it is by the nature of the world itself. James Clerk Maxwell's interpretation of his famous thermodynamic sorting demon had carried a comparable implication.[29]

Einstein's Natural Philosophy

Einstein's dislike of quantum mechanics manifested itself in other attempts to discredit it, similarly focused on intuitive implausibility (what one might call metaphysical implausibility). His approach was to identify what he took to be unacceptable consequences of its basic concepts and technical procedures. A paper published in 1935 attacked another aspect of the new theory; the paper bore Einstein's name alongside those of two collaborators.[30] One of its central concerns was the issue of "completeness" of the theory of quantum mechanics. The authors were committed to the position that a satisfactory physical theory should refer to all properties of the physical reality that it represents; it should allow for the measurement of all such features to an arbitrarily high degree of precision. But a central feature of quantum mechanics was the Heisenberg uncertainty principle, which showed that for each of certain conjoined, or conjugate, pairs of measurable quantities, such as the position and momentum of a particle, the more precise the measurement of one quantity, the more imprecise the simultaneous measurement of the other—not just that the measurement itself would be imprecise, but that the property being measured would be inherently indeterminate. Yet a "complete" theory, according to the authors, should allow both members of such a pair to have precise values, regardless of the circumstances of their measurement; precise values should exist in reality whether or not they happened to be measured. And any theory that was not complete in this sense would need to be supplemented with "hidden" variables that would govern (to an arbitrary degree of precision) the properties of the physical system under consideration.

The most striking consequence of this analysis concerned what came to be called nonlocality. Nonlocality was a property whereby the outcome of a measurement (of momentum or position) made on a particle in one place and time had consequences for a measurement made on another particle at another place and time, even though no communication between those particles that might causally link the two measurements was possible (no signal passing between the particles in an allowable time). The value of a property of one particle could be inferred from the value of the same property of the other because, before separation, they had been interrelated. So a precise measurement of the momentum of one would allow similarly precise inference of that of the other, but the corresponding *uncertainty*, that is, *indeterminacy*, of the conjugate quantity, its position, for the first particle would also immediately apply to the second, distant particle: the latter would display an inherently indeterminate position even though it had remained undisturbed by any local measuring process. Einstein later branded such a thing "spooky [*spukhafte*] action at a distance" and deemed it out of bounds for a physical theory.[31] For Einstein, the only satisfactory way to understand the situation would be to accept that particles always possess precise, determinate properties, but quantum mechanics as currently formulated could not measure or infer them all simultaneously. And, he argued, since it seemed to be a consequence of quantum mechanics, so much the worse for quantum mechanics, at least in the form promoted by Bohr and his allies. "We are thus forced to conclude that the quantum-mechanical description of physical reality given by wave functions [i.e., Schrödinger's version of QM] is not complete."[32] Einstein reckoned that, because of this inadequacy, quantum mechanics was seriously flawed as an account of reality. There must, he thought, be hidden variables involved that, if discoverable and measurable, would preserve the assumption of normal causal relationships. They might not all be measurable in practice (because they depended on inaccessible, hidden values of some kind), but that would be a fault of the knower, not a shortcoming in nature itself. The paper concludes, "While we have thus shown that the wave function does not provide a complete description of the physical reality, we left out the question of whether or not such a description exists. We believe, however, that such a theory is possible."[33]

By the close of the 1930s Einstein had adopted the view that the fundamental mistake in Bohr's ideas lay in their retention of so-called classical concepts like position, momentum, and energy. What Einstein saw as the problems of quantum mechanics would ultimately have to be solved by rejecting those concepts for completely new ones, although he was unable to suggest what those might look like. Bohr, on the other hand, argued that these classical concepts had to be used because they were the only way of talking about what was measured in the laboratory. They represented the way that human beings interact with the physical world, so they could never be eliminated.

Attempts have been made since then, on occasion, to show that a deterministic level of understanding of quantum phenomena could, conceivably, exist behind the probabilistic calculations of orthodox quantum mechanics. In the 1950s serious efforts were made by the American physicist David Bohm to develop the kind of "hidden variables" interpretation of quantum mechanics that Einstein had hoped for, although Bohm's ideas did not meet with any great degree of acceptance, even by Einstein.[34] For some years, a proposed proof by the mathematician John von Neumann of the impossibility of a hidden variables version of quantum mechanics was generally accepted, but its rejection soon followed, in 1964, after a demonstration by John Bell in the United Kingdom of particular implications of hidden variables versions of quantum mechanics that failed in the face of empirical testing.[35] So perhaps the world really is indeterminate.

Nonetheless, most physical scientists have continued to calculate and predict according to the well-established procedures, and the Copenhagen interpretation has remained essentially unchallenged. The natural-philosophical enterprise that spawned Bohr's ideas ground almost to a halt in the face of various theoretical challenges arising from particle physics: thanks to the development of particle accelerators since the late 1930s with the work of E. O. Lawrence at Berkeley, many new elementary particles were found, whether predicted or not, giving rise to the "standard model" of elementary particles in the 1960s, which invoked quarks as explanatory constituents of most other particles.[36] Attempts in the 1970s by Roger Penrose and Stephen Hawking, among others, to reconcile quantum mechanics and the general theory of relativity continue to generate cosmological speculation on the grandest scale. The notoriety of these two theories, and their problematic inconsistencies with one another, have been popularized through the charismatic focus of black holes, cosmological superstars for which a curious and misleading antecedent appears in the work of John Michell in the mid-eighteenth century.[37]

Bohr's and Einstein's concerns are now questions that are mostly discussed by philosophers of science rather than by physicists themselves. The marginalization of these sorts of questions amounts to the demise of natural philosophy, at least in physics, in the twentieth century. Quantum mechanics "works," and that's good enough for most scientists who encounter it.

CONCLUSION

The World We Have Gained... and Lost

THIS BOOK has tracked, over the course of a couple of centuries, aspects of the intellectual enterprise of natural philosophy—the "understanding" part of modern science—up into the twentieth century. The coherence of scientific endeavor throughout has rested on institutional and cultural continuities among scientific societies, publications (especially serial publications), and educational practices. These are what enabled the continuity of science, allowing scientists to see themselves as part of a progressive tradition stretching back centuries.[1]

Science as a Form of Understanding

The institutional continuity of science has been overwhelmingly European and, developing immensely in the twentieth century, American (the United States). Although science is nowadays a global enterprise, the dominance of so-called Western scientific centers remains, and leading scientists from around the world are disproportionately trained (and often employed) at European and American universities. During the formative period from 1700 until the early twentieth century, Western Europe tended to dominate the new enterprise; hence the frequent appearance in this book of French, British, and German figures alongside others from elsewhere in Europe. But the Dane Oersted studied in

Paris, the Russian Mendeleev (who had studied in Germany) became a Foreign Member of the Royal Society (an acknowledgment from a core of scientific culture), and the Pole Curie spent most of her career in Paris. That geographical concentration marked the presence of the institutionally defined endeavor called "science" as a specific cultural practice pursued in particular communities. It was a practice subsequently emulated much further afield, most notably in the United States, and one of its products has been the creation of natural-philosophical images of the world that inform the ways in which large segments of the world's population perceive the world and their place within it: the world, in other words, as we know it.

Of course, the hegemony of what is sometimes called the "scientific worldview" is not, and never has been, complete. Today in the United States many people refuse to accept the evolutionary perspective associated with Darwin, despite its practically universal acceptance among life scientists. On the other hand, some version of a cosmological picture featuring an unbounded universe, with its unimaginable dimensions and age, is unquestioningly acknowledged by those in modern literate societies who pay any attention to such matters. Overall, and despite some doubts and rejections that will spring readily to mind concerning anthropogenic climate change or the reliability of vaccines, the worldview within which scientists pursue their trade is, with all its inevitable inconsistencies, broadly accepted. That is why the state apparatus of most democratic nations (and not just those) includes scientific advisory panels and support for scientific research.

The world we think of ourselves as inhabiting is often that of past, sometimes out-of-date, science. Space extends around us, stretching to infinity, rather like an eighteenth-century Newtonian universe, except when we think of the expanding universe of Hubble and Einstein, and the Big Bang. These are different pictures, but we can operate with either because, perhaps, it doesn't much matter what we choose. We are the products of evolution and natural selection when we are not the biological equivalent of elaborate *Matrix*-style computer simulations (biological processes as information systems); we can think of ourselves in different ways, all validated by the prestige of various scientific specialties. In the nineteenth century James Clerk Maxwell juggled the contrasting models of the kinetic theory of gases and the physical universe being an immense space-filling aether; the contrast did not seem to bother him, and such apparent inconsistencies between different theoretical constructs, each facilitating understanding, do not bother most of us now. But theoretical physicists concerned with reconciling relativity and quantum mechanics do aim at finding underlying consistency. Some seek at the most totalizing level to embrace everything under a grand theory of superstrings, a reductionist attempt to distill everything. All will be derivable, in this view, from those most basic concepts; although, as the eminent physicist Steven Weinberg wrote, "A final theory would not end scientific research, not even pure scientific research, nor even pure research in physics. Wonderful phenomena, from turbulence to thought, will still need explanation whatever final theory is discovered.... A final theory will be final in only one sense—it will bring to an end a certain sort of science, the ancient search for those principles that cannot be explained in terms of deeper principles."[2] And we tend to nod in bemused fashion at the ambition.

Weinberg's stress on "explanation" appeals to the idea of science in its guise as natural philosophy: a final theory would be the grandest achievement of natural philosophy, but not of an instrumentally applied system of techniques. In fact, the notion of a "final theory" would make little sense in the latter case; the massive energies generated by the CERN particle accelerator, which produced the Higgs Boson, are not dedicated to practical ends.

Science: The Fate of a Nineteenth-Century Project

Science in the twenty-first century shapes cultural images of humanity and of the world at large. Many developments of the last hundred years have contributed to those images—far too many to detail in a book like this—but the fundamental structures of the world picture created by the middle of the twentieth century remain in place. In that sense we still live in a world made in the nineteenth century, just as science itself is a nineteenth-century enterprise. Among the most prominent cases of striking new scientific ideas that inform our sense of place in the world are the so-called Big Bang account of the origins of the universe, perhaps also plate tectonics ("continental drift"), and especially genetics—more specifically genomics.[3]

Despite the increasing prominence of biological sciences in more recent decades, the idea persists of physics as the fundamental explanatory discipline to which all natural phenomena can in principle be reduced. This is a view with deep historical roots, appearing in a variety of approaches: Laplace's forces-and-particles ontology; the view, shared by many Victorian British physicists, of thermodynamics ("energetics") as a unifying framework (effectively replacing forces and particles with energy and particles, as in Helmholtz's work); or the concern in the years around 1900 with unifying electromagnetism and mechanics, exemplified by ongoing British attempts at reducing both to mechanics and Einstein's relativistic encompassment of the two domains under a common set of principles.

Such approaches have as their focus a concern with natural philosophy, a desire to create a picture of what the world is *really like*—the world as God knows it—rather than simply having instrumental or operational control over phenomena. The dispute between Bohr and Einstein over quantum mechanics can be seen in a similar light, perhaps contrasting Bohr's sometimes instrumentalist conceptions (seeing knowledge as a tool for accomplishing tasks, including making predictions) with Einstein's focus on natural philosophy (a project of intellectual understanding).

The professionalization of science has served, in effect, to distance science from its philosophical pretensions. Instrumentally oriented science still informs, nonetheless, a representation of the world the credibility of which rests on technical scientific concepts developed for the purpose of managing experimental measurements: thus quantum mechanics, designed to describe microscale phenomena, gives license to speculate about the origins of the universe and its ultimate fate.[4] Genomics has displaced traditional taxonomy in biology, for example, so that what was once chiefly a matter of morphological comparison (comparative anatomy) has increasingly been displaced by techniques that focus on DNA sequences (with paleontology being the broadest exception to this generalization, except when DNA samples are in some cases recoverable).

When, many decades ago, Princeton University Press published *The Edge of Objectivity*, by the eminent historian of science Charles Gillispie, things seemed simpler.[5] Gillispie's book—still in print—covers similar ground to that of the present work, but illustrates the propensity for historical overviews to differ widely from one another despite drawing on comparable materials. *Edge of Objectivity* is an accomplished work of literature that takes the form of tragedy: as Gillispie put it elsewhere, science "dooms to conquer," accomplishing its cold ends by disenchanting the world that it comes to understand.[6] Gillispie in effect lamented the failure of Romanticism and what he saw as the inevitability of science's harsh analytical realism in dissolving a world of affect and wonder, the worlds of Blake or Goethe.[7]

I have tried, by contrast, to tell a story about the world as it appears in the dominant imaginaries of what is known as the Western world, which might better be described, tautologically, as the world in which the ideology of modern science holds sway. The cosmology (in the broad sense) of that world has been the subject of this book, laid out in a partly chronological and partly topical format, integrated by recurrent thematic considerations. At the center, however, lies a worldview and an associated set of assumptions that comprise an architecture of the universe, one managed by a kind of metaphorical operating system. The features of that operating system are deeply rooted in a particular historical passage: assumptions about what an explanation should look like and what it should contain; whether explanations should appeal to intuitions about intelligibility or simply coordinate phenomena in formalistic ways; and, intimately bound up with such matters, assumptions about God as the ontological guarantor of the universe.

God was an explicit part of the worldviews of Joule, Faraday, and Maxwell in nineteenth-century Britain, or, in the twentieth, of the devout Quaker Arthur Eddington (leader of the 1919 eclipse expedition that tested general relativity), just as He had been for Newton.[8] Comparably, the almost mystical views (albeit distanced from organized religion) of such as the American anthropologist and naturalist Loren Eiseley have continued to carry much extra-empirical freight in the work of theoreticians. It has never been easy to separate such scientific worldviews from broader cosmological commitments, at least in the work of those who address large questions of natural philosophy. But it is not a large component of the professional lives of most people today called "scientists," and has not been for a long time, despite the popular idea of an estrangement or conflict between science and religion; frequent attempts to counter that view by historians and religious apologists seem not to hold the same attraction as the older, agonistic model—and no wonder.[9] Science, or some conception of science, offers a powerful frame within which to generate cosmologies that offer senses of meaning to go along with the literally unimaginable computer-manipulated images from the marvelous space telescopes.

In the last few years, in a world dominated by the COVID-19 pandemic, the slogan "Science Is Real" frequently appeared on yard signs or bumper stickers in the United States, usually alongside others such as "No Human Is Illegal" or "Women's Rights Are Human Rights." "Science Is Real" has a longer and more complicated genealogy than this, but taken in itself it seems to express (beyond an implied COVID-era message about vaccines) the idea that beliefs or conclusions contrary to scientific orthodoxy are in some way unreal, fantastic, or "made up."[10] This creates a useful binary for political purposes, but obscures the ways in which science is always a work in progress; its pronouncements are seldom final, at least in the short run, and are always understood to be, in principle, up for debate.[11] Of course science as a human enterprise is real, but on this issue there seems to be little more to be said. And if the phrase means "everything that scientists say is, ipso facto, true," then it can safely be said to be hugely unlikely, to the extent of being false.

The moral weight of science, however, surely rests on the view not that science necessarily speaks the truth, but that science is always *honest*. The circular conclusion must then be that if some scientific work is managed deceitfully or dishonestly, then it must not be truly scientific. The only true form of science would be natural-philosophical, because only disinterested intellectual knowledge could be morally pure; utilitarian knowledge of nature would always and necessarily be prey to corrupting interests.[12] In fact, of course, science being a human enterprise, we may be sure that even natural philosophy will never be free from potential corruption, but even this perspective assumes an idealized view of science in which its threatened purity must be protected.[13]

It is a genuinely historical question as to how science could ever have come to be seen in such a light in the first place: it is not the only way to understand the nature of science, as Marxist conceptions have displayed particularly starkly.[14] In that sense, the "pure" view of science as natural philosophy is a fallacious ideology that distorts reality. But the alternative, Marxist understanding of science as fundamentally utilitarian, the product of emergent capitalism and its attempts to command the material world, is no less ideological in this sense of being systematically distorting. Science can be seen as either of these two ideologies (at least); no doubt there has always been a bit of both, depending on the parallels adduced in any particular case. Where science is imagined according to a theological model of pure understanding, natural philosophy is foremost; where science is imagined according to a craft model of practical manipulation, instrumentality or utility is foremost.[15] The world as we know it is mediated by these differing models; to escape from one is to be caught up in the arms of the other. Our modern scientific worldviews are the result of this restless process.

ACKNOWLEDGMENTS

Because of the breadth and scope of the topics and arguments informing this book, specific acknowledgments are difficult to make. But by the same token, it is surely the case that any thing or person that helped inform my understanding of and enthusiasm for the history of science warrants grateful mention. Thus, inter alia, John Schuster, Simon Schaffer, Steven Shapin, and Robert Westman, together with the late Michael Mahoney, Gerald Geison, and Owen Hannaway, are preeminent (whether they realize it or not) among my senior mentors, while among my peers and contemporaries should be numbered Thomas Broman, Michael Dennis, John Carson, Lissa Roberts, Jan Golinski, and a whole generation of historians of which it was a privilege to be a part. And of course I have learned much from my own students. Anna Märker, Adelheid Voskuhl, Carin Berkowitz, Bill Lynch, Shelley Costa, and the seminars and study groups in which they and others participated, taught me much, and haunt these pages as much as the undergraduates of Cornell's HIST/S&TS 282 course and its cognates, from which this book developed. Thanks also to my editor at Princeton University Press, Eric Crahan, and to the press's indefatigable Becca Binnie, without whose aid I could not have managed.

PRD

NOTES

Introduction

1. Three more recent examples are Stephen Gaukroger's series of volumes starting with *The Emergence of a Scientific Culture: Science and the Shaping of Modernity, 1210–1685* (Oxford: Clarendon Press, 2006); Floris Cohen, *How Modern Science Came into the World: Four Civilizations, One 17th-Century Breakthrough* (Amsterdam: Amsterdam University Press, 2010); David Wootton, *The Invention of Science: A New History of the Scientific Revolution* (New York: Harper, 2015).

2. One can, of course, argue this both ways: see Marwa Elshakry, "When Science Became Western: Historiographical Reflections," *Isis* 101 (2010), pp. 98–109.

3. As Bruno Latour famously argued in *Science in Action: How to Follow Scientists and Engineers through Society* (Cambridge, MA: Harvard University Press, 1988).

4. See Richard Yeo, *Defining Science: William Whewell, Natural Knowledge and Public Debate in Early Victorian Britain* (Cambridge: Cambridge University Press, 1993), pp. 110–11.

5. Exemplified in Paul White, *Thomas Huxley: Making the "Man of Science"* (Cambridge: Cambridge University Press, 2002).

6. A useful fuller discussion of the terms *science*, *scientific*, and *scientist* appears on pages 23–32 of Wootton, *Invention of Science*, which also touches on related labels and usages in other European languages. See also Raymond Williams, *Keywords: A Vocabulary of Culture and Society* (Oxford: Oxford University Press, 2014), q.v. "Science," pp. 215 ff. The meaning of *scientia* in the early modern period is also examined in some depth in T. Demeter, B. Láng, and D. Schmal, "Scientia in the Renaissance, Concept of," in Sgarbi, M. (eds.), *Encyclopedia of Renaissance Philosophy*, https://link.springer.com/referenceworkentry/10.1007/978-3-319-14169-5_266.

7. See Edward Grant, *A History of Natural Philosophy: From the Ancient World to the Nineteenth Century* (Cambridge: Cambridge University Press, 2007). For a broad overview, see Peter Dear, "The Natural Philosopher," in Bernard Lightman (ed.), *Blackwell Companion to the History of Science* (Oxford: Blackwell, 2016), pp. 71–83.

8. This is the basic argument of Peter Dear, *Discipline and Experience: The Mathematical Way in the Scientific Revolution* (University of Chicago Press, 1995); for more on utility, see also Dear, *Revolutionizing the Sciences: European Knowledge in Transition, 1500–1700*, 3rd ed. (Princeton: Princeton University Press, 2019), chap. 3. Note that I capitalize Science here to denote its emergence as a broadly encompassing modern term that took root in the nineteenth century, as opposed to the specificities of individual sciences.

9. Of the many treatments of the Scientific Revolution, one might mention Wootton, *The Invention of Science*, and for a briefer account, Dear, *Revolutionizing the Sciences*.

10. See essays in Daniel Garber and Sophie Roux (eds.), *The Mechanization of Natural Philosophy* (Dordrecht: Springer, 2013), for an indication of these complexities.

11. Alexandre Koyré, *From the Closed World to the Infinite Universe* (Baltimore: Johns Hopkin's University Press, 1957), remains an excellent treatment of these issues.

12. Another development of the preceding century: see, e.g., Dear, *Revolutionizing the Sciences*, chap. 7.

13. See the classic discussion in G.A.J. Rogers, "Descartes and the Method of English Science," *Annals of Science* 29 (1972), pp. 237–55.

14. A view most forcefully argued by Andrew Cunningham, e.g. in "Getting the Game Right: Some Plain Words on the Identity and Invention of Science," *Studies in History and Philosophy of Science* 19 (1998), pp. 365–89.

15. This is a good point at which to address the role of women scientists in the enterprises that form the subject of this book (and the considerable historical scholarship devoted to this issue). Women were routinely excluded from institutional scientific activities throughout the period that this book covers. This is unquestionably true despite a good number of accomplished individual women scientists and mathematicians, some of whom appear in references in this book. It would be easy to make lists: Émilie du Châtelet, Laura Bassi, Caroline Herschel, Mary Sommerville, Marie Curie, Henrietta Leavitt, and so on. However, my book attempts to provide a history of scientific ideas rather than a sociocultural history of science; it would misrepresent the misogynistic character of much scientific activity and institutional structures by presenting such lists as if they were typical. Indeed, the chief reason for such enumerations would seem to be to refute suggestions that women are not typically capable of high-level scientific activity—a foolishness that much more recent history has in any case definitively disproved. For the periods and places considered in this book, the relative paucity of women in the sciences is socially overdetermined, regardless of individual notables.

16. Medicine may furnish a partial exception to this generalization, but only insofar as a university professor of medicine would typically also be a practicing physician who might thereby acquire experiential knowledge, sometimes to be publicized.

17. For recent essays on this topic, see Mordechai Feingold and Giulia Giannini (eds.), *The Institutionalization of Science in Early Modern Europe* (Leiden: Brill, 2019).

18. Broad studies on this theme include Londa Schiebinger, *The Mind has No Sex? Women in the Origins of Modern Science* (Cambridge, MA: Harvard University Press, 1989); and Patricia Fara, *Pandora's Breeches: Women, Science and Power in the Enlightenment* (London: Pimlico, 2004). In seventeenth-century England, a prominent example of an excluded woman is the Duchess of Newcastle, Margaret Cavendish, and her relationship with the Royal Society; instructive exceptions are examined in Massimo Mazzotti, *The World of Maria Gaetana Agnesi: Mathematician of God* (Baltimore: Johns Hopkins University Press, 2007), and in Paula Findlen's classic article "Science as a Career in Enlightenment Italy: The Strategies of Laura Bassi," *Isis* 84 (1993), pp. 441–69.

Chapter 1. Divine Order

1. For an examination of this point, see Robert E. Schofield, "An Evolutionary Taxonomy of Eighteenth-Century Newtonianisms," *Studies in Eighteenth-Century Culture* 7 (1978), pp. 175–92.

2. Alexander Pope, "Epitaph Intended for Sir Isaac Newton" (1735).

3. A detailed description of the bas-relief can be found at the website of Westminster Abbey, https://www.westminster-abbey.org/abbey-commemorations/commemorations/sir-isaac-newton. Newton's reflecting telescope, his optical studies (a cherub peering through a prism) and his work as Master of the Mint are also represented, while the smelting furnace on the right may, besides coinage, also allude to his chemical and alchemical investigations.

4. See Alexandre Koyré, *From the Closed World to the Infinite Universe* (Baltimore: Johns Hopkins University Press, 1957), chap. 7, which quotes some of the relevant texts.

5. See Tad Schmaltz, "Nicolas Malebranche," in *The Stanford Encyclopedia of Philosophy* (Spring 2022), Edward N. Zalta (ed.), https://plato.stanford.edu/archives/spr2022/entries/malebranche/.

6. Descartes had originally outlined his world-system in an unpublished work of the 1630s called *Le Monde*: see René Descartes, *The World and Other Writings*, trans. Stephen Gaukroger (Cambridge: Cambridge University Press, 1998). I discuss Descartes and Cartesianism at greater length in Peter Dear, *Revolutionizing the Sciences: European Knowledge in Transition, 1500–1700*, 3rd ed. (Princeton: Princeton University Press, 2019), chap. 5.

7. The classic treatment of these issue is John Henry, "Occult Qualities and the Experimental Philosophy. Active Principles in Pre-Newtonian Matter Theory," *History of Science* 24 (1986), pp. 335–81.

8. Margaret C. Jacob, *The Newtonians and the English Revolution 1689–1720* (Ithaca: Cornell University Press, 1976), p. 144.

9. Newton to Bentley, December 10, 1692, The Newton Project, https://www.newtonproject.ox.ac.uk/view/texts/normalized/THEM00254. Irregular spelling and punctuation follow the original.

10. On the early Boyle lectures and their political salience, see Jacob, *Newtonians*, chap. 5.

11. Jacob, *Newtonians*.

12. Quoted in Jacob, *Newtonians*, pp. 192–93.

13. That is, "Lord of All," from the "General Scholium" of the *Principia*'s second edition, 1713. See Isaac Newton, *The Principia*, 940.

14. Koyré, *Closed World*, pp. 176–189.

15. John Toland, *Letters to Serena* (London, 1704), Letter V, "Motion essential to Matter."

16. Henry Guerlac, *Newton on the Continent* (Ithaca: Cornell University Press, 1981). For a comprehensive overview, see essays in Helmut Pulte and Scott Mandelbrote (eds.), *The Reception of Isaac Newton in Europe*, 3 vols. (London: Bloomsbury Academic, 2019).

17. J. B. Shank, *The Newton Wars and the Beginning of the French Enlightenment* (Chicago: University of Chicago Press, 2008), chaps. 4 and 5.

18. Two biographical treatments of Du Châtelet, which reflect increasing historical attention to her contemporary significance and reputation, are Judith Zinsser, *Dame d'Ésprit: A Biography of the Marquise du Châtelet* (New York: Viking, 2006) and Robyn Arianrhod, *Seduced by Logic: Émilie du Châtelet, Mary Somerville, and the Newtonian Revolution* (Oxford: Oxford University Press, 2012).

19. See Thomas L. Hankins, *Jean d'Alembert: Science and the Enlightenment* (Oxford: Clarendon Press, 1970); more broadly, Shank, *Newton Wars*, esp. chap. 7. Outside France, see also essays in Elizabethanne Boran and Mordechai Feingold (eds.), *Reading Newton in Early Modern Europe* (Leiden: Brill, 2017).

20. *Encyclopédie ou Dictionnaire raisonné des sciences, des arts et des métiers*. The best available edition of this work is the online ARTFL text edited by Robert Morrissey, https://encyclopedie.uchicago.edu/.

21. Locke's relevant work here is John Locke, *An Essay Concerning Human Understanding* (London, 1691); the standard modern edition is edited by Peter Nidditch (Oxford: Clarendon Press, 1979).

22. See, on the Dutch, Steffen Ducheyne and Jip van Besouw, "Newton and the Dutch 'Newtonians': 1713–1750," in Eric Schliesser and Chris Smeenk (eds), *The Oxford Handbook of Newton* (online edition, Oxford Academic, March 2017), https://doi.org/10.1093/oxfordhb/9780199930418.013.20.

23. I. Bernard Cohen promoted this characterization in his classic book *Franklin and Newton: An Inquiry into Speculative Newtonian Science and Franklin's Work in Electricity as an Example Thereof* (Philadelphia: American Philosophical Society, 1956), chaps. 5 and 6.

24. For a detailed study of chemical work largely independent of Newton, see Lawrence M. Principe, *The Transmutations of Chemistry: Wilhem Homberg and the Académie Royale des Sciences* (Chicago: University of Chicago Press, 2020).

25. Newton, *Opticks*, 378. For the best modern scholarly edition, see *The Optical Papers of Isaac Newton*, ed. Alan E. Shapiro, vol. 2 (Cambridge: Cambridge University Press, 2021).

26. Newton, *The Principia*, 940. A more detailed account of the unnecessary features of the solar system that revealed God's hand were shortly afterward provided by Newton's acolyte William Whiston in his *Astronomical Principles of Religion, Natural & Reveal'd* (London, 1717), pp. 81–90.

27. For *Physico-Theology* I use the third edition, published in 1714. For *Astro-Theology* I use the second edition, published in 1715.

28. Derham, *Physico-Theology*, "To the Reader," A4v. However, see the editors' introduction to Ann Blair and Kaspar von Greyerz (eds.), *Physico-Theology: Religion and Science in Europe, 1650–1750* (Baltimore: Johns Hopkins University Press, 2020), pp. 1–20, noting that there was much more to natural theology than simply the argument from design.

29. On the concept of the terraqueous globe, see Grant, *Planets, Stars, and Orbs*, 630–37; also Wootton, *The Invention of Science*, 110–43.

30. Derham, *Physico-Theology*, 111–12.

31. Derham, *Physico-Theology*, 332–33, n. 1.

32. Marjorie Grene makes an elegant case for the primacy of living beings in shaping Aristotle's philosophy in *A Portrait of Aristotle* (Chicago: University of Chicago Press, 1963).

33. See Katharine Calloway, "'Rather Theological than Philosophical': John Ray's Seminal *Wisdom of God Manifested in the Works of Creation*," in Blair and von Greyerz (eds.), *Physico-Theology*, 115–26.

34. Quoted in Paul Lawrence Farber, *The Naturalist Tradition from Linnaeus to E. O. Wilson* (Baltimore: Johns Hopkins University Press, 2000), p. 13.

Chapter 2. Celestial Order and Universal Gravity

1. René Descartes, *The World and Other Writings*, trans. Stephen Gaukroger (Cambridge: Cambridge University Press, 1998). See figure 1.2, this volume.

2. See, for an overview of relevant ideas on matter after Newton, Robert E. Schofield, *Mechanism and Materialism: British Natural Philosophy in an Age of Reason* (Princeton: Princeton University Press, 1970).

3. See Imre Lakatos, *The Methodology of Scientific Research Programmes*, ed. John Worrall and Gregory Currie (Cambridge: Cambridge University Press, 1978).

4. See Michael A. Hoskin, "Newton, Providence, and the Universe of Stars," *Journal for the History of Astronomy* 8 (1977), pp. 77–101.

5. Primarily a catalogue by the astronomer Vincent Wing. See also the related discussions of Newton's astronomical-historical labors by Jed Buchwald and Mordechai Feingold, *Newton and the Origin of Civilization* (Princeton: Princeton University Press, 2012).

6. Isaac Newton, *The Principia: Mathematical Principles of Natural Philosophy*, trans. I. Bernard Cohen and Anne Whitman (Berkeley: University of California Press, 1999), "General Scholium," p. 940. David Gregory suggested, on the basis of discussions with Newton, that Newton had earlier contemplated stellar immobility as a "continual miracle," where God chooses to suspend His ordinary laws of gravitational attraction. See Michael Hoskin, "Newton, Providence, and the Universe of Stars," *Journal for the History of Astronomy* 8 (1977), p. 77, quoting David Gregory on discussions with Newton. For further important remarks on Newton, Newtonians, and miracles see James E. Force, *William Whiston: Honest Newtonian* (Cambridge: Cambridge University Press, 1985), esp. pp. 124–26.

7. Edmund Halley, "Of the Infinity of the Sphere of Fix'd Stars," *Philosophical Transactions* 31 (1720–21), pp. 22–24; also Halley, "Of the Number, Order, and Light of the Fixed Stars," *Philosophical Transactions* 31 (1720–21), pp. 24–26. The two together effectively rehearse the same argument as Newton's; see also Hoskin, "Newton, Providence."

8. William Derham, *Astro-Theology: Or, A Demonstration of the Being and Attributes of God, From a Survey of the Heavens*, 2nd ed. (London, 1715), pp. 56–61.

9. John Keill, *An Introduction to the True Astronomy* (London, 1721), p. 47.

10. Roger Cotes's preface to the second edition (1713) of the *Principia* made the requirement of universality very clear.

11. William Whiston, *Astronomical Principles of Religion, Natural & Reveal'd* (London, 1717).

12. Whiston, *Astronomical Principles*, p. 88.

13. Whiston, *Astronomical Principles*, pp. 88–89.

14. Whiston, *Astronomical Principles*, p. 89. As with others of Whiston's views, this one went beyond what Newton himself was prepared to assert publicly.

15. John D. North, *The Norton History of Astronomy and Cosmology* (New York: Norton, 1994), pp. 395–96.

16. Immanuel Kant, *Allgemeine Naturgeschichte und Theorie des Himmels* (Königsberg, 1755).

17. Thomas Wright [of Durham], *An Original Theory or New Hypothesis of the Universe*, ed. Michael A. Hoskin (London: Macdonald, 1971); Michael Hoskin, "The Cosmology of Thomas Wright of Durham," *Journal for the History of Astronomy* 1 (1970), 44–52; Kenneth Glyn Jones, "The Observational Basis for Kant's *Cosmogony*: A Critical Analysis," *Journal for the History of Astronomy* 2 (1971), 29–34.

18. Jones, "Observational Basis."

19. See, for an interesting discussion of the relationship of Laplace's and Herschel's ideas to biological evolution in the nineteenth century, Stephen G. Brush, "The Nebular Hypothesis and the Evolutionary Worldview," *History of Science* 25 (1987), 245–78.

20. Ironically, Herschel was himself of German origin, like the Georgian kings of England in the eighteenth century.

21. This general idea was also used by other natural philosophers in the eighteenth century, including Joseph Priestley. See Arnold Thackray, *Atoms and Powers: An Essay on Newtonian Matter-Theory and the Development of Chemistry* (Cambridge, MA: Harvard University Press, 1970), esp. chap. 5. Boscovich's first name appears in various spellings, sometimes Serbo-Croat, sometimes Italian, sometimes Latin, but all are equivalents of the English name Roger. His surname is usually written as here.

22. Roger Joseph Boscovich, *A Theory of Natural Philosophy* (Chicago, 1922), a translation by J. M. Child of Rogerius Boscovich, *Theoria philosophiae naturalis* (Venice, 1763), 291; cf. also 105–7.

23. Boscovich, *Theory of Natural Philosophy*, 291.

24. Boscovich, *Theory of Natural Philosophy*, 291.

25. John Michell, "An Inquiry into the Probable Parallax, and Magnitude of the Fixed Stars, from the Quantity of Light which They Afford us, and the Particular Circumstances of their Situation," *Philosophical Transactions* 57 (1767), pp. 234–64; see, for analysis of Michell's work in this paper, Russell McCormmach, *Weighing the World: The Reverend John Michell of Thornhill* (Dordrecht: Springer, 2012), pp. 145–51; also McCormmach, "John Michell and Henry Cavendish: Weighing the Stars," *British Journal for the History of Science* 4 (1968), 126–55.

26. The term *double star* refers simply to two stars appearing close together in the sky, whether binary or line-of-sight doubles.

27. Michell, "Parallax," 243.

28. Michell, "Parallax," 249.

29. Isaac Newton, *Opticks*, 2nd (English) edition (London, W. and J. Innis, 1717/18), 350–82.

30. See McCormmach, *Weighing the World*, 216–18. For Michell's ideas about stars so massive that light could not escape their gravitational pull, see ibid., 210–16; attention was drawn to Michell's role from the usual ascription to Laplace's *Exposition du système du monde* of 1796 in Simon Schaffer, "John Michell and Black Holes," *Journal for the History of Astronomy* 10 (1979), 42–43; and Jean Eisenstaedt, "De l'influence de la gravitation sur la propagation de la lumière en théorie newtonienne: L'archéologie des trous noirs," *Archive for History of Exact Sciences* 42 (1991), 315–86.

31. Michael Hoskin, *The Construction of the Heavens: William Herschel's Cosmology* (Cambridge: Cambridge University Press, 2012); Hoskin, *Discoverers of the Universe: William and Caroline Herschel* (Princeton: Princeton University Press, 2011); for a concise overview, see Hoskin, "William Herschel and the Construction of the Heavens," *Proceedings of the American Philosophical Society*, 133 (1989), pp. 427–33. Caroline gained a considerable reputation in her own right as an observer, systematically recording nebulae and discovering several comets. Her cataloguing work led in 1828 to her being awarded a gold medal by the recently founded Astronomical Society (soon to become the Royal Astronomical Society), an unusual honor for a woman at that time.

32. The existence of a variety of distance forces was an idea shared by Michell and many others. On Herschel's natural history of the heavens, see Simon Schaffer, "Herschel in Bedlam: Natural History and Stellar Astronomy," *British Journal for the History of Science* 13 (1980), pp. 211–39; Steven J. Dick, *Discovery and Classification in Astronomy: Controversy and Consensus* (Cambridge: Cambridge University Press, 2013), chap. 3, "In Herschel's Gardens."

33. Newton's theology is examined at length in Rob Iliffe, *Priest of Nature: The Religious Worlds of Isaac Newton* (Oxford: Oxford University Press, 2017).

Chapter 3. Mixed Mathematics and Probability

1. Peter Dear, "Mixed Mathematics," in Peter Harrison, Michael Shank, and Ronald Numbers (eds.), *Wrestling with Nature: From Omens to Science* (Chicago: University of Chicago Press, 2011), pp. 149–72.

2. See esp. Lorraine Daston, *Classical Probability in the Enlightenment* (Princeton: Princeton University Press, 1988).

3. Robert Smith, *A Compleat Systeme of Opticks*, 2 vols. (Cambridge, 1738); Smith, *Harmonics, or the Philosophy of Musical Sounds* (Cambridge, 1749). Smith was Plumian Professor of Astronomy and master of Trinity College, Cambridge, as well as a cousin of Roger Cotes, the editor of the second edition of Newton's *Principia*.

4. Daston, *Classical Probability*, pp. 230–41; Ian Hacking, *The Emergence of Probability: A Philosophical Study of Early Ideas about Probability, Induction and Statistical Inference* (Cambridge: Cambridge University Press, 1975), chaps. 16 and 17. Bernoulli's book is translated by Edith Dudley Sylla as Jacob Bernoulli, *The Art of Conjecturing* (Baltimore: Johns Hopkins University Press, 2005).

5. John Arbuthnot, "An Argument for Divine Providence," *Philosophical Transactions* (January 1710), pp. 186–90; Arbuthnot's paper is also discussed in Hacking, *The Emergence of Probability*, chap. 18.

6. See discussion in James Franklin, *The Science of Conjecture: Evidence and Probability Before Pascal* (Baltimore: Johns Hopkins University Press, 2001), chap. 11.

7. Arbuthnot, "An Argument," 187.

8. Arbuthnot, "An Argument," 188.

9. It's worth considering why we would not regard Arbuthnot's mathematical argument as sound. Clearly the difficulty is that he provides no way of calculating the chances involved in having an extreme distribution, far from 50:50, in any particular year or set of coin tosses. He only says that it is "very improbable" that you will not sometimes get one of these extreme distributions; he provides no reason to accept that assertion.

10. Arbuthnot, "An Argument," p. 176 n. 8. See also the discussion in Hacking, *The Emergence of Probability*, pp. 169–70.

11. See Stephen M. Stigler, *The History of Statistics: The Measurement of Uncertainty before 1900* (Cambridge, MA: Harvard University Press, 1986), p. 65.

12. Abraham De Moivre, *The Doctrine of Chances: or, A Method of Calculating the Probability of Events in Play* (London, 1718), dedication (n.p.); see Daston, *Classical Probability*, pp. 250–53; also Stigler, *History of Statistics*, pp. 70–88.

13. Hacking, *The Emergence of Probability*, pp. 170–71.

14. De Moivre, *Doctrine*, 3rd ed. (London, 1756), pp. 252–53.

15. De Moivre, *Doctrine*, p. 253.

16. De Moivre, *Doctrine*, p. 252.

17. See, for a full discussion, Daston, *Classical Probability*, pp. 68–81.

18. Daston, *Classical Probability*, pp. 70–76.

19. Daston, *Classical Probability*, pp. 76–79.

20. "Sur le calcul des probabilités," in Jean D'Alembert, *Opuscules mathématiques*, vol. 7 (Paris, 1780), p. 48 (my translation).

21. "Sur le calcul des probabilités," p. 48.

Chapter 4. Inventories of Electricity

1. The significance of scientific training is the focus of essays in David Kaiser (ed.), *Pedagogy and the Practice of Science: Historical and Contemporary Perspectives* (Cambridge, MA: MIT Press, 2005).

2. Much the best comprehensive account of early modern electricity remains John L. Heilbron, *Electricity in the 17th and 18th Centuries: A Study of Early Modern Physics* (Berkeley: University of California Press, 1979; reprint with new preface, Mineola, NY: Dover, 1999); on Gilbert, see chap. 3 in the same volume. Thomas L. Hankins, *Science in the Enlightenment* (Cambridge: Cambridge University Press, 1983), pp. 53–71, remains a good overview of eighteenth-century electrical ideas and research.

3. On the use of a checklist of "trials" as a means, in effect, of defining a natural object, see Bruno Latour, *Science in Action: How to Follow Scientists and Engineers through Society* (Cambridge, MA: Harvard University Press, 1988), esp. pp. 86–94.

4. Further details on the English work from Hauksbee to Gray may be found in Heilbron, *Electricity*, chap. 8.

5. Isaac Newton, *Opticks*, 2nd English edition (London: W. and J. Innys, 1717/18), pp. 324–27.

6. Isaac Newton, *The Principia: Mathematical Principles of Natural Philosophy*, trans. I. Bernard Cohen and Anne Whitman (Berkeley: University of California Press, 1999), pp. 943–44 (in the "General Scholium" of the 1713 edition).

7. Stephen Gray, "A Letter to Cromwell Mortimer . . . Containing Several Experiments Concerning Electricity," in *Philosophical Transactions of the Royal Society* 37 (1731), pp. 18–44. Besides Heilbron, see also I. Bernard Cohen, *Franklin and Newton: An Inquiry into Speculative Newtonian Science* (Philadelphia: American Philosophical Society, 1956), pp. 368–70.

8. Gray, "Letter to Mortimer," p. 20.

9. Gray, "Letter to Mortimer," p. 20.
10. Some details were more salient than others; see n.13 below on human subjects.
11. Gray, "Letter to Mortimer," p. 26.
12. On Dufay, see Heilbron, *Electricity*, chap. 9, esp. pp. 252–57. See also translations in Peter Dear (ed.), *Scientific Practices in European History, 1200–1800: A Book of Texts* (London: Routledge, 2018), pp. 111–15, from Dufay's papers in the irregularly numbered volume of the *Mémoires de l'Académie Royale des Sciences* for 1733.
13. For discussion of Dufay's philosophical style, see Michael Bycroft, "Wonders in the Academy: The Value of Strange Facts in the Experimental Research of Charles Dufay," *Historical Studies in the Natural Sciences* 43 (2013), pp. 334–70. The association of electricity with natural wonders was a perennial theme in the eighteenth century (intersecting also with ideas about the animal or human body); a classic study is Simon Schaffer, "Natural Philosophy and Public Spectacle in the Eighteenth Century," *History of Science* 21 (1983), pp. 1–43. My focus in this chapter is on ideas about the physical, natural-philosophical nature of electricity, but physiological and medical perspectives on electricity were active throughout the period, playing an important role in the late eighteenth-century invention of current electricity discussed in chapter 13. See for such themes Paola Bertucci, "Shocking Subjects: Human Experiments and the Material Culture of Medical Electricity in Eighteenth-Century England," in Erika Dick and Larry Stewart (eds.), *The Uses of Humans in Experiment: Perspectives from the 17th to the 20th Century* (Leiden: Brill, 2016), pp. 111–38; essays in Bertucci and Giuliano Pancaldi (eds.), *Electric Bodies. Episodes in the History of Medical Electricity* (Bologna: CIS, University of Bologna, 2001); James Delbourgo, *A Most Amazing Scene of Wonders: Electricity and Enlightenment in Early America* (Cambridge, MA: Harvard University Press, 2006).
14. Jean Theophile Desaguliers, "Experiments Made before the Royal Society, Feb. 2. 1737–8," *Philosophical Transactions of the Royal Society* 41 (1739), pp. 193–99.
15. Heilbron, *Electricity*, pp. 264–70.
16. Heilbron, *Electricity*, pp. 312–18.
17. Heilbron, *Electricity*, p. 317.
18. Quoted in Heilbron, *Electricity*, p. 318.
19. Cohen, *Franklin and Newton*, p. 412.
20. William Watson, *A Sequel to the Experiments and Observations Tending to Illustrate the Nature and Properties of Electricity* (London, 1746), p. 35; Cohen, *Franklin and Newton*, pp. 402–3.
21. See I. Bernard Cohen (ed.), *Benjamin Franklin's Experiments: A New Edition of Benjamin Franklin's* Experiments and Observations on Electricity (Cambridge, MA: Harvard University Press, 1941), pp. 174–76; also extracted in Dear, *Scientific Practices*, pp. 116–18.
22. In Cohen, *Franklin's Experiments*, pp. 174–76; Dear, *Scientific Practices*, pp. 116–18.
23. For an examination of this dimension of Franklin's work, see Jessica Riskin, "Poor Richard's Leyden Jar: Electricity and Economy in Franklinist France," *Historical Studies in the Physical and Biological Sciences* 28 (1998), pp. 301–36.
24. Following the account in Heilbron, *Electricity*, pp. 330–33. See the use of this material by Thomas S. Kuhn, *The Structure of Scientific Revolutions*, 2nd ed. (Chicago: University of Chicago Press, 1970), pp. 17–18, 61–62, 117–18; and a more elaborate reading of it in J. A. Schuster and G. Watchirs, "Natural Philosophy, Experiment and Discourse in the Eighteenth Century: Beyond the Kuhn/Bachelard Problematic," in Homer E. LeGrand (ed.), *Experimental Inquiries: Historical, Philosophical and Social Studies of Experiment* (Dordrecht: Reidel, 1990), pp. 1–48.
25. Heilbron, *Electricity*, , pp. 337–38. For a close account, see Roderick W. Home, "Franklin's Electrical Atmospheres," *British Journal for the History of Science* 6 (1972), pp. 131–51; this and other relevant articles are reprinted in R. W. Home, *Electricity and Experimental Physics in Eighteenth-Century Europe* (Brookfield, VT: Variorum, 1992).

26. E.g., Heilbron, *Electricity*, pp. 333-34.

27. For Priestley, see Robert E. Schofield, *The Enlightenment of Joseph Priestley: A Study of His Life and Work, 1733-1773* (University Park: Pennsylvania State University Press, 1997), p. 150. Heilbron notes the imprecision of Coulomb's technique to confirm an inverse square relationship and the merely suggestive experimental technique used by Priestley to the same end: Heilbron, *Electricity*, pp. 464-65, 468-77. At around the same period, the German philosopher Immanuel Kant (among others) argued that the inverse square law for gravity was necessarily true; it could not be otherwise, given the fundamental status of the laws of mechanics: see Michael Friedman, "Causal Laws and the Foundations of Natural Science," in Paul Guyer (ed.), *The Cambridge Companion to Kant* (Cambridge: Cambridge University Press, 1992), pp. 161-99, esp. pp. 176-78.

28. On two-fluid versus one-fluid theories, see Heilbron, *Electricity*, chap. 18.

29. See Heilbron, *Electricity*, pp. 376-77.

Chapter 5. Organization

1. As persuasively argued in Marjorie Grene, *A Portrait of Aristotle* (Chicago: University of Chicago Press, 1962).

2. For a general overview, see Paul Lawrence Farber, *Finding Order in Nature: The Naturalist Tradition from Linnaeus to E. O. Wilson* (Baltimore: Johns Hopkins University Press, 2000). On the character of the classificatory enterprise in this period, see Michel Foucault, *The Order of Things: An Archaeology of the Human Sciences* (New York: Pantheon, 1970), chap. 5, esp. pp. 145-50; and Simon Schaffer, "Herschel in Bedlam: Natural History and Stellar Astronomy," *British Journal for the History of Science* 13 (1980), pp. 211-39. See for more discussion Peter Dear, *The Intelligibility of Nature: How Science Makes Sense of the World* (Chicago: University of Chicago Press, 2006), chap. 2. A valuable discussion of classificatory practices as means of insight for historians of science may be found in Ursula Klein and Wolfgang Lefèvre, *Materials in Eighteenth-Century Science: A Historical Ontology* (Cambridge, MA: MIT Press, 2007), chap. 3.

3. John Gascoigne, *Science in the Service of Empire: Joseph Banks, the British State and the Uses of Science in the Age of Revolution* (Cambridge: Cambridge University Press, 1998); see also Patricia Fara, *Sex, Botany and Empire: The Story of Carl Linnaeus and Joseph Banks* (Cambridge: Icon, 2003).

4. See Brian W. Ogilvie, *The Science of Describing: Natural History in Renaissance Europe* (Chicago: University of Chicago Press, 2006), esp. chap. 5.

5. I cheat slightly here: the category "family," located between order and genus, was not generally used by Linnaeus himself, although its use soon became standard among later botanists in the eighteenth century, such as Michel Adanson.

6. For a classic examination of the cultural meanings of Linnaeus's category "mammal," see Londa Schiebinger, "Why Mammals Are Called Mammals: Gender Politics in Eighteenth-Century Natural History," *American Historical Review* 98 (1993), pp. 382-411.

7. John E. Lesch, *Science and Medicine in France: The Emergence of Experimental Physiology, 1790-1855* (Cambridge, MA: Harvard University Press, 1984), p. 20.

8. Lesch, *Science and Medicine in France*, pp. 66-67.

9. Compare Farber, *Finding Order in Nature*, p. 13.

10. Phillip R. Sloan, "John Locke, John Ray, and the Problem of the Natural System," *Journal of the History of Biology* 5 (1972), p. 37. Sloan's article remains an essential guide to understanding these issues.

11. See Sloan, "John Locke," p. 38.

12. Translated in Sloan, "John Locke," p. 48.

13. Jacques Roger, *Buffon: A Life in Natural History* (Ithaca: Cornell University Press, 1997).

14. Published over the course of forty years and running to thirty-six volumes.

15. Buffon's arguments can be seen as a variety of philosophical skepticism, with the crucial argumentative element being that of trying to place the burden of proof on his opponents.

16. Following the translation by John Lyon, "Initial Discourse" to vol. 1 of Buffon's *Histoire naturelle* (1749) in John Lyon and Phillip R. Sloan (eds. and trans.), *From Natural History to the History of Nature: Readings from Buffon and His Critics* (Notre Dame: University of Notre Dame Press, 1981), p. 111.

17. Roger, *Buffon*, chap. 19, esp. pp. 313–20.

18. These developments in Buffon's views over time are chronicled in John H. Eddy, "Buffon's *Histoire naturelle*: History? A Critique of Recent Interpretations," *Isis* 85 (1994), pp. 644–61.

Chapter 6. Cleaning up Chemistry

1. It is crucial to recognize the restricted selection of materials that came to constitute the academic chemist's toolbox in the eighteenth century. These materials, whether subsequently regarded as simple or compound, were not prepackaged as potential elements or chemical compounds (as opposed to mixtures), but had to be given such statuses as a result of chemical work at both the experimental and theoretical levels. This produced the list of "simple substances" famous from Lavoisier's 1789 *Elements of Chemistry* and informing the great *Méthode de nomenclature chimique, proposée par MM. de Morveau, Lavoisier, Bertholet* [sic], *et de Fourcroy* (Paris, 1787). See the important article by Wolfgang Lefèvre, "The *Méthode de nomenclature chimique* (1787): A Document of Transition," *Ambix* 65 (2018), pp. 9–29, and Ursula Klein and Lefèvre, *Materials in Eighteenth-Century Science: A Historical Ontology* (Cambridge, MA: MIT Press, 2007).

2. Thomas S. Kuhn, "Robert Boyle and Structural Chemistry in the Seventeenth Century," *Isis* 43 (1952), pp. 12–36. However, William R. Newman counters Kuhn's claim that Boyle's work had very little impact on subsequent chemistry: Newman, "Robert Boyle, Transmutation and the History of Chemistry before Lavoisier: A Response to Kuhn," *Osiris* 29 (2014), pp. 63–77.

3. Denis Diderot, *Encyclopédie ou Dictionnaire raisonné des sciences, des arts et des métiers*, online ARTFL text ed. by Robert Morrissey, https://encyclopedie.uchicago.edu/, vol. 3, p. 408. Notice that Venel takes it for granted that chemistry is a disciplinary natural kind, rather than just a term of convenience.

4. Diderot, *Encyclopédie*, vol. 3, p. 408. On Stahl, see Ku-ming Kevin Chang, "Communications of Chemical Knowledge: Georg Ernst Stahl and the Chemists at the French Academy of Sciences in the First Half of the Eighteenth Century," *Osiris* 29 (2014), pp. 135–57; Chang, "Phlogiston and Chemical Principles: The Development and Formulation of Georg Ernst Stahl's Principle of Inflammability," in Karen Hunger Parshall, Michael T. Walton, and Bruce T. Moran (eds.), *Bridging Traditions: Alchemy, Chemistry, and Paracelsian Practices in the Early Modern Era* (Kirksville, MO: Truman State University Press, 2015), pp. 101–31.

5. Georg Ernst Stahl, *Philosophical Principles of Universal Chemistry*, trans. Peter Shaw of Stahl, *Fundamenta chymiae* (London, 1730), p. 4.

6. John C. Powers, *Inventing Chemistry: Herman Boerhaave and the Reform of the Chemical Arts* (Chicago: University of Chicago Press, 2012), chap. 3; the idea goes back to Daniel Sennert in the mid-seventeenth century.

7. For a synoptic account of alchemy, see Lawrence Principe, *The Secrets of Alchemy* (Chicago: University of Chicago Press, 2013).

8. Isaac Newton, *Opticks*, 2nd English ed. (London: W. and J. Innys, 1717/18), pp. 355–56.

9. The best overall treatment of this topic is Mi Gyung Kim, *Affinity, That Elusive Dream: A Genealogy of the Chemical Revolution* (Cambridge, MA: MIT Press, 2003); see also Klein and

Lefèvre, *Materials*, pp. 147–53; Wolfgang Lefèvre, "*Méthode*," on the restricted choice of materials in Geoffroy's table.

10. Treated in Arnold Thackray, *Atoms and Powers: An Essay on Newtonian Matter-Theory and the Development of Chemistry* (Cambridge, MA: Harvard University Press, 1970).

11. Lefèvre, "*Méthode*," p. 18.

12. Stephen Hales, *Vegetable Staticks, or, An Account of some Statical Experiments*... (London, 1727). There is also a modern reprint with a foreword by Michael Hoskin (London: Scientific Book Guild, 1961).

13. Thomas L. Hankins, *Science in the Enlightenment* (Cambridge: Cambridge University Press, 1983), pp. 89–91, provides a useful account of Black's work that tracks the account I give here.

14. The foundational study is Henry Guerlac, *Lavoisier—The Crucial Year: The Background and Origin of his First Experiments on Combustion in 1772* (Ithaca: Cornell University Press, 1961). See also J. B. Gough, "The Origins of Lavoisier's Theory of the Gaseous State," in Harry Woolf (ed.), *The Analytic Spirit: Essays in the History of Science in Honor of Henry Guerlac* (Ithaca: Cornell University Press, 1981), pp. 15–39.

15. This narrative can be usefully compared with the somewhat fuller account in William Brock, *The Fontana History of Chemistry* (London: Fontana, 1992), chap. 3. See also C. E. Perrin, "Research Traditions, Lavoisier, and the Chemical Revolution," *Osiris* 4 (1988), pp. 53–81.

16. Note the assumption that one seeks a material component, a substance, to provide the property under consideration—what the historian Robert Schofield called "materialism" in the matter theory of this period. See Robert E. Schofield, *Mechanism and Materialism: British Natural Philosophy in an Age of Reason* (Princeton: Princeton University Press, 1970).

17. *Méthode de nomenclature chimique, proposée par MM. de Morveau, Lavoisier, Bertholet, et de Fourcroy* (Paris, 1787); see Maurice P. Crosland, *Historical Studies in the Language of Chemistry* (London: Heinemann, 1962), esp. Pt. 3, chap. VI; also Lefèvre, "The *Méthode de nomenclature chimique*."

18. As quoted in Antoine Lavoisier, *Elements of Chemistry*, trans. Robert Kerr (Edinburgh, 1790), pp. xiii–xiv.

19. Lavoisier, *Elements of Chemistry*, p. xiv.

20. Although Lavoisier apologizes for sidelining issues to do with chemical affinities (related to the chemistry of salts), his excuse is that this area is already being addressed in an authoritative article being written by his colleague Guyton de Morveau, one of Lavoisier's collaborators on the new nomenclature: Lavoisier, *Elements of Chemistry*, pp. xxi–xxii.

21. Lavoisier, *Elements of Chemistry*, p. xv.

22. Lavoisier, *Elements of Chemistry*, pp. xxvii–xxviii.

23. On the new nomenclature, see Maurice P. Crosland, *Historical Studies in the Language of Chemistry* (London: Heinemann, 1962), chap. 6, and Lefèvre, "Méthode de nomenclature chimique."

Chapter 7. Laplace, Revolutionary Order, and the Invention of Mathematical Physics

1. On French education immediately before the Revolution, see Charles Coulston Gillispie, *Science and Polity in France: The Revolutionary and Napoleonic Years* (Princeton: Princeton University Press, 2004), pp. 129–36.

2. Gillispie, *Science and Polity*, pp. 146–64; L. Pearce Williams, "Science, Education and Napoleon I," *Isis* 47 (1956), pp. 369–82.

3. Gillispie, *Science and Polity*, pp. 520–40.

4. Gillispie, *Science and Polity*, pp. 520–40.

5. Laplace's rise under Napoleon is the subject of chapter 20 in Charles Coulston Gillispie, *Pierre-Simon Laplace, 1749–1827: A Life in Exact Science* (Princeton: Princeton University Press, 1997). On Napoleon and the sciences, see Maurice Crosland, *The Society of Arcueil: A View of French Science at the Time of Napoleon I* (London: Heinemann, 1967), esp. chap. 1; Roger Hahn, *Pierre Simon Laplace, 1749–1827: A Determined Scientist* (Cambridge, MA: Harvard University Press, 2005). See also for a synoptic overview Maurice Crosland, "A Science Empire in Napoleonic France," *History of Science* 44 (2006), pp. 29–48.

6. Crosland, *Society of Arcueil*, pp. 63–64.

7. Lavoisier's demise is detailed in Gillispie, *Science and Polity*, pp. 318–26.

8. Gillispie, *Science and Polity*, pp. 186–94, 210–22, 448–51.

9. See the classic treatment by Robert Fox, "The Rise and Fall of Laplacian Physics," *Historical Studies in the Physical Sciences* 4 (1974), pp. 89–136; also the valuable account in Gillispie, *Science and Polity*, pp. 675–95.

10. Crosland, *Society of Arcueil*.

11. Quoted in Fox, "Rise and Fall," p. 100 n. 35.

12. Crosland, *Society of Arcueil*, pp. 245–48.

13. Although the authors make a point of noting that the significance of their work is the same regardless of whether heat is regarded as a substance or the consequence of particulate motion. See Henry Guerlac, "Chemistry as a Branch of Physics: Laplace's Collaboration with Lavoisier," *Historical Studies in the Physical Sciences* 7 (1976), esp. pp. 249–58; see also Robert Fox, *The Caloric Theory of Gases: From Lavoisier to Regnault* (Oxford: Clarendon, 1971); Gillispie, *Laplace*, chap. 14. For a facsimile and translation of the original 1783 publication, *Mémoire sur la chaleur*, see Antoine Lavoisier and Pierre Simon de Laplace, *Memoir on Heat*, trans. and ed. Henry Guerlac (New York: Neal Watson Academic Publications, 1982).

14. Lavoisier and Laplace, *Memoir on Heat*, p. 32.

15. Eugene Frankel, "J. B. Biot and the Mathematization of Experimental Physics in Napoleonic France," *Historical Studies in the Physical Sciences* 8 (1977), pp. 33–72.

16. Translated in Frankel, "Biot," pp. 62–63.

17. The conclusion drawn in Fox, "Rise and Fall."

Entr'acte

1. A term more familiar in reference to the early modern period, as applied to Isaac Newton in Rob Iliffe, *Priest of Nature: The Religious Worlds of Isaac Newton* (Oxford: Oxford University Press, 2017), e.g., pp. 16–17; Harold Fisch, "The Scientist as Priest: A Note on Robert Boyle's Natural Theology," *Isis* 44 (1953), pp. 252–65.

2. This is an image promoted in his work by Francis Bacon in the early seventeenth century. See especially Sophie Weeks, "The Role of Mechanics in Francis Bacon's Great Instauration," in Claus Zittel, Gisela Engel, Romano Nanni, and Nicole C. Karafyllis (eds.), *Philosophies of Technology: Francis Bacon and His Contemporaries* (Leiden, 2008), pp. 133–95.

3. Owen Hannaway, *The Chemists and the Word: The Didactic Origins of Chemistry* (Baltimore: Johns Hopkins University Press, 1975). Hannaway's insight was pursued by J. R. R. Christie and J. V. Golinski, "The Spreading of the Word: New Directions in the History of Chemistry, 1600–1800," *History of Science* 20 (1982), pp. 235–66; Golinski, "Chemistry in the Scientific Revolution: Problems of Language and Communication," in David C. Lindberg and Robert S. Westman (eds.), *Reappraisals of the Scientific Revolution* (Cambridge: Cambridge University Press, 1990), pp. 367–96.

4. Kathryn M. Olesko, *Physics as a Calling: Discipline and Practice in the Königsberg Seminar for Physics* (Ithaca: Cornell University Press, 1991); Andrew Warwick, *Masters of Theory: Cambridge and the Rise of Mathematical Physics* (Chicago: University of Chicago Press, 2003).

5. William Clark, *Academic Charisma and the Origins of the Research University* (Chicago: University of Chicago Press, 2006).

6. For a good overview see Bas van Bommel, "Between 'Bildung' and 'Wissenschaft': The 19th-Century German Ideal of Scientific Education," *Brewminate* (blog), August 6, 2019, https://brewminate.com/between-bildung-and-wissenschaft-the-19th-century-german-ideal-of-scientific-education/.

7. See the classic article by R. Steven Turner, "The Growth of Professorial Research in Prussia, 1818–1848: Causes and Context," *Historical Studies in the Physical Sciences* 3 (1971), pp. 137–82. Later in the nineteenth century, in the United States, the label "pure science" was frequently used to signal a similar impression of moral purity for the natural sciences, understood as a disinterested search for the truth. See Laurence R. Veysey, *The Emergence of the American University* (Chicago: University of Chicago Press, 1965).

8. The originating discussion is Jack Morell, "The Chemist Breeders: The Research Schools of Liebig and Thomas Thomson," *Ambix* 19 (1972), pp. 1–46.

9. The moral ideal is indicated well in the fictional form of Sinclair Lewis's novel *Arrowsmith* (New York, 1925).

10. Lawrence Scientific School at Harvard (1846); Sheffield Scientific School at Yale (1847), with focuses on applied science.

11. See, e.g., Martin J. S. Rudwick, *Worlds Before Adam: The Reconstruction of Geohistory in the Age of Reform* (Chicago: University of Chicago Press, 2008).

12. See Jack Morrell and Arnold Thackray, *Gentlemen of Science: Early Years of the British Association for the Advancement of Science* (Oxford: Clarendon Press, 1981). One exception to these generalizations about Britain is the Royal Institution, founded in London in 1799. It combined public lectures on a fairly popular level with research and had laboratories on the premises. The chemist Humphry Davy ran it at first, and then in the '30s and '40s Michael Faraday. Faraday worked on his own and founded no school of researchers around him; in that regard he differed little from the brewer Joule, except for the source of financial support.

13. Steven Shapin, *The Scientific Life: A Moral History of a Late Modern Vocation* (Chicago: University of Chicago Press, 2008).

14. See John Gascoigne, *Science and the State: From the Scientific Revolution to World War II* (Cambridge: Cambridge University Press, 2019), esp. chap. 6, which addresses the Russian case as well as the four principals discussed here.

Chapter 8. Classification and Extinction

1. This rapid ascent later prompted rumors by Cuvier's enemies of blackmail, although less sensational explanations seem more probable. For a biographical introduction to Cuvier's early career, see Martin J. S. Rudwick, *Bursting the Limits of Time: The Reconstruction of Geohistory in the Age of Revolution* (Chicago: University of Chicago Press, 2005), pp. 353–56; more extensively, Dorinda Outram, *Georges Cuvier: Vocation, Science, and Authority in Post-Revolutionary France* (Manchester: Manchester University Press, 1984), and the interesting discussion in Dorinda Outram, "Scientific Biography and the Case of Georges Cuvier," *History of Science* 14 (1976), pp. 101–37.

2. Botanists' attempts to develop more justifiably "natural" systems for botany in the later eighteenth century are discussed in Peter Dear, *The Intelligibility of Nature: How Science Makes Sense of the World*, rev. ed. (Chicago: University of Chicago Press, 2008), pp. 54–60.

3. Other aspects of Cuvier's intellectual context are investigated in Dorinda Outram, "Uncertain Legislator: Cuvier's Laws of Nature in Their Intellectual Context," *History of Science* 19 (1986), pp. 323–68.

4. Xavier Bichat, *Anatomie générale, appliquée à la physiologie et la médecine, première partie, tome premier* (Paris, 1801), pp. liv–lv. On Bichat's ambitions to rival Newton, see John E. Lesch,

Science and Medicine in France: The Emergence of Experimental Physiology, 1790–1855 (Cambridge, MA: Harvard University Press, 1984), esp. p. 68.

5. Lesch represents Bichat as pursuing two kinds of physiology: one observational/experimental, the other a taxonomy of vital properties. Lesch, *Science and Medicine*, chap. 3; the latter is of particular interest here.

6. For general biographical detail, see Lesch, *Science and Medicine*, pp. 89–92.

7. François Magendie, "Quelques idées générales sur les phénomènes particuliers aux corps vivans," *Bulletin des Sciences Medicales* 4 (1809), p. 145, my translation.

8. See the discussion in Lesch, *Science and Medicine*, pp. 92–96; Bruno Belhoste, "From Quarry to Paper: Cuvier's Three Epistemological Cultures," in Karine Chemla and Evelyn Fox Keller (eds.), *Cultures without Culturalism: The Making of Scientific Knowledge* (Raleigh: Duke University Press, 2017), pp. 250–77, esp. p. 253.

9. The best account of Cuvier's anatomical rules may be found in William Coleman, *Georges Cuvier, Zoologist* (Cambridge, MA: Harvard University Press, 1964). Valuable texts in translation appear in Martin J. S. Rudwick (ed. and trans.), *Georges Cuvier, Fossil Bones, and Geological Catastrophes* (Chicago: University of Chicago Press, 1997).

10. Trans. Rudwick, *Georges Cuvier*, p. 217, from Cuvier's *Discours sur les révolutions de la surface du globe* (3rd ed., 1826). Versions of these remarks can also be found elsewhere in Cuvier's writings.

11. Rudwick, *Bursting the Limits*, pp. 409–10; Rudwick, *The Meaning of Fossils: Episodes in the History of Paleontology* (Chicago: University of Chicago Press, 1985 [1972]), pp. 113–16; this was in 1804. See also the discussion in Belhoste, "From Quarry to Paper," esp. pp. 263–65.

12. See Cuvier's further discussion from *Discours*, in Rudwick, *Georges Cuvier*, pp. 217–19.

13. See Coleman, *Cuvier*, pp. 67–72, for discussion.

14. See Coleman, *Cuvier*, chap. 4.

15. Bert Theunissen suggests a more effective, less post hoc role for Cuvier's "laws." Whether the rules were "really" effective or simply after-the-fact rationalizations of judgments made on other grounds, the result is the same: Cuvier's work changed the image of natural history. Theunissen, "The Relevance of Cuvier's *lois zoologiques* for His Paleontological Work," *Annals of Science* 43 (1986), pp. 543–56.

16. For these aspects of Cuvier's career, see Outram, *Georges Cuvier*.

17. Rudwick, *Bursting the Limits*, pp. 364–88; Rudwick, *Georges Cuvier*, chap. 12; Belhoste, "From Quarry to Paper."

18. Rudwick, *Meaning of Fossils*, pp. 107–9; Rudwick, *Bursting the Limits*, pp. 361–63.

19. Georges Cuvier, "Discours préliminaire," pp. 1–116 of *Recherches sur les ossemens fossiles de quadrupèdes*, vol. 1 (Paris, 1812); see Rudwick, *Georges Cuvier*, pp. 183–252, for English translation. See also Rudwick, *Meaning of Fossils*, pp. 149–50; Rudwick, *Bursting the Limits*, pp. 499–512.

20. See Rudwick, *Bursting the Limits*, pp. 590–91.

21. Rudwick, *Meaning of Fossils*, pp. 124–34.

22. See Martin J. S. Rudwick, *Worlds Before Adam: The Reconstruction of Geohistory in the Age of Reform* (Chicago: University of Chicago Press, 2008), esp. pp. 16–34, for an account of Cuvier's role in British geology and paleontology in the early nineteenth century.

Chapter 9. Darwin's Taxonomy

1. "Darwin on marriage," Darwin Correspondence Project, https://www.darwinproject.ac.uk/tags/about-darwin/family-life/darwin-marriage. Besides the correspondence project, the massive online compilation the Complete Work of Charles Darwin Online, http://darwin-online.org.uk/, edited by John van Wyhe, is an immensely valuable resource for all of Darwin's writings, in print and manuscript, and much other Darwiniana besides.

2. Janet Browne, *Charles Darwin: Voyaging* (Princeton: Princeton University Press, 1996), pp. 436–47; printed texts in Darwin Online F1556.

3. Martin Rudwick, "Darwin and Glen Roy: A 'Great Failure' in Scientific Method?" *Studies in History and Philosophy of Science* 5 (1974), pp. 97–185. Alistair Sponsel traces the path of Darwin's career anxieties about appearing overly speculative, whether in geology or in natural history and evolution, sparked by his early success with coral reefs. Sponsel, *Darwin's Evolving Identity: Adventure, Ambition, and the Sin of Speculation* (Chicago: University of Chicago Press, 2018).

4. Charles Darwin, *Variation of Animals and Plants under Domestication*, 2 vols. (London, 1868) published the earlier material from the draft "Big Species Book" (as Darwin called it); the rest has been published as R. C. Stauffer (ed.), *Charles Darwin's Natural Selection: Being the Second Part of His Big Species Book Written from 1856 to 1858* (Cambridge: Cambridge University Press, 1975).

5. Charles Darwin, *The Autobiography of Charles Darwin, 1809–1882, with Original Omissions Restored*, ed. Nora Barlow (New York: Norton, 1969 [1958]), p. 59.

6. See Darwin, *Origin of Species*, p. 402. On the Galápagos visit, see Browne, *Voyaging*, pp. 296–305.

7. Charles Darwin, *Voyages of the Adventure and Beagle*, 3 vols., vol. 3, Charles Darwin, "Journal and Remarks, 1832–1836" (London, 1839), p. 462, to be found under Darwin's "Publications: Journal of Researches" in Darwin Online, http://darwin-online.org.uk/converted/pdf/1839 _voyage_F10.3.pdf. This is the original version of Darwin's famous *Voyage of the Beagle* of 1845.

8. William Whewell promoted the term in the 1830s: see Martin J. S. Rudwick, *Worlds Before Adam: The Reconstruction of Geohistory in the Age of Reform* (Chicago: University of Chicago Press, 2008), p. 358.

9. See Charles Coulston Gillispie, *Pierre-Simon Laplace, 1749–1827: A Life in Exact Science* (Princeton: Princeton University Press, 1997), p. 47.

10. Charles Lyell, *Principles of Geology*, vol. 1 (London: John Murray, 1830), p. 152. For discussion of Lyell's first volume, see Rudwick, *Worlds Before Adam*, chap. 21.

11. Lyell, *Principles of Geology*, p. 153.

12. Lyell, *Principles of Geology*, pp. 153–55. Much later, in 1863, several years after the appearance of Darwin's *Origin of Species*, Lyell published a book called *The Antiquity of Man*. This assembled the rather scanty direct evidence that indicated the lowly animalistic (hence continuist) antecedents of humans in historical time.

13. Lyell, *Principles of Geology*, pp. 155–56.

14. Lyell, *Principles of Geology*, p. 157.

15. Lyell, *Principles of Geology*, p. 162.

16. Lyell, *Principles of Geology*, p. 164.

17. For an exposition of the relationship of Darwin's ideas on species to his reading of Malthus (directed against attempts to downplay its significance), see Peter Vorzimmer, "Darwin, Malthus, and the Theory of Natural Selection," *Journal of the History of Ideas* 30 (1969), pp. 527–42.

Chapter 10. Evolution and Scientific Naturalism

1. Darwin's underlying assumptions about the "goal" of evolution are examined in Robert J. Richards, "Darwin's Theory of Natural Selection and Its Moral Purpose," in Michael Ruse and Robert J. Richards (eds.), *The Cambridge Companion to the* Origin of Species (Cambridge University Press, 2009), pp. 47–66.

2. This is Darwin's own oft-used term in the *Origin of Species*.

3. Derek Partridge, "Darwin's Two Theories, 1844 and 1859," *Journal of the History of Biology* 51 (2018), pp. 563–92. I am grateful to an anonymous referee for bringing this important article to my attention.

4. A similar theme is considered in Michael Ruse, *The Darwinian Revolution: Science Red in Tooth and Claw*, 2nd ed. (Chicago: University of Chicago Press, 1999), chap. 6.

5. The best study of Owen is Nicolaas A. Rupke, *Richard Owen: Biology without Darwin*, rev. ed. (Chicago: University of Chicago Press, 2009).

6. Nicolaas A. Rupke, "Richard Owen's Vertebrate Archetype," *Isis* 84 (1993), pp. 231–51, p. 241.

7. James A. Secord, *Victorian Sensation: The Extraordinary Publication, Reception, and Secret Authorship of* Vestiges of the Natural History of Creation (Chicago: University of Chicago Press, 2000).

8. One of the most intransigent opponents on this point was a Scottish engineer, Fleeming Jenkin. See Janet Browne, *Charles Darwin: The Power of Place* (Princeton: Princeton University Press, 2002), p. 282.

9. As implied by the title of Iwan Rhys Morus, *When Physics Became King* (Chicago: University of Chicago Press, 2005).

10. Chronicled in Joe D. Burchfield, *Lord Kelvin and the Age of the Earth* (Chicago: University of Chicago Press, 1990). Not until the twentieth century did new evidence appear to extend Thomson's figure, initially from the discovery of radioactive decay and the heat it produced. But at the time it was a problem for Darwin and natural selection, if not one for evolution in general.

11. Charles Darwin, review of Henry Walter Bates, "Contributions to an Insect Fauna of the Amazon Valley," *Natural History Review* 3 (1863), pp. 219–24, http://darwin-online.org.uk/converted/pdf/1863_mimetic_F1725.pdf.

12. As, for example: "These cases of relationship, without identity, of the inhabitants of seas now disjoined, and likewise of the past and present inhabitants of the temperate lands of North America and Europe, are inexplicable on the theory of creation." Charles Darwin, *On the Origin of Species* (London, 1859), p. 372.

13. See George W. Stocking Jr., *Victorian Anthropology* (New York: Free Press, 1987), p. 15, for a definition.

14. Besides Stocking, see especially Adrian Desmond and James Moore, *Darwin's Sacred Cause: How a Hatred of Slavery Shaped Darwin's Views on Human Evolution* (Boston: Houghton Mifflin Harcourt, 2009); Cuvier quoted in George Coleman, *Georges Cuvier, Zoologist* (Cambridge, MA: Harvard University Press, 1964), pp. 165–66; Charles Darwin, *The Descent of Man, and Selection in Relation to Sex*, 2 vols. (London, 1871), esp. vol. 1, p. 235.

15. Darwin discusses sexual selection in the *Origin of Species*, pp. 87–90, but the major statement comes in *Descent of Man*. For a comprehensive study, see Evelleen Richards, *Darwin and the Making of Sexual Selection* (Chicago: University of Chicago Press, 2017).

16. Richards, *Sexual Selection*, esp. pp. 437–40.

17. First seriously investigated in Frank M. Turner, *Between Science and Religion: The Reaction to Scientific Naturalism in Late Victorian England* (New Haven: Yale University Press, 1974); see also Gowan Dawson and Bernard Lightman (eds.), *Victorian Scientific Naturalism: Community, Identity, Continuity* (Chicago: University of Chicago Press, 2014).

18. Darwin's reclusiveness meant that he was never a member, despite his sympathy with the group's aims. On the X Club, see Ruth Barton, *The X Club: Power and Authority in Victorian Science* (Chicago: University of Chicago Press, 2018); on its beginnings, Barton, "'Huxley, Lubbock, and Half a Dozen Others': Professionals and Gentlemen in the Formation of the X Club," 1851–1864, *Isis* 89 (1998), pp. 410–44.

19. Various instances of this point appear repeatedly in Darwin, *Origin of Species*, e.g., pp. 129, 185, 203, 372, 406, 470, 471, 478, 487, and elsewhere.

20. Darwin, *Origin of Species*, p. 471.

21. Darwin, review of Bates, p. 222.

22. Darwin, review of Bates, p. 222.

23. Matthew Stanley, *Huxley's Church and Maxwell's Demon: From Theistic Science to Naturalistic Science* (Chicago: University of Chicago Press, 2015).

24. *Molecules* referred to particles defined by weight, thus making no distinction with atoms. See chap. 14.

25. James Clerk Maxwell, "Molecules," in Maxwell, *Scientific Papers*, ed. W. D. Niven, 2 vols. (Cambridge: Cambridge University Press, 1890), vol. 2, p. 376.

26. Maxwell, "Molecules," 2:376.

27. Maxwell, "Molecules," 2:376.

28. Richard Noakes, *Physics and Psychics: The Occult and the Sciences in Modern Britain* (Cambridge: Cambridge University Press, 2019). John Tyndall, by contrast, scorned spiritualism and its performances, and enjoyed exposing it as fraudulent. Noakes, *Physics and Psychics*, p. 119.

29. The potential problems with naturalistic accounts of the origins of life ("spontaneous generation") for Huxley and X Club orthodoxy is valuably investigated in James E. Strick, "Darwinism and the Origin of Life: The Role of H. C. Bastian in the British Spontaneous Generation Debates, 1868–1873," *Journal of the History of Biology* 32 (1999), pp. 51–92, and in Strick, *Sparks of Life: Darwinism and the Victorian Debates over Spontaneous Generation* (Cambridge, MA: Harvard University Press, 2000), esp. chaps. 3 and 4.

30. John Tyndall, *Address Delivered before the British Association Assembled at Belfast* (London: Longmans, Green, 1874). Tyndall combats the charge of "material atheism" in the preface, p. viii. See Bernard Lightman, "Scientists as Materialists in the Periodical Press: Tyndall's Belfast Address," in Geoffrey Cantor and Sally Shuttleworth (eds.), *Science Serialized: Representations of the Sciences in Nineteenth-Century Periodicals* (Cambridge, MA: MIT Press, 2004), pp. 199–238.

31. Tyndall, *Address*, p. 55.

32. Tyndall, *Address*, pp. 55–56

33. Tyndall, *Address*, p. 56.

Chapter 11. Thermodynamics and Modern Physics

1. On Darwin and consilience, see the discussion in Michael Ruse, "Darwin's Debt to Philosophy: An Examination of the Influence of the Philosophical Ideas of John F. W. Herschel and William Whewell on the Development of Charles Darwin's Theory of Evolution," *Studies in History and Philosophy of Science* 6 (1975), pp. 159–81.

2. For a helpful overview, see George Smith, "The *vis viva* Dispute: A Controversy at the Dawn of Dynamics," *Physics Today* 59, no. 10 (2006), p. 31.

3. A still useful overview is Yehuda Elkana, *The Discovery of the Conservation of Energy* (Cambridge, MA: Harvard University Press, 1974).

4. David Philip Miller, "A New Perspective on the Natural Philosophy of Steams and Its Relation to the Steam Engine," *Technology and Culture* 61 (2020), pp. 1129–48, moves beyond the more teleological framing of the history of thermodynamics to be found in such (still excellent) treatments as D.S.L. Cardwell, *From Watt to Clausius: The Rise of Thermodynamics in the Early Industrial Age* (Ithaca: Cornell University Press, 1971).

5. See Charles C. Gillispie, *Lazare Carnot, Savant: A Monograph Treating Carnot's Scientific Work* (Princeton: Princeton University Press, 1971).

6. There is an English translation: Sadi Carnot, *Reflections on the Motive Power of Fire* (New York: Dover, 1960). Note that fire, *feu*, in Carnot's title is the heat substance also called *caloric*.

7. J.-A.-C. Charles was a public lecturer on the physical sciences in late eighteenth-century Paris. Robert Boyle, a leading fellow of the Royal Society, had performed his work on pressure and volume in mid-seventeenth-century England.

8. Carnot, *Reflections*, pp. 17–19; see also Cardwell, *Watt to Clausius*, pp. 193–201.

9. For a biographical treatment, see D.S.L. Cardwell, *James Joule: A Biography* (Manchester: Manchester University Press, 1989). See also Crosbie Smith, *The Science of Energy: A Cultural History of Energy Physics in Victorian Britain* (Chicago: University of Chicago Press, 1998), chap. 4.

10. J. P. Joule, "On the Calorific Effects of Magneto-Electricity, and on the Mechanical Value of Heat," in *The Scientific Papers of James Prescott Joule* (Cambridge: Cambridge University Press, 2011), vol. 1, pp. 123–59 (1843). The first part of the paper is concerned with establishing that the heating effect of electrical current results from the generation of heat by a current-carrying wire rather than from the transfer of preexisting heat from elsewhere in the apparatus; the second part (pp. 149–59) attempts to measure the mechanical equivalent of the generated heat. The historian Otto Sibum made an illuminating attempt to replicate Joule's experimental work and to learn more about the practical skills that it must have involved. Heinz Otto Sibum, "Reworking the Mechanical Value of Heat: Instruments of Precision and Gestures of Accuracy in Early Victorian England," *Studies in History and Philosophy of Science* 26 (1995), pp. 73–106. The virtues of Système International (SI) units for scientific measurement, which were introduced later in the nineteenth century with subsequent revisions until the present day, are clear. . . . See chapter 13 for more on metrology in the nineteenth century.

11. J. P. Joule, "On the Changes of Temperature Produced by the Rarefaction and Condensation of Air," in *Scientific Papers of James Prescott Joule* (Cambridge: Cambridge University Press, 2011), vol. 1, pp. 172–89.

12. Joule, "Changes of Temperature," p. 188.

13. Joule, "Changes of Temperature," p. 189.

14. "Über die Erhaltung der Kraft," translated into English as H. Helmholtz, "On the Conservation of Force: A Physical Memoir," trans. John Tyndall, in Tyndall and William Francis (eds.), *Scientific Memoirs: Natural Philosophy* (London, 1853), pp.114–162.

15. David Cahan, "Helmholtz and the British Scientific Elite: From Force Conservation to Energy Conservation," *Notes and Records of the Royal Society* 66 (2012), pp. 55–68; Kenneth L. Caneva, *Helmholtz and the Conservation of Energy: Contexts of Creation and Reception* (Cambridge, MA: MIT Press, 2021); David Cahan, *Helmholtz: A Life in Science* (Chicago: University of Chicago Press, 2018).

16. Helmholtz, "Conservation of Force," p. 119.

17. See, for a thorough intellectual contextualization, Ken Caneva, "Helmholtz, the Conservation of Force and the Conservation of *vis viva*," *Annals of Science* 76 (2019), pp. 17–57. On the reception of this work in the ten years following its publication, especially in Britain, see Caneva, *Helmholtz*, chap. 7. (Scientific ideas do not propagate themselves, as the case of Gregor Mendel reminds us.)

18. H. Helmholtz, "On the Interaction of Natural Forces," trans. John Tyndall, in Helmholtz, *Popular Lectures on Scientific Subjects* (New York, 1873), pp. 185–86.

19. John Tyndall, *Heat Considered as a Mode of Motion* (London: Longmans, Green, 1863), pp. 431–32. These particularly colorful passages do not appear in later editions of the book; they parallel, as Tyndall notes, similar, but much earlier, observations made by John Herschel in his *Outlines of Astronomy* of 1833. See also for a fuller discussion Ted Underwood, *The Work of the Sun: Literature, Science, and Economy, 1760–1860* (New York: Palgrave Macmillan, 2005), chap. 8.

20. See the text in Harry M. Geduld (ed.), *The Definitive Time Machine: A Critical Edition of H. G. Wells's Scientific Romance* (Bloomington: Indiana University Press, 1987), esp. pp. 84–86; see also for a similar point Iwan Rhys Morus, *When Physics Became King* (Chicago: University of Chicago Press, 2005), pp. 140–42.

21. Smith, *Science of Energy*, chap. 10.

22. Crosbie Smith and M. Norton Wise, *Energy and Empire: A Biographical Study of Lord Kelvin* (Cambridge: Cambridge University Press, 1989), pp. 312–15.

23. See Smith, *Science of Energy*, pp. 166–69.

24. Smith, *Science of Energy*, pp. 256–57.

25. Relevant extracts from Clausius (in English translation) may be found in Steven Brush, *Kinetic Theory*, vol. 1, "The Nature of Gases and of Heat" (Oxford: Pergamon, 1965).

Chapter 12. Chance and Determinism

1. Peter Dear, "Darwin's Sleepwalkers: Naturalists, Nature, and the Practices of Classification," *Historical Studies in the Natural Sciences* 44 (2014), pp. 297–318; Mary Pickard Winsor, "Darwin's Dark Matter: Utter Extinction," *Annals of Science* (2023), https://doi.org/10.1080/00033790.2023.2194889, pp. 1–33.

2. Theodore M. Porter, *The Rise of Statistical Thinking, 1820–1900* (Princeton: Princeton University Press, 1986).

3. A convenient English translation is Pierre Simon Laplace, *A Philosophical Essay on Probabilities*, trans. Frederick Wilson Truscott and Frederick Lincoln Emory (New York: Dover, 1952); see Charles Coulston Gillispie, *Science and Polity in France: The Revolutionary and Napoleonic Years* (Princeton: Princeton University Press, 2004), chaps. 25 and 26.

4. Laplace, *Philosophical Essay*, p. 62. See Lorenz Krüger, "The Slow Rise of Probabilism: Philosophical Arguments in the Nineteenth Century," in Lorenz Krüger, Lorraine J. Daston, and Michael Heidelberger (eds.), *The Probabilistic Revolution*, 2 vols. (Cambridge, MA: MIT Press, 1887), vol. 1, pp. 50–90.

5. Jean D'Alembert had made a similar point in "Sur le calcul des probabilités," in Jean D'Alembert, *Opuscules mathématiques*, vol. 7 (Paris, 1780), p.48.

6. Gerd Gigerenzer et al., *The Empire of Chance: How Probability Changed Science and Everyday Life* (Cambridge: Cambridge University Press, 1989), chap. 2.

7. Porter, Theodore M., *The Rise of Statistical Thinking, 1820–1900* (Princeton: Princeton University Press, 1986), chap. 1.

8. Theodore M. Porter, "The Mathematics of Society: Variation and Error in Quetelet's Statisics," *British Journal for the History of Science* 18 (1985), pp. 51–69.

9. Quoted in Krüger, "Slow Rise of Probabilism," p. 76.

10. John Herschel, "Quetelet on Probabilities," *Edinburgh Review* 92 (1850), pp. 1–57. Herschel also weighed in on John Michell's 1767 arguments concerning star groupings and their evidence for gravitational associations of stars. See Porter, *Rise of Statistical Thinking*, p. 79.

11. Marx worked on his great treatise *Das Kapital*, largely written in the 1860s, while living in London. On Buckle, see Porter, *Rise of Statistical Thinking*, pp. 60–65.

12. For Victorian debates about formal quantitative data on risk, as opposed to trust in professional actuaries, see Theodore M. Porter, "Precision and Trust: Early Victorian Insurance and the Politics of Calculation," in M. Norton Wise, *The Values of Precision* (Princeton: Princeton University Press, 1995), pp. 173–97.

13. Charles Coulston Gillispie, "Intellectual Factors in the Background of Analysis by Probabilities," in A. C. Crombie (ed.), *Scientific Change* (London: Heinemann, 1963), pp. 431–53.

14. Matthew Stanley, *Huxley's Church and Maxwell's Demon: From Theistic Science to Naturalistic Science* (Chicago: University of Chicago Press, 2015), p. 224, chap. 6 passim.

15. Relevant primary sources may be found in Stephen G. Brush, *Kinetic Theory*, vol. 1, "The Nature of Gases and of Heat" (London: Pergamon, 1965).

16. A valuable study is Matthew Stanley, "The Pointsman: Maxwell's Demon, Victorian Free Will, and the Boundaries of Science," *Journal of the History of Ideas* 69 (2008), pp. 467–91.

17. P. M. Harman, *The Natural Philosophy of James Clerk Maxwell* (Cambridge: Cambridge University Press, 1998), pp. 197–208; Crosbie Smith, *The Science of Energy: A Cultural History of Energy Physics in Victorian Britain* (Chicago: University of Chicago Press, 1998), pp. 247–53.

18. See the discussion in Smith, *The Science of Energy*, pp. 249–51. A good exposition of the demon thought experiment, with useful diagrams, is W. Ehrenberg, "Maxwell's Demon," *Scientific American* 217, no. 5 (Nov. 1967), pp. 103–11.

19. Stanley, "Pointsman."

20. Francis Galton, *Hereditary Genius: An Inquiry into Its Laws and Consequences* (London: Macmillan, 1869), the Darwins on pp. 209–10; similar sentiments in Charles Darwin, *The Descent of Man*, 2 vols. (London, 1871), vol. 2, pp. 402–4.

21. Galton, *Hereditary Genius*, p. 67.

22. The view is later clearly expressed in Francis Galton, *Natural Inheritance* (London: Macmillan, 1889), pp. 32–33. For fuller discussion, see also Nicholas Wright Gillham, *A Life of Sir Francis Galton: From African Exploration to the Birth of Eugenics* (Oxford: Oxford University Press, 2001), chap. 20.

23. Reflected in one of his book titles: Francis Galton, *English Men of Science: Their Nature and Nurture* (London, 1874).

24. See material in Karl Pearson, *The Life, Letters and Labours of Francis Galton*, 3 vols. (Cambridge: Cambridge University Press, 1914–1930), vol. 3A, pp. 87–93. The term *agnostic* was coined by Huxley and seems to have become the professed norm for members of the X Club.

25. See Stephen M. Stigler, *The History of Statistics: The Measurement of Uncertainty before 1900* (Cambridge, MA: Harvard University Press, 1986), pp. 281–83.

26. Galton, *Natural Inheritance*, chap. 6.

27. Galton, *Natural Inheritance*, esp. chaps. 5 and 7.

28. See Stigler, *History of Statistics*, pp. 281–99; Donald A. MacKenzie, *Statistics in Britain 1865–1930: The Social Construction of Scientific Knowledge* (Edinburgh: Edinburgh University Press, 1981), pp. 51–68. MacKenzie stresses throughout the role of eugenics in motivating Galton's work on statistics.

29. Galton, *Natural Inheritance*, p. 106.

30. Stigler, *The History of Statistics*, pp. 281–99; on Pearson, see the superb study by Theodore M. Porter, *Karl Pearson: The Scientific Life in a Statistical Age* (Princeton: Princeton University Press, 2004), p. 258; see also MacKenzie, *Statistics in Britain*, p. 91.

31. A significant overview of the past century or so is Jon Agar, *Science in the Twentieth Century and Beyond* (Cambridge: Polity, 2012), which gives some idea of the fate of natural philosophy.

Chapter 13. Electromagnetism, Action at a Distance, and Aether

1. Marcello Pera, *The Ambiguous Frog: The Galvani-Volta Controversy on Animal Electricity*, trans. Jonathan Mandelbaum (Princeton: Princeton University Press, 1992); cf. Mesmer's "animal magnetism" in the late eighteenth century: Robert Darnton, *Mesmerism and the End of the Enlightenment in France* (Cambridge, MA: Harvard University Press, 1968).

2. Giuliano Pancaldi, *Volta: Science and Culture in the Age of Enlightenment* (Princeton: Princeton University Press, 2003), chap. 6; see also on electric eels James Delbourgo, *A Most Amazing Scene of Wonders: Electricity and Enlightenment in Early America* (Cambridge, MA: Harvard University Press, 2006), chap. 5.

3. For annotated extracts from Volta's presentation to the Royal Society, see Peter Dear (ed.), *Scientific Practices in European History, 1200–1800: A Book of Texts* (London: Routledge, 2018), chap. 23.

4. The body of the experimenter was often an integral part of experimentation in this period; see Simon Schaffer, "Self Evidence," *Critical Inquiry* 18 (1992), pp. 327–62; Stuart Strickland, "The Ideology of Self-Knowledge and the Practice of Self-Experimentation," *Eighteenth-Century Studies* 31 (1998), pp. 453–71.

5. The principal scientific biography of Faraday remains L. Pearce Williams, *Michael Faraday: A Biography* (London: Chapman and Hall, 1965).

6. Robert Richards discusses *Naturphilosophie* in its central arena of the sciences of life. Richards, *The Romantic Conception of Life: Science and Philosophy in the Age of Goethe* (Chicago: University of Chicago Press, 2002).

7. See H.A.M. Snelders, "Oersted's Discovery of Electromagnetism," in Andrew Cunningham and Nicholas Jardine (eds.), *Romanticism and the Sciences* (Cambridge: Cambridge University Press, 1990), pp. 228–40.

8. Williams, *Faraday*, pp. 142–43.

9. On the early reception, see Christine Blondel and Abdelmadjid Benseghir, "The Key Role of Oersted's and Ampère's 1820 Electromagnetic Experiments in the Construction of the Concept of Electric Current," *American Journal of Physics* 85, no. 5 (2017), pp. 369–80.

10. Williams, *Faraday*, chap. 2, esp. pp. 156–57.

11. On Ampère: see James R. Hofmann, *André-Marie Ampère: Enlightenment and Electrodynamics* (Cambridge: Cambridge University Press, 2006).

12. Which is how Joule could do his first study of the mechanical value of heat (see chap. 11).

13. Faraday's work on electrical induction is helpfully presented in Williams, *Faraday*, chap. 5. Faraday's work was published serially as "Experimental Researches in Electricity," initially as an extended sequence of numbered papers published mostly by the Royal Society and subsequently collected in book form. Michael Faraday, *Experimental Researches in Electricity*, 3 vols. (Cambridge: Cambridge University Press, 2012). This material on induction first appears as "V. Experimental Researches in Electricity," *Philosophical Transactions of the Royal Society* 122 (1832), pp. 125–62.

14. This perspective on Faraday's work resonates with that of David Gooding, *Experiment and the Making of Meaning: Human Agency in Scientific Observation and Experiment* (Dordrecht: Kluwer, 1990).

15. Discussed in Williams, *Faraday*, pp. 241–57, 296–306 (on series XI of Faraday's sequence of "Electrical Researches").

16. See diagram in Williams, *Faraday*, p. 296.

17. These particle fragments Faraday named, with William Whewell's linguistic blessing, "ions" (Oxford English Dictionary [1834], *q.v.*).

18. Williams, *Faraday*, pp. 283–99.

19. See David Gooding, "Conceptual and Experimental Bases of Faraday's Denial of Electrostatic Action at a Distance," *Studies in History and Philosophy of Science* 9 (1978), pp. 117–49.

20. On Boscovich, see chap. 2, this volume; P. M. Heimann, "Faraday's Theories of Matter and Electricity," *British Journal for the History of Science* 5 (1971), pp. 235–57.

21. The effect subsequently known as magneto-optic rotation. See for further discussion Peter Dear, *The Intelligibility of Nature: How Science Makes Sense of the World* (Chicago: University of Chicago Press, 2006), pp. 122–23.

22. "On the Physical Character of the Lines of Force," in Michael Faraday, *Experimental Researches in Electricity*, 3 vols. (Cambridge: Cambridge University Press, 2012), vol. 3, p. 436.

23. For a particularly valuable and accessible study of Maxwell, see P. M. Harman, *The Natural Philosophy of James Clerk Maxwell* (Cambridge: Cambridge University Press, 1998); also John Hendry, *James Clerk Maxwell and the Theory of the Electromagnetic Field* (Boston: Hilger, 1986).

24. See Gooding, *Experiment and the Making of Meaning*.

25. Faraday refers to the "strong conviction expressed by Sir Isaac Newton, that even gravity cannot be carried on to produce a distant effect except by some interposed agent," footnoting one of Newton's letters. Faraday, *Experimental Researches in Electricity*, vol. 3, p. 532. See also James Clerk Maxwell, "On Action at a Distance," *Proceedings of the Royal Institution of Great Britain* 7 (1873–1875), pp. 44–54, esp. p. 48.

26. Maxwell, "Action at a Distance," p. 45.

27. With famous mathematical names such as Green, McCullough, and Stokes: see the classic account in E. T. Whittaker, *A History of the Theories of Aether and Electricity*, 2 vols. (rev. ed., London: Nelson, 1951), vol. 1, esp. chap. 5.

28. Thomson's role in the history of electromagnetism in the nineteenth century was immense, and encompassed, besides the theoretical work highlighted here, much practical engagement and innovation, most notably his involvement with the great project of the transatlantic cable. The best and most comprehensive account of Thomson (later Lord Kelvin) and his work is Crosbie Smith and M. Norton Wise, *Energy and Empire: A Biographical Study of Lord Kelvin* (Cambridge: Cambridge University Press, 1989).

29. Maxwell, "Action at a Distance," p. 52.

30. See Dear, *Intelligibility*, chap. 5.

31. Maxwell, "Action at a Distance." Maxwell more intimately analogizes action at a distance to the action of muscles. Maxwell, "Action at a Distance," pp. 52–53. See on the muscle analogy Jordi Cat, "On Understanding: Maxwell on the Methods of Illustration and Scientific Metaphor," *Studies in History and Philosophy of Modern Physics* 32 (2001), pp. 395–441, esp. pp. 413–15. This seems, perhaps, to be more than mere analogy; see also Dear, *Intelligibility of Nature*, chap. 5.

32. It is worth noting at this point that Faraday himself had not been clear on the nature of an electric current. He tended to regard it as a particular condition of the circuit rather than something flowing through the circuit. See, e.g., Williams, *Faraday*, pp. 311–13.

33. See Bruce J. Hunt, *Imperial Science: Cable Telegraphy and Electrical Physics in the Victorian British Empire* (Cambridge: Cambridge University Press, 2020), chap. 5, esp. pp. 189–92.

34. See especially Simon Schaffer, "Accurate Measurement is an English Science," in M. Norton Wise (ed.), *The Values of Precision* (Princeton: Princeton University Press, 1995), pp. 135–72.

35. John Herschel, "The Yard, the Pendulum, and the Metre," in Herschel, *Familiar Lectures on Scientific Subjects* (London and New York: Strahan, 1866), p. 445.

36. See Simon Schaffer, "Metrology, Metrication, and Victorian Values," in Bernard Lightman (ed.), *Victorian Science in Context* (Chicago: University of Chicago Press, 1997), pp. 438–74.

37. Bruce Hunt argues for Maxwell's engagement with an engineering tradition at this time (1863–64) that somewhat drew him from his natural-philosophical project. Hunt, "Maxwell, Measurement, and the Modes of Electromagnetic Theory," *Historical Studies in the Natural Sciences* 45 (2015), pp. 303–39, and Hunt, *Imperial Science*, chap. 5.

38. Maxwell, "On Action at a Distance," p. 53. Mr. Huggins will appear again in chapter 15.

39. Bruce J. Hunt, *The Maxwellians* (Ithaca: Cornell University Press, 1991).

40. See Jed Z. Buchwald, *The Creation of Scientific Effects: Heinrich Hertz and Electric Waves* (Chicago: University of Chicago Press, 1994), esp. chap. 17.

Chapter 14. The Chemical Use of Atoms

1. Arnold Thackray, *John Dalton: Critical Assessments of His Life and Science* (Cambridge, MA: Harvard University Press, 1972), pp. 47–55. Modern scholarship on Dalton has had to cope with the loss of the greater part of his manuscripts caused by German bombing in the Second World War: Thackray, *John Dalton*, p. 175.

2. On the origins of Dalton's atomic ideas and their character, see Thackray, *John Dalton*, chap. 5; Alan J. Rocke, *Chemical Atomism in the Nineteenth Century: From Dalton to Cannizzaro* (Columbus: Ohio State University Press, 1984), chap. 2; also William Brock, *The Fontana History of Chemistry* (London: Fontana, 1992), esp. pp. 135–43. Dalton had published on his meteorological work in John Dalton, *Meteorological Observations and Essays* (London, 1793).

3. Thackray, *John Dalton*, pp. 66–75. All this is related to Dalton's law of partial pressures.

4. John Dalton, *A New System of Chemical Philosophy* (Manchester, 1808), part 1.

5. Dalton, *New System*, p. 1.

6. The modern distinction between the meanings of the words *atom* and *molecule* became standard only in the 1860s. Before that, the word *molecule*, because it refers to a weight unit, could mean both a single atom and a chemically combined group of atoms. On Dalton, Gay-Lussac, and Dumas, a useful overview is Mary Jo Nye, *Before Big Science: The Pursuit of Modern Chemistry and Physics, 1800–1940* (New York: Twayne, 1996), chap. 2.

7. William Wollaston, "A Synoptic Scale of Chemical Equivalents," *Philosophical Transactions of the Royal Society* 104 (1814), pp. 1–22. See Alan J. Rocke, "Atoms and Equivalents: The Early Development of the Chemical Atomic Theory," *Historical Studies in the Physical Sciences* 9 (1978), pp. 225–63, esp. pp. 228–29; see also, and more extensively, Rocke, *Chemical Atomism*.

8. Such considerations mean that our familiar description of water as H_2O, not canonically accepted for most of the nineteenth century, is less self-evident than we often think. See Hasok Chang, *Is Water H_2O? Evidence, Pluralism and Realism* (Dordrecht: Springer, 2012).

9. See Rocke, *Chemical Atomism*, pp. 52–56.

10. It seems fairer to call the idea Avogadro's conjecture, to emphasize its empirical ungroundedness, but calling it a hypothesis is a time-honored usage.

11. Discussed in detail in Maurice Crosland, *Gay-Lussac, Scientist and Bourgeois* (Cambridge University Press, 1978), chap. 5; also Rocke, *Chemical Atomism*, pp. 40–42.

12. Rocke, *Chemical Atomism*, pp. 40–42.

13. J. R. Partington, *A History of Chemistry*, 4 vols. (London: Macmillan, 1962–70), vol. 4, chap. 15; Alan Chalmers, *The Scientist's Atom and the Philosopher's Stone: How Science Succeeded and Philosophy Failed to Gain Knowledge of Atoms* (Dordrecht: Springer, 2009), chap. 10.

14. Rocke, *Chemical Atomism*, pp. 101–3. Avogadro avoided concrete images of diatomic molecules, as Rocke notes, although this is how his views were later represented.

15. John Hedley Brooke, "Avogadro's Hypothesis and Its Fate: A Case-Study in the Failure of Case-Studies," *History of Science* 19 (1981), pp. 235–73.

16. See Rocke, *Chemical Atomism*, pp. 115–18; Crosland, *Gay-Lussac*, pp. 111–12.

17. See Chalmers, *The Scientist's Atom*, pp. 218–19.

18. Rocke, *Chemical Atomism*, chap. 10; Partington, *A History of Chemistry* vol. 4, chap. 15; Chalmers, *The Scientist's Atom*, chap. 10.

19. Chalmers, *The Scientist's Atom*, p. 223.

20. Michael D. Gordin, *A Well-Ordered Thing: Dimitrii Mendeleev and the Shadow of the Periodic Table*, rev. ed. (Princeton: Princeton University Press, 2019).

21. Rocke, *Chemical Atomism*, pp. 313–21.

22. Quoted in Alan J. Rocke, *Image and Reality: Kekulé, Kopp, and the Scientific Imagination* (Chicago: University of Chicago Press, 2010), p. 225; discussed in Chalmers, *The Scientist's Atom*, p. 189. An excellent collection of primary sources in translation on atomism, both chemical and physical, is Mary Jo Nye (ed.), *The Question of the Atom: From the Karlsruhe Congress to the First Solvay Conference, 1860–1911* (Los Angeles: Tomash, 1984).

23. Organic chemists in living memory used to joke about how many pedagogical generations each of them was removed from Liebig: Liebig trained *A*, who trained *B*, who trained *C*, and so on, down to themselves.

24. J. B. Morrell, "The Chemist Breeders: The Research Schools of Liebig and Thomas Thomson," *Ambix* 19 (1972), pp. 1–46; Alan J. Rocke, "Origins and Spread of the 'Giessen Model' in University Science," *Ambix* 50 (2003), pp. 90–115.

25. Simon Werrett, "From the Grand Whim to the Gasworks: 'Philosophical Fireworks' in Georgian England," in Lissa Roberts, Simon Schaffer, and Peter Dear (eds.), *The Mindful Hand: Inquiry and Invention from the Late Renaissance to Early Industrialisation* (Amsterdam: Edita, 2007), pp. 325–48.

26. Anthony S. Travis, *The Rainbow Makers: The Origins of the Synthetic Dyestuffs Industry in Western Europe* (Bethlehem: Lehigh University Press, 1993).

27. For a thorough investigation, see Rocke, *Image and Reality*.

28. For a discussion of whether the position is better described as positivist or instrumentalist, see Chalmers, *The Scientist's Atom*, pp. 228–29.

29. A useful discussion is Martin J. Klein, "The Physics of J. Willard Gibbs in His Time," *Physics Today* 43 (1990), pp. 40–48.

30. Chalmers, *The Scientist's Atom*, p. 333 and chap. 12; also Charlotte Bigg, "Evident Atoms: Visuality in Jean Perrin's Brownian Motion Research," *Studies in History and Philosophy of Science* 39 (2008), pp. 312–22; Mary Jo Nye, *Molecular Reality: A Perspective on the Scientific Work of Jean Perrin* (London: Macdonald, 1972).

31. The big study remains Thomas S. Kuhn, *Black-Body Theory and the Quantum Discontinuity, 1894–1912* (Oxford: Clarendon, 1978). A useful synoptic discussion is Helge Kragh, *Quantum Generations: A History of Physics in the Twentieth Century* (Princeton: Princeton University Press, 1999), chap. 5.

Chapter 15. Laboratories of the Heavens

1. Isaac Todhunter (ed.), *William Whewell, Master of Trinity College, Cambridge: An Account of his Writings*, 2 vols. (Cambridge: Cambridge University Press, 2011 [1876]), vol. 1, p. 199.

2. See the discussion in John North, *The Norton History of Astronomy and Cosmology* (New York: Norton, 1994), pp. 427–30.

3. In the previous century, John Michell had devoted much of his philosophical career to this problem. See chapter 2, this volume.

4. Bradley was the Savilian Professor of Astronomy at Oxford from 1721, and Astronomer Royal from 1742. A second of arc is a sixtieth of a sixtieth of a degree, where the disks of sun and moon each subtend about half a degree (closer to thirty-one minutes) of arc. On Bradley and others, see Michael A. Hoskin, "Hooke, Bradley and the Aberration of Light," in Hoskin, *Stellar Astronomy: Historical Studies* (New York: Science History Publications, 1982), pp. 29–36.

5. M. A. Hoskin, "The English Background to the Cosmology of Wright and Herschel," in Wolfgang Yourgrau and Allen D. Breck (eds.), *Cosmology, History, and Theology* (New York: Plenum, 1977), pp. 219–31; Mari Williams, "James Bradley and the Eighteenth Century 'Gap' in Attempts to Measure Annual Stellar Parallax," *Notes and Records of the Royal Society of London* 37 (1982), pp. 83–100.

6. See Theodore M. Porter, *The Rise of Statistical Thinking, 1820–1900* (Princeton: Princeton University Press, 1986), pp. 93–96; Christoph Hoffmann, "Constant Differences: Friedrich Wilhelm Bessel, the Concept of the Observer in Early Nineteenth-Century Practical Astronomy and the History of the Personal Equation," *British Journal for the History of Science* 40 (2007), pp. 333–65.

7. See Michael J. Crowe, *Modern Theories of the Universe from Herschel to Hubble* (New York: Dover, 1994), pp. 152–59 (the volume as a whole is an excellent source for reprinted primary material); and North's discussion, especially focused on Bessel, in *Norton History of Astronomy and Cosmology*, pp. 414–20.

8. Michael Hoskin, *The Construction of the Heavens: William Herschel's Cosmology* (Cambridge: Cambridge University Press, 2012).

9. Simon Schaffer, "The Leviathan of Parsonstown: Literary Technology and Scientific Representation," in Timothy Lenoir (ed.), *Inscribing Science: Scientific Texts and the Materiality of Communication* (Stanford: Stanford University Press, 1998), pp. 182–222; Omar W. Nasim, *Observing by Hand: Sketching the Nebulae in the Nineteenth Century* (Chicago: University of Chicago Press, 2013), chap. 2.

10. Crowe, *Modern Theories*, chap. 4.

11. Michael A. Hoskin, "The Nebulae from Herschel to Huggins," in Hoskin, *Stellar Astronomy: Historical Studies* (New York: Science History Publications, 1982), pp. 137–53.

12. The best treatment of Fraunhofer as an instrument maker is Myles Jackson, *Spectrum of Belief: Joseph von Fraunhofer and the Craft of Precision Optics* (Cambridge, MA: MIT Press, 2000).

13. On this work of Kirchhoff and Bunsen, see A. J. Meadows, "The Origins of Astrophysics," in Owen Gingerich (ed.), *The General History of Astronomy*, vol. 4, "Astrophysics and Twentieth-Century Astronomy to 1950: Part A" (Cambridge: Cambridge University Press, 1984), pp. 3–15, esp. pp. 5–6.

14. Meadows, "The Origins of Astrophysics," p. 9.

15. Barbara J. Becker, *Unravelling Starlight: William and Margaret Huggins and the Rise of the New Astronomy* (Cambridge: Cambridge University Press, 2011), chap. 4.

16. William Huggins, "On the Spectra of Some of the Nebulae," *Philosophical Transactions of the Royal Society* 154 (1864), p. 438.

17. Huggins, "On the Spectra," p. 438.

18. Huggins, "On the Spectra," p. 438.

19. Huggins, "On the Spectra," p. 438. This nebula was the one that had convinced William Herschel in the 1790s that there was true nebulosity, on the basis of its general appearance. See Becker, *Unravelling Starlight*, chap. 5; Steven J. Dick, *Discovery and Classification in Astronomy: Controversy and Consensus* (Cambridge: Cambridge University Press, 2013), esp. pp.75–77. Huggins, like Parsons before him, was to serve as president of the Royal Society.

20. Meadows, "The Origins of Astrophysics."

21. The historian Derek J. de Solla Price seems to have coined the term "Big Science," although it is sufficiently generic as to render the question somewhat obscure. De Solla Price, *Little Science, Big Science* (New York: Columbia University Press, 1963).

22. Points interrogated by Schaffer, "Leviathan of Parsonstown"; Nasim, *Observing by Hand*.

23. John Lankford, "The Impact of Photography on Astronomy," in Owen Gingerich (ed.), *The General History of Astronomy*, vol. 4, "Astrophysics and Twentieth-Century Astronomy to 1950: Part A" (Cambridge: Cambridge University Press, 1984), pp. 16–39, esp. pp. 22–23.

24. Quoted in John Lankford, *American Astronomy: Community, Careers, and Power, 1859–1940* (Chicago: University of Chicago Press, 1997), p. 55.

25. Simon Newcomb, "The Place of Astronomy among the Sciences," *Sidereal Messenger* 7 (1888), p. 69.

26. Becker, *Unravelling Starlight*, chap. 7. For extracts from Huggins on the discoveries, see Crowe, *Modern Theories*, pp. 183–94.

27. On Clark and his telescopes, see Albert van Helden, "Telescope Building, 1850–1900," in Owen Gingerich (ed.), *The General History of Astronomy*, vol. 4, "Astrophysics and Twentieth-Century Astronomy to 1950: Part A" (Cambridge: Cambridge University Press, 1984), pp. 44–51.

28. Trudy E. Bell, "The Lick Observatory," in Owen Gingerich (ed.), *The General History of Astronomy*, vol. 4, "Astrophysics and Twentieth-Century Astronomy to 1950: Part A" (Cambridge: Cambridge University Press, 1984), pp. 127–30.

29. The University of Chicago was established in 1892 with Rockefeller money. See Van Helden, "Telescope Building," pp. 135–44.

30. On Hale and Mount Wilson, see, as well as Van Helden, North, *Norton History of Astronomy and Cosmology*, chap. 16.

31. Steven G. Brush, *Nebulous Earth: The Origin of the Solar System and the Core of the Earth from Laplace to Jeffreys* (Cambridge: Cambridge University Press, 1996). Also see chap. 2, this volume.

32. Michael Hoskin, *Discoverers of the Universe: William and Caroline Herschel* (Princeton: Princeton University Press, 2011).

33. Margaret W. Rossiter, *Women Scientists in America: Struggles and Strategies to 1940* (Baltimore: Johns Hopkins University Press, 1982), pp. 53–57; Dava Sobel, *The Glass Universe: How the Ladies of the Harvard Observatory Took the Measure of the Stars* (New York: Viking, 2016). Another area of astronomical endeavor that was treated as women's work can also be seen at Mount Wilson. While there were no women astronomers among the scientific staff, a lot of female labor went into the production of results there. Just as today, in order to have quantitative observational data processed, computing power was needed. In that time and place, a computer designated a *human* calculator: observational data was sent down the mountain to an office in Pasadena to be computed by human workers, who were exclusively female. Female computers were employed at the Harvard Observatory, too; the social model was that of the now-defunct company typing pool. See David Alan Grier, *When Computers Were Human* (Princeton: Princeton University Press, 2005); generally, Naomi Oreskes, "Objectivity or Heroism? On the Invisibility of Women in Science," *Osiris* 11 (1996), pp. 87–113.

34. George Johnson, *Miss Leavitt's Stars: The Untold Story of the Woman Who Discovered How to Measure the Universe* (New York: Norton, 2005); Crowe, *Modern Theories*, chap. 6 with a reprint of Leavitt's 1912 paper.

35. North, *Norton History of Astronomy and Cosmology*, pp. 488–89.

36. There was an initial error by a factor of ten: see North, *Norton History of Astronomy and Cosmology*, p. 489.

37. The so-called Great Debate: see Crowe, *Modern Theories*, chaps. 7 and 8.

38. Crowe, *Modern Theories*, chap. 9.

39. Van Helden, "Telescope Building," pp. 142–43.

40. The instrument sits atop Mount Palomar in Southern California. See Van Helden, "Telescope Building," esp. pp. 144–52.

41. On the human genome project, see Stephen Hilgartner, *Reordering Life: Knowledge and Control in the Genomics Revolution* (Cambridge, MA: MIT Press, 2017).

Chapter 16. New Modes of Natural Philosophy

1. Dennis Overbye, science correspondent for the *New York Times,* exemplifies this point nicely in the subtitle of his book *Lonely Hearts of the Cosmos: The Story of the Scientific Quest for the Secret of the Universe* (New York: Little, Brown, 2021). For the natural-philosophical ambitions of particle physics, see also Nasser Zakariya, *A Final Theory: Science, Myth, and Beginnings* (Chicago: University of Chicago Press, 2017), chap. 11, dealing with Steven Weinberg and the canceled project of the Superconducting Super Collider.

2. See chapter 2, this volume.

3. See the discussion in Richard Staley, *Einstein's Generation: The Origins of the Quantum Revolution* (Chicago: University of Chicago Press, 2008), chap. 2.

4. On interferometry and Michelson the experimentalist, see Staley, *Einstein's Generation*, chap. 3.

5. A convenient partial version of the paper in English may be found in L. Pearce Williams (ed.), *Relativity Theory: Its Origins and Impact on Modern Thought* (New York: Wiley, 1968), pp. 49–55. Einstein's other two papers of 1905 concerned the photoelectric effect and Brownian motion.

6. The device of the elevator is used as a general expository device in Albert Einstein and Leopold Infeld, *The Evolution of Physics: From Early Concepts to Relativity and Quanta* (New York: Simon & Schuster, 1938), pp. 214–22. The case of the candle is discussed in A. Zee, "An Old Man's Toy," in Donald Goldsmith and Marcia Bartusiak (eds.), *E = Einstein: His Life, His*

Thought and His Influence on Our Culture (New York: Sterling, 2006), pp. 229–30. The expository choice of a candle is reminiscent of Faraday's celebrated Royal Institution lectures: Michael Faraday, *A Course of Six Lectures on the Chemical History of a Candle* (London, 1861).

7. Einstein laid out his views in the wake of his new celebrity following Eddington's eclipse expedition, although they had informed his work since the beginning of the century. See Albert Einstein, "What Is the Theory of Relativity?," *The Times*, London, November 28, 1919, in *German History in Documents and Images*, vol. 6, *Weimar Germany, 1918/19–1933*, https://germanhistorydocs.ghi-dc.org/pdf/eng/EDU_Einstein_ENGLISH.pdf; Arthur I. Miller, *Albert Einstein's Special Theory of Relativity: Emergence (1905) and Early Interpretation (1905–1911)* (Reading, MA: Addison-Wesley, 1981), pp. 133–37; Suman Seth, *Crafting the Quantum: Arnold Sommerfeld and the Practice of Theory, 1890–1926* (Cambridge, MA: MIT Press, 2010), pp. 189–94, discusses in particular "theories of principle" as understood by Einstein and Max Planck.

8. Quoting from *The Collected Papers of Albert Einstein*, vol. 2, "The Swiss Years: Writings, 1900–1909," English translation supplement, trans. Anna Beck (Princeton: Princeton University Press, 1989), p. 141.

9. *Collected Papers of Albert Einstein*, vol. 2, pp. 166–67.

10. Miller, *Albert Einstein's Special Theory of Relativity*, pp. 29–32. Lorentz's conjectured explanation for the contraction effect was explicitly hypothetical, and hence, in Einstein's later terminology, "constructive" in nature.

11. The matter is discussed, at possibly inconsequential length, in Gerald Holton, "Einstein, Michelson, and the 'Crucial' Experiment," in Holton, *Thematic Origins of Scientific Thought: Kepler to Einstein* (Cambridge, MA: Harvard University Press, 1975), pp. 261–352; on Einstein's 1907 review, see Staley, *Einstein's Generation*, p. 11.

12. See Gerald Holton, "Mach, Einstein, and the Search for Reality," in Holton, *Thematic Origins*, pp. 219–59.

13. Albert Einstein, "Autobiographical Notes," in Paul Arthur Schilpp, ed., *Albert Einstein, Philosopher-Scientist* (Evanston, IL: Library of Living Philosophers, 1949), p. 5.

14. See Nathan Rotenstreich, "Relativity and Relativism," in Gerald Holton and Yehuda Elkana (eds.), *Albert Einstein: Historical and Cultural Perspectives* (Princeton: Princeton University Press, 1982), pp. 175–204.

15. As detailed in Iwan Rhys Morus, *When Physics Became King* (Chicago: University of Chicago Press, 2005).

16. Quoted in Loren R. Graham, "The Reception of Einstein's Ideas: Two Examples from Contrasting Political Cultures," in Holton and Yehuda Elkana, *Albert Einstein: Historical and Cultural Perspectives* (Princeton: Princeton University Press, 1982), p. 107.

17. Quoted in Gerald Holton, "Introduction: Einstein and the Shaping of Our Imagination," in Holton and Yehuda Elkana, *Albert Einstein: Historical and Cultural Perspectives* (Princeton: Princeton University Press, 1982), p. xiv.

18. Holton, "Introduction," p. xv.

19. On Eddington and the expedition, see Matthew Stanley, *Practical Mystic: Religion, Science, and A. S. Eddington* (Chicago: University of Chicago Press, 1907), chap. 3; see also Gerard Gilmore and Gudrun Tausch-Pebody, "The 1919 Eclipse Results that Verified General Relativity and Their Later Detractors: A Story Retold," in *Notes and Records of the Royal Society* 76 (2022), pp. 155–80.

20. Quoted in Holton, "Mach, Einstein," pp. 236–37, from a manuscript by Einstein's student Ilse Rosenthal-Schneider.

21. One thinks of the philosopher Karl Popper and his "falsification" criterion of scientificity.

22. Alas, Newton's idea of "imponderable" matter, matter that has no weight, slightly spoils the neatness of this picture.

23. See Einstein's introductory account of the equivalence principle in Albert Einstein, *The Meaning of Relativity: Four Lectures Delivered at Princeton University, May, 1921* (Princeton: Princeton University Press, 1922), pp. 60–65.

24. Thomas S. Kuhn, *Black-Body Theory and the Quantum Discontinuity, 1894–1912* (Oxford: Clarendon, 1978), pp. 130–34.

25. Albert Einstein, "On a Heuristic Point of View Concerning the Production and Transformation of Light," in *The Collected Papers of Albert Einstein*, vol. 2, "The Swiss Years: Writings, 1900–1909," English translation supplement, trans. Anna Beck (Princeton: Princeton University Press, 1989), pp. 86–103.

26. John L. Heilbron and Thomas S. Kuhn, "The Genesis of the Bohr Atom," *Historical Studies in the Physical Sciences* 1 (1969), pp. 211–90; Helge Kragh, *Niels Bohr and the Quantum Atom* (Oxford: Oxford University Press, 2012).

27. I give a basic account in Peter Dear, *The Intelligibility of Nature: How Science Makes Sense of the World*, rev. ed. (Chicago: University of Chicago Press, 2008), chap. 6.

28. An excellent study of all these issues is Arthur Fine, *The Shaky Game: Einstein, Realism, and the Quantum Theory*, 2nd ed. (Chicago: University of Chicago Press, 1996).

29. On Maxwell's demon, see chap. 12, this volume.

30. A. Einstein, B. Podolsky, and N. Rosen, "Can Quantum-Mechanical Description of Physical Reality Be Considered Complete?," *Physical Review* 47 (1935), pp. 777–80. For a thorough exposition, see Arthur Fine and Thomas A. Ryckman, "The Einstein-Podolsky-Rosen Argument in Quantum Theory," *The Stanford Encyclopedia of Philosophy* (2020), Edward N. Zalta (ed.), https://plato.stanford.edu/archives/sum2020/entries/qt-epr/.

31. Newtonian action at a distance in its classical (eighteenth-century) sense would also have been "spooky" because of the instantaneousness of its action; the quantum paradox relies on the Einsteinian disallowance of superluminal velocities.

32. Einstein, Podolsky, and Rosen, "Quantum Mechanical Description," p. 780.

33. Einstein, Podolsky, and Rosen, "Quantum Mechanical Description," p. 780. Einstein had not entirely approved of the final text of this paper (written by Podolsky), which he had not reviewed before publication (he thought that, as written, its central points were obscure), but these aspects of it seem to have reflected his own views.

34. James T. Cushing, *Quantum Mechanics: Historical Contingency and the Copenhagen Hegemony* (Chicago: University of Chicago Press, 1994), esp. chap. 9; see also David Bohm, *Causality and Chance in Modern Physics* (London: Routledge and Kegan Paul, 1957).

35. Arthur Fine, "Bell's theorem," in *The Routledge Encyclopedia of Philosophy* (London: Taylor and Francis, 1998), https://www.rep.routledge.com/articles/thematic/bells-theorem/v-1.

36. See, for example, on the Superconducting Super Collider, Peter Galison, *Image and Logic: A Material Culture of Microphysics* (Chicago: University of Chicago Press, 1997), pp. 671–78. Steven Weinberg, *The Discovery of Subatomic Particles*, rev. ed. (Cambridge: Cambridge University Press, 2003) is an accessible account of particle physics over the course of the twentieth century by one of the creators of the so-called standard model governing subatomic particles. A useful short overview of particle physics from 1930s to 1960s may be found in J. L. Heilbron, *Physics: A Short History from Quintessence to Quarks* (Oxford: Oxford University Press, 2015), pp. 165–88.

37. See chapter 2, this volume.

Conclusion

1. Rachel Laudan, "Histories of the Sciences and their Uses: A Review to 1913," *History of Science* 31 (1993), pp. 1–34; for the suggestion that scientists have generally abandoned their former engagement with histories of their own fields, see Peter Dear, "The History of Science

and the History of the Sciences: Sarton, *Isis*, and the Two Cultures," *Isis* 100 (2009), pp. 89–93. The title of this chapter is a nod to Peter Laslett, *The World We Have Lost* (London: Routledge, 2021 [1965]).

2. Steven Weinberg, *Dreams of a Final Theory* (New York: Pantheon, 1992), p. 18.

3. Helge Kragh, *Cosmology and Controversy: The Historical Development of Two Theories of the Universe* (Princeton: Princeton University Press, 1999); Martin J. S. Rudwick, *Earth's Deep History: How It Was Discovered and Why it Matters* (Chicago: University of Chicago Press, 2014); Stephen Hilgartner, *Reordering Life: Knowledge and Control in the Genomics Revolution* (Cambridge, MA: MIT Press, 2017).

4. See, on the historical development of such grand narratives, Nasser Zakariya, *A Final Theory: Science, Myth, and Beginnings* (Chicago: University of Chicago Press, 2017).

5. Charles C. Gillispie, *The Edge of Objectivity: An Essay on the History of Scientific Ideas* (Princeton: Princeton University Press, 1960).

6. Charles C. Gillispie, "The *Encyclopédie* and the Jacobin Philosophy of Science: A Study in Ideas and Consequences," in Marshall Clagett (ed.), *Critical Problems in the History of Science* (Madison: University of Wisconsin Press, 1959), p. 255; the figure of disenchantment was used by the early twentieth-century sociologist Max Weber in describing the decline of religion.

7. Gillispie expounds this theme powerfully in *The Edge of Objectivity*, particularly in chapter 5, "Science and the Enlightenment." William Clark treats Gillispie's book as an instance of epic romance: Clark, "Narratology and the History of Science," *Studies in History and Philosophy of Science* 26 (1995), pp. 1–71. But tragedy captures the book's affect equally directly, where the principal actor is "science." Perhaps in my account "natural philosophy" encounters a similar inevitable downfall that leaves it as an impoverished and instrumentalized "science," although I hope to have made it clear that the "cosmological" project of natural philosophy is still robust.

8. Geoffrey Cantor, *Michael Faraday, Sandemanian and Scientist: A Study of Science and Religion in the Nineteenth Century* (New York: St. Martin's, 1991); Matthew Stanley, *Huxley's Church and Maxwell's Demon: From Theistic Science to Naturalistic Science* (Chicago: University of Chicago Press, 2015); Stanley, *Practical Mystic: Religion, Science, and A. S. Eddington* (Chicago: University of Chicago Press, 2007); see also Zakariya, *A Final Theory*, chap. 8.

9. The crusading atheism (so to speak) of the evolutionary biologist Richard Dawkins, as in *The God Delusion* (London: Bantam Press, 2006), keeps this view very much current, despite the vigorous attempts of historians and others to delineate a much more complex role for topics of "religion and science" in the cultural matrix of modernity. Jeff Hardin, Ronald L. Numbers, and Ronald A. Binzley (eds.), *The Warfare between Science and Religion: The Idea that Wouldn't Die* (Baltimore: Johns Hopkins University Press, 2018) is a good recent collection investigating these issues in various settings; Ronald Numbers has coedited several volumes on such themes.

10. "Science Is Real" is a song on the 2009 album *Here Comes Science* by the band They Might Be Giants.

11. Of course there will always be practical limits to what can be questioned, and investigating the location and character of those limits is a legitimate task for the historian or sociologist of science.

12. However, the sociologist Robert K. Merton might well have disagreed with this conclusion, seeing self-correcting mechanisms in the conduct of scientific knowledge making that might preclude self-defeating dishonesty in the longer run. See, for example, the classic papers collected in Robert K. Merton, *The Sociology of Science: Theoretical and Empirical Investigations* (Chicago: University of Chicago Press, 1973).

13. Steven Shapin, "Rarely Pure and Never Simple: Talking About Truth," *Configurations* 7 (1999), pp. 1–14. One of the most celebrated cases of scientific malfeasance is detailed in

Daniel J. Kevles, *The Baltimore Case: A Trial of Politics, Science, and Character* (New York: Norton, 1998).

14. The classic example from the 1930s is Boris Hessen, *The Social and Economic Roots of Newton's* Principia (New York: H. Fertig, 1971), originally published in Russian in 1932.

15. Argued at greater length in Peter Dear, "What Is the History of Science the History *of*? Early Modern Roots of the Ideology of Modern Science," *Isis* 96 (2005), pp. 390–406, and Dear, "Science Is Dead; Long Live Science," *Osiris*, n.s. 27 (2012), pp. 37–55.

BIBLIOGRAPHY

Agar, Jon, *Science in the Twentieth Century and Beyond* (Cambridge: Polity, 2012).
Arbuthnot, John, "An Argument for Divine Providence," *Philosophical Transactions* (January 1710), pp. 186–90.
Arianrhod, Robyn, *Seduced by Logic: Émilie du Châtelet, Mary Somerville, and the Newtonian Revolution* (Oxford: Oxford University Press, 2012).
Barton, Ruth, "'Huxley, Lubbock, and Half a Dozen Others': Professionals and Gentlemen in the Formation of the X Club," 1851–1864, *Isis* 89 (1998), pp. 410–44.
Barton, Ruth, *The X Club: Power and Authority in Victorian Science* (Chicago: University of Chicago Press, 2018).
Becker, Barbara J., *Unravelling Starlight: William and Margaret Huggins and the Rise of the New Astronomy* (Cambridge: Cambridge University Press, 2011).
Belhoste, Bruno, "From Quarry to Paper: Cuvier's Three Epistemological Cultures," in Karine Chemla and Evelyn Fox Keller (eds.), *Cultures without Culturalism: The Making of Scientific Knowledge* (Raleigh: Duke University Press, 2017), pp. 250–77.
Bell, Trudy E., "The Lick Observatory," in Owen Gingerich (ed.), *The General History of Astronomy*, vol. 4, "Astrophysics and Twentieth-Century Astronomy to 1950: Part A" (Cambridge: Cambridge University Press, 1984), pp. 127–30.
Bernoulli, Jacob, *The Art of Conjecturing*, trans. Edith Dudley Sylla (Baltimore: Johns Hopkins University Press, 2005).
Bertucci, Paola, "Shocking Subjects: Human Experiments and the Material Culture of Medical Electricity in Eighteenth-Century England," in Erika Dick and Larry Stewart (eds.), *The Uses of Humans in Experiment: Perspectives from the 17th to the 20th Century* (Leiden: Brill, 2016), pp. 111–38.
Bertucci, Paola, and Giuliano Pancaldi (eds.), *Electric Bodies: Episodes in the History of Medical Electricity* (Bologna: CIS, University of Bologna, 2001).
Bichat, Xavier, *Anatomie générale, appliquée à la physiologie et la médecine, première partie, tome premier* (Paris, 1801).
Bigg, Charlotte, "Evident Atoms: Visuality in Jean Perrin's Brownian Motion Research," *Studies in History and Philosophy of Science* 39 (2008), pp. 312–22.
Blair, Ann, and Kaspar von Greyerz, eds., *Physico-Theology: Religion and Science in Europe, 1650–1750* (Baltimore: Johns Hopkins University Press, 2020).
Blondel, Christine, and Abdelmadjid Benseghir, "The Key Role of Oersted's and Ampère's 1820 Electromagnetic Experiments in the Construction of the Concept of Electric Current," *American Journal of Physics* 85, no. 5 (2017), pp. 369–80.
Bohm, David, *Causality and Chance in Modern Physics* (London: Routledge and Kegan Paul, 1957).
Boran, Elizabethanne, and Mordechai Feingold (eds.), *Reading Newton in Early Modern Europe* (Leiden: Brill, 2017).

Boscovich, Roger Joseph, *A Theory of Natural Philosophy*, trans. J. M. Child (Chicago, 1922).
Brock, William, *The Fontana History of Chemistry* (London: Fontana, 1992).
Brooke, John Hedley, "Avogadro's Hypothesis and Its Fate: A Case-Study in the Failure of Case-Studies," *History of Science* 19 (1981), pp. 235-73.
Browne, Janet, *Charles Darwin: The Power of Place* (Princeton: Princeton University Press, 2002).
Browne, Janet, *Charles Darwin: Voyaging* (Princeton: Princeton University Press, 1996).
Brush, Stephen G., *Kinetic Theory*, vol. 1, "The Nature of Gases and of Heat" (London: Pergamon, 1965).
Brush, Stephen G., "The Nebular Hypothesis and the Evolutionary Worldview," *History of Science* 25 (1987), pp. 245-78.
Brush, Steven G., *Nebulous Earth: The Origin of the Solar System and the Core of the Earth from Laplace to Jeffreys* (Cambridge: Cambridge University Press, 1996).
Buchwald, Jed Z., *The Creation of Scientific Effects: Heinrich Hertz and Electric Waves* (Chicago: University of Chicago Press, 1994).
Buchwald, Jed Z., and Mordechai Feingold, *Newton and the Origin of Civilization* (Princeton: Princeton University Press, 2012).
Burchfield, Joe D., *Lord Kelvin and the Age of the Earth* (Chicago: University of Chicago Press, 1990).
Bycroft, Michael, "Wonders in the Academy: The Value of Strange Facts in the Experimental Research of Charles Dufay," *Historical Studies in the Natural Sciences* 43 (2013), pp. 334-70.
Cahan, David, *Helmholtz: A Life in Science* (Chicago: University of Chicago Press, 2018).
Cahan, David, "Helmholtz and the British Scientific Elite: From Force Conservation to Energy Conservation," *Notes and Records of the Royal Society* 66 (2012), pp. 55-68.
Calloway, Katharine, "'Rather Theological than Philosophical': John Ray's Seminal *Wisdom of God Manifested in the Works of Creation*," in Blair and von Greyerz, *Physico-Theology*, pp. 115-26.
Caneva, Kenneth L., "Helmholtz, the Conservation of Force and the Conservation of *vis viva*," *Annals of Science* 76 (2019), pp. 17-57.
Caneva, Kenneth L., *Helmholtz and the Conservation of Energy: Contexts of Creation and Reception* (Cambridge, MA: MIT Press, 2021).
Cantor, Geoffrey, *Michael Faraday, Sandemanian and Scientist: A Study of Science and Religion in the Nineteenth Century* (New York: St. Martin's, 1991).
Cardwell, D.S.L., *From Watt to Clausius: The Rise of Thermodynamics in the Early Industrial Age* (Ithaca: Cornell University Press, 1971).
Cardwell, D.S.L., *James Joule: A Biography* (Manchester: Manchester University Press, 1989).
Carnot, Sadi, *Reflections on the Motive Power of Fire* (New York: Dover, 1960).
Cat, Jordi, "On Understanding: Maxwell on the Methods of Illustration and Scientific Metaphor," *Studies in History and Philosophy of Modern Physics* 32 (2001), pp. 395-441.
Chalmers, Alan, *The Scientist's Atom and the Philosopher's Stone: How Science Succeeded and Philosophy Failed to Gain Knowledge of Atoms* (Dordrecht: Springer, 2009).
Chang, Hasok, *Is Water H_2O? Evidence, Pluralism and Realism* (Dordrecht: Springer, 2012).
Chang, Ku-ming Kevin, "Communications of Chemical Knowledge: Georg Ernst Stahl and the Chemists at the French Academy of Sciences in the First Half of the Eighteenth Century," *Osiris* 29 (2014), pp. 135-57.
Chang, Ku-ming Kevin, "Phlogiston and Chemical Principles: The Development and Formulation of Georg Ernst Stahl's Principle of Inflammability," in Karen Hunger Parshall, Michael T. Walton, and Bruce T. Moran, *Bridging Traditions: Alchemy, Chemistry, and Paracelsian Practices in the Early Modern Era* (Kirksville, MO: Truman State University Press, 2015), pp. 101-31.

Christie, J.R.R., and J. V. Golinski, "The Spreading of the Word: New Directions in the History of Chemistry, 1600–1800," *History of Science* 20 (1982), pp. 235–66.
Clark, William, *Academic Charisma and the Origins of the Research University* (Chicago: University of Chicago Press, 2006).
Clark, William, "Narratology and the History of Science," *Studies in History and Philosophy of Science* 26 (1995), pp. 1–71.
Cohen, Floris, *How Modern Science Came into the World: Four Civilizations, One 17th-Century Breakthrough* (Amsterdam: Amsterdam University Press, 2010).
Cohen, I. Bernard (ed.), *Benjamin Franklin's Experiments: A New Edition of Benjamin Franklin's Experiments and Observations on Electricity* (Cambridge, MA: Harvard University Press, 1941).
Cohen, I. Bernard, *Franklin and Newton: An Inquiry into Speculative Newtonian Experimental Science and Franklin's Work in Electricity as an Example Thereof* (Philadelphia: American Philosophical Society, 1956).
Coleman, William, *Georges Cuvier, Zoologist* (Cambridge, MA: Harvard University Press, 1964).
Crosland, Maurice, *Gay-Lussac, Scientist and Bourgeois* (Cambridge: Cambridge University Press, 1978).
Crosland, Maurice P., *Historical Studies in the Language of Chemistry* (London: Heinemann, 1962).
Crosland, Maurice, "A Science Empire in Napoleonic France," *History of Science* 44 (2006), pp. 29–48.
Crosland, Maurice, *The Society of Arcueil: A View of French Science at the Time of Napoleon I* (London: Heinemann, 1967).
Crowe, Michael J., *Modern Theories of the Universe from Herschel to Hubble* (New York: Dover, 1994).
Cunningham, Andrew, "Getting the Game Right: Some Plain Words on the Identity and Invention of Science," *Studies in History and Philosophy of Science* 19 (1998), pp. 365–89.
Cushing, James T., *Quantum Mechanics: Historical Contingency and the Copenhagen Hegemony* (Chicago: University of Chicago Press, 1994).
Cuvier, Georges, "Discours préliminaire," *Recherches sur les ossemens fossiles de quadrupèdes* (Paris, 1812), vol. 1, pp. 1–116.
D'Alembert, Jean, "Sur le calcul des probabilités," in D'Alembert, *Opuscules mathématiques*, vol. 7 (Paris, 1780), pp. 39–60.
Dalton, John, *Meteorological Observations and Essays* (London, 1793).
Dalton, John, *A New System of Chemical Philosophy*, part 1 (Manchester, 1808).
Darnton, Robert, *Mesmerism and the End of the Enlightenment in France* (Cambridge, MA: Harvard University Press, 1968).
Darwin, Charles, *The Autobiography of Charles Darwin, 1809–1882, with Original Omissions Restored*, ed. Nora Barlow (New York: Norton, 1969 [1958]).
Darwin, Charles, *Charles Darwin's Natural Selection: Being the Second Part of His Big Species Book Written from 1856 to 1858*, ed. R. C. Stauffer (Cambridge: Cambridge University Press, 1975).
Darwin, Charles, *The Descent of Man, and Selection in Relation to Sex*, 2 vols. (London, 1871).
Darwin, Charles, "Journal and Remarks, 1832–1836," vol. 3 of *Voyages of the Adventure and Beagle*, 3 vols. (London, 1839).
Darwin, Charles, *On the Origin of Species* (London, 1859).
Darwin, Charles, review of Henry Walter Bates, "Contributions to an Insect Fauna of the Amazon Valley," *Natural History Review* 3 (1863), pp. 219–24.
Darwin, Charles, *Variation of Animals and Plants under Domestication*, 2 vols. (London, 1868).
Daston, Lorraine, *Classical Probability in the Enlightenment* (Princeton: Princeton University Press, 1988).

Dawkins, Richard, *The God Delusion* (London: Bantam, 2006).
Dawson, Gowan, and Bernard Lightman (eds.), *Victorian Scientific Naturalism: Community, Identity, Continuity* (Chicago: University of Chicago Press, 2014).
De Moivre, Abraham, *The Doctrine of Chances: or, A Method of Calculating the Probability of Events in Play* (London, 1718).
De Moivre, Abraham, *The Doctrine of Chances: or, A Method of Calculating the Probability of Events in Play*, 3rd ed. (London, 1756).
Dear, Peter, "Darwin's Sleepwalkers: Naturalists, Nature, and the Practices of Classification," *Historical Studies in the Natural Sciences* 44 (2014), pp. 297–318.
Dear, Peter, *Discipline and Experience: The Mathematical Way in the Scientific Revolution* (University of Chicago Press, 1995).
Dear, Peter, "The History of Science and the History of the Sciences: Sarton, *Isis*, and the Two Cultures," *Isis* 100 (2009), pp. 89–93.
Dear, Peter, *The Intelligibility of Nature: How Science Makes Sense of the World*, rev. ed. (Chicago: University of Chicago Press, 2008).
Dear, Peter, "Mixed Mathematics," in Peter Harrison, Michael Shank, and Ronald Numbers (eds.), *Wrestling with Nature: From Omens to Science* (Chicago: University of Chicago Press, 2011), pp. 149–72.
Dear, Peter, "The Natural Philosopher," in Bernard Lightman (ed.), *Blackwell Companion to the History of Science* (Oxford: Blackwell, 2016), pp. 71–83.
Dear, Peter, *Revolutionizing the Sciences: European Knowledge in Transition, 1500–1700* (Princeton: Princeton University Press, 2019).
Dear, Peter, "Science Is Dead; Long Live Science," *Osiris*, n.s. 27 (2012), pp. 37–55.
Dear, Peter (ed.), *Scientific Practices in European History, 1200–1800: A Book of Texts* (London: Routledge, 2018).
Dear, Peter, "What Is the History of Science the History of? Early Modern Roots of the Ideology of Modern Science," *Isis* 96 (2005), pp. 390–406.
Delbourgo, James, *A Most Amazing Scene of Wonders: Electricity and Enlightenment in Early America* (Cambridge, MA: Harvard University Press, 2006).
Demeter, T., B. Láng, and D. Schmal "Scientia in the Renaissance, Concept of," in M. Sgarbi, *Encyclopedia of Renaissance Philosophy* (Springer, 2020), https://doi.org/10.1007/978-3-319-02848-4_266-2.
Derham, William, *Astro-Theology: or, A Demonstration of the Being and Attributes of God, from a Survey of the Heavens* (London, 1715 [1714]).
Derham, William, *Physico-Theology: or, A Demonstration of the Being and Attributes of God, from His Works of Creation* (London, 1714 [1713]).
Desaguliers, Jean Theophile, "Experiments Made before the Royal Society, Feb. 2. 1737–8," *Philosophical Transactions of the Royal Society* 41 (1739), pp. 193–99.
Descartes, René, *The World and Other Writings*, trans. Stephen Gaukroger (Cambridge: Cambridge University Press, 1998).
Desmond, Adrian, and James Moore, *Darwin's Sacred Cause: How a Hatred of Slavery Shaped Darwin's Views on Human Evolution* (Boston: Houghton Mifflin Harcourt, 2009).
Dick, Steven J., *Discovery and Classification in Astronomy: Controversy and Consensus* (Cambridge: Cambridge University Press, 2013).
Ducheyne, Steffen, and Jip van Besouw, "Newton and the Dutch 'Newtonians': 1713–1750," in Eric Schliesser and Chris Smeenk (eds.), *The Oxford Handbook of Newton* (online edition, Oxford Academic, March 6, 2017, https://doi.org/10.1093/oxfordhb/9780199930418.013.20.
Eddy, John H., "Buffon's *Histoire naturelle*: History? A Critique of Recent Interpretations," *Isis* 85 (1994), pp. 644–61.

Ehrenberg, W., "Maxwell's Demon," *Scientific American* 217, no. 5 (1967), pp. 103–11.
Einstein, Albert, "Autobiographical Notes," in Paul Arthur Schilpp, ed., *Albert Einstein, Philosopher-Scientist* (Evanston, IL: Library of Living Philosophers, 1949), pp. 2–95.
Einstein, Albert, *The Collected Papers of Albert Einstein*, vol. 2, *The Swiss Years: Writings, 1900–1909*, English translation supplement, trans. Anna Beck (Princeton: Princeton University Press, 1989).
Einstein, Albert, *The Meaning of Relativity: Four Lectures Delivered at Princeton University, May, 1921* (Princeton: Princeton University Press, 1922).
Einstein, Albert, "On a Heuristic Point of View Concerning the Production and Transformation of Light," in *Collected Papers of Albert Einstein*, vol. 2, "The Swiss Years: Writings, 1900–1909," trans. Anna Beck (Princeton: Princeton University Press, 1989), pp. 86–103.
Einstein, Albert, "What Is the Theory of Relativity?," *The Times*, London, November 28, 1919, in *German History in Documents and Images*, vol. 6, *Weimar Germany, 1918/19–1933*, https://germanhistorydocs.ghi-dc.org/pdf/eng/EDU_Einstein_ENGLISH.pdf.
Einstein, Albert, and Leopold Infeld, *The Evolution of Physics: From Early Concepts to Relativity and Quanta* (New York: Simon & Schuster, 1938).
Einstein, Albert, Boris Podolsky, and Nathan Rosen, 1935, "Can Quantum-Mechanical Description of Physical Reality be Considered Complete?," *Physical Review* 47 (1935), pp. 777–80.
Eisenstaedt, Jean, "De l'influence de la gravitation sur la propagation de la lumière en théorie newtonienne: L'archéologie des trous noirs," *Archive for History of Exact Sciences* 42 (1991), pp. 315–86.
Elkana, Yehuda, *The Discovery of the Conservation of Energy* (Cambridge, MA: Harvard University Press, 1974).
Elshakry, Marwa, "When Science Became Western: Historiographical Reflections," *Isis* 101 (2010), pp. 98–109.
Fara, Patricia *Pandora's Breeches: Women, Science and Power in the Enlightenment* (London: Pimlico, 2004).
Fara, Patricia, *Sex, Botany and Empire: The Story of Carl Linnaeus and Joseph Banks* (Cambridge: Icon, 2003).
Faraday, Michael, *A Course of Six Lectures on the Chemical History of a Candle* (London, 1861).
Faraday, Michael, "On the Physical Character of the Lines of Force" in Michael Faraday, *Experimental Researches in Electricity*, 3 vols. (Cambridge: Cambridge University Press, 2012), vol. 3, pp. 407–37.
Faraday, Michael, "V. Experimental Researches in Electricity," *Philosophical Transactions of the Royal Society* 122 (1832), pp. 125–62.
Farber, Paul Lawrence, *Finding Order in Nature: The Naturalist Tradition from Linnaeus to E. O. Wilson* (Baltimore: Johns Hopkins University Press, 2000).
Feingold, Mordechai, and Giulia Giannini (eds.), *The Institutionalization of Science in Early Modern Europe* (Leiden: Brill, 2019).
Findlen, Paula, "Science as a Career in Enlightenment Italy: The Strategies of Laura Bassi," *Isis* 84 (1993), pp. 441–69.
Fine, Arthur, "Bell's Theorem," in *The Routledge Encyclopedia of Philosophy* (London: Taylor and Francis, 1998), https://www.rep.routledge.com/articles/thematic/bells-theorem/v-1.
Fine, Arthur, *The Shaky Game: Einstein, Realism, and the Quantum Theory*, 2nd ed. (Chicago: University of Chicago Press, 1996).
Fine, Arthur, and Thomas A. Ryckman, "The Einstein-Podolsky-Rosen Argument in Quantum Theory," *Stanford Encyclopedia of Philosophy* (2020), Edward N. Zalta (ed.), https://plato.stanford.edu/archives/sum2020/entries/qt-epr/.
Fisch, Harold, "The Scientist as Priest: A Note on Robert Boyle's Natural Theology," *Isis* 44 (1953), pp. 252–65.

Force, James E., *William Whiston: Honest Newtonian* (Cambridge: Cambridge University Press, 1985).
Foucault, Michel, *The Order of Things: An Archaeology of the Human Sciences* (New York: Pantheon, 1970).
Fox, Robert, *The Caloric Theory of Gases: From Lavoisier to Regnault* (Oxford: Clarendon, 1971).
Fox, Robert, "The Rise and Fall of Laplacian Physics," *Historical Studies in the Physical Sciences* 4 (1974), pp. 89–136.
Frankel, Eugene, "J. B. Biot and the Mathematization of Experimental Physics in Napoleonic France," *Historical Studies in the Physical Sciences* 8 (1977), pp. 33–72.
Franklin, James, *The Science of Conjecture: Evidence and Probability Before Pascal* (Baltimore: Johns Hopkins University Press, 2001).
Friedman, Michael, "Causal Laws and the Foundations of Natural Science," in Paul Guyer (ed.), *The Cambridge Companion to Kant* (Cambridge: Cambridge University Press, 1992), pp. 161–99.
Galison, Peter, *Image and Logic: A Material Culture of Microphysics* (Chicago: University of Chicago Press, 1997).
Galton, Francis, *English Men of Science: Their Nature and Nurture* (London, 1874).
Galton, Francis, *Hereditary Genius: An Inquiry into Its Laws and Consequences* (London: Macmillan, 1869).
Galton, Francis, *Natural Inheritance* (London: Macmillan, 1889).
Garber, Daniel, and Sophie Roux (eds.), *The Mechanization of Natural Philosophy* (Dordrecht: Springer, 2013).
Gascoigne, John, *Science and the State: From the Scientific Revolution to World War II* (Cambridge: Cambridge University Press, 2019).
Gascoigne, John, *Science in the Service of Empire: Joseph Banks, the British State and the Uses of Science in the Age of Revolution* (Cambridge: Cambridge University Press, 1998).
Gaukroger, Stephen, *The Emergence of a Scientific Culture: Science and the Shaping of Modernity, 1210–1685* (Oxford: Clarendon, 2006).
Geduld, Harry M. (ed.), *The Definitive Time Machine: A Critical Edition of H. G. Wells's Scientific Romance* (Bloomington: Indiana University Press, 1987).
Gigerenzer, Gerd, Zeno Swijtink, Theodore Porter, Lorraine Daston, John Beatty, and Lorenz Kruger, *The Empire of Chance: How Probability Changed Science and Everyday Life* (Cambridge: Cambridge University Press, 1989).
Gillham, Nicholas Wright, *A Life of Sir Francis Galton: From African Exploration to the Birth of Eugenics* (Oxford: Oxford University Press, 2001).
Gillispie, Charles Coulston, *The Edge of Objectivity: An Essay on the History of Scientific Ideas* (Princeton: Princeton University Press, 1960).
Gillispie, Charles Coulston, "The *Encyclopédie* and the Jacobin Philosophy of Science: A Study in Ideas and Consequences," in Marshall Clagett (ed.), *Critical Problems in the History of Science* (Madison: University of Wisconsin Press, 1959), pp. 255–89.
Gillispie, Charles Coulston, "Intellectual Factors in the Background of Analysis by Probabilities," in A. C. Crombie (ed.), *Scientific Change* (London: Heinemann, 1963), pp. 431–53.
Gillispie, Charles Coulston, *Lazare Carnot, Savant: A Monograph Treating Carnot's Scientific Work* (Princeton: Princeton University Press, 1971).
Gillispie, Charles Coulston, *Pierre-Simon Laplace, 1749–1827: A Life in Exact Science* (Princeton: Princeton University Press, 1997).
Gillispie, Charles Coulston, *Science and Polity in France: The Revolutionary and Napoleonic Years* (Princeton: Princeton University Press, 2004).

Gilmore, Gerard, and Gudrun Tausch-Pebody, "The 1919 Eclipse Results that Verified General Relativity and Their Later Detractors: A Story Retold," in *Notes and Records of the Royal Society* 76 (2022), pp. 155-80.

Golinski, Jan, "Chemistry in the Scientific Revolution: Problems of Language and Communication," in David C. Lindberg and Robert S. Westman (eds.), *Reappraisals of the Scientific Revolution* (Cambridge: Cambridge University Press, 1990), pp. 367-96.

Gooding, David, "Conceptual and Experimental Bases of Faraday's Denial of Electrostatic Action at a Distance," *Studies in History and Philosophy of Science* 9 (1978), pp. 117-49.

Gooding, David, *Experiment and the Making of Meaning: Human Agency in Scientific Observation and Experiment* (Dordrecht: Kluwer, 1990).

Gordin, Michael D., *A Well-Ordered Thing: Dimitrii Mendeleev and the Shadow of the Periodic Table*, rev. ed. (Princeton: Princeton University Press, 2019).

Graham, Loren R. "The Reception of Einstein's Ideas: Two Examples from Contrasting Political Cultures," in Gerald Holton and Yehuda Elkana, *Albert Einstein: Historical and Cultural Perspectives* (Princeton: Princeton University Press, 1982), pp. 107-36.

Grant, Edward, *A History of Natural Philosophy: From the Ancient World to the Nineteenth Century* (Cambridge: Cambridge University Press, 2007).

Grant, Edward, *Planets, Stars, and Orbs: The Medieval Cosmos, 1200-1687* (Cambridge: Cambridge University Press, 1994).

Gray, Stephen, "A Letter to Cromwell Mortimer ... Containing Several Experiments Concerning Electricity," in *Philosophical Transactions of the Royal Society* 37 (1731), pp. 18-44.

Grene, Marjorie, *A Portrait of Aristotle* (Chicago: University of Chicago Press, 1963).

Grier, David Alan, *When Computers Were Human* (Princeton: Princeton University Press, 2005).

Guerlac, Henry, "Chemistry as a Branch of Physics: Laplace's Collaboration with Lavoisier," *Historical Studies in the Physical Sciences* 7 (1976), pp. 193-276.

Hacking, Ian, *The Emergence of Probability: A Philosophical Study of Early Ideas about Probability, Induction and Statistical Inference* (Cambridge: Cambridge University Press, 1975).

Hahn, Roger, *Pierre Simon Laplace, 1749-1827: A Determined Scientist* (Cambridge, MA: Harvard University Press, 2005).

Hales, Stephen, *Vegetable Staticks, or, An Account of Some Statical Experiments on the Sap in Vegetables* (London, 1727).

Halley, Edmund, "Of the Infinity of the Sphere of Fix'd Stars," *Philosophical Transactions* 31 (1720-21), pp. 22-24.

Halley, Edmund, "Of the Number, Order, and Light of the Fix'd Stars," *Philosophical Transactions* 31 (1720-21), pp. 24-26.

Hankins, Thomas L., *Jean d'Alembert: Science and the Enlightenment* (Oxford: Clarendon Press, 1970).

Hankins, Thomas L., *Science in the Enlightenment* (Cambridge: Cambridge University Press, 1983).

Hannaway, Owen, *The Chemists and the Word: The Didactic Origins of Chemistry* (Baltimore: Johns Hopkins University Press, 1975).

Hardin, Jeff, Ronald L. Numbers, and Ronald A. Binzley (eds.), *The Warfare between Science and Religion: The Idea that Wouldn't Die* (Baltimore: Johns Hopkins University Press, 2018).

Harman, P. M., *The Natural Philosophy of James Clerk Maxwell* (Cambridge: Cambridge University Press, 1998).

Heilbron, John L., *Electricity in the 17th and 18th Centuries: A Study of Early Modern Physics* (Berkeley: University of California Press, 1979; repr. with new preface, Mineola, NY: Dover, 1999).

Heilbron, John L., *Physics: A Short History from Quintessence to Quarks* (Oxford: Oxford University Press, 2015).

Heilbron, John L., and Thomas S. Kuhn, "The Genesis of the Bohr Atom," *Historical Studies in the Physical Sciences* 1 (1969), pp. 211–90.

Heimann, P. M., "Faraday's Theories of Matter and Electricity," *British Journal for the History of Science* 5 (1971), pp. 235–57.

Helmholtz, H., "On the Conservation of Force: A Physical Memoir," trans. John Tyndall, in John Tyndall and William Francis (eds.), *Scientific Memoirs: Natural Philosophy* (London, 1853), pp. 114–62.

Helmholtz, H., "On the Interaction of Natural Forces," trans. John Tyndall, in Helmholtz, *Popular Lectures on Scientific Subjects* (New York, 1873), pp. 153–96.

Hendry, John, *James Clerk Maxwell and the Theory of the Electromagnetic Field* (Boston: Hilger, 1986).

Henry, John, "Occult Qualities and the Experimental Philosophy: Active Principles in Pre-Newtonian Matter Theory," *History of Science* 24 (1986), pp. 335–81.

Herschel, John, "Quetelet on Probabilities," *Edinburgh Review* 92 (1850), pp. 1–57.

Herschel, John, "The Yard, the Pendulum, and the Metre," in Herschel, *Familiar Lectures on Scientific Subjects* (London and New York: Strahan, 1866), pp. 419–51.

Hilgartner, Stephen, *Reordering Life: Knowledge and Control in the Genomics Revolution* (Cambridge, MA: MIT Press, 2017).

Hessen, Boris, *The Social and Economic Roots of Newton's Principia* (New York: H. Fertig, 1971).

Hoffmann, Christoph, "Constant Differences: Friedrich Wilhelm Bessel, the Concept of the Observer in Early Nineteenth-Century Practical Astronomy and the History of the Personal Equation," *British Journal for the History of Science* 40 (2007), pp. 333–65.

Hofmann, James R., *André-Marie Ampère: Enlightenment and Electrodynamics* (Cambridge: Cambridge University Press, 2006).

Holton, Gerald, "Introduction: Einstein and the Shaping of Our Imagination," in Holton and Yehuda Elkana, *Albert Einstein: Historical and Cultural Perspectives* (Princeton: Princeton University Press, 1982), pp. viii–xxxii.

Holton, Gerald, *Thematic Origins of Scientific Thought: Kepler to Einstein* (Cambridge, MA: Harvard University Press, 1975).

Home, Roderick W., *Electricity and Experimental Physics in Eighteenth-Century Europe* (Brookfield, VT: Variorum, 1992).

Home, Roderick W., "Franklin's Electrical Atmospheres," *British Journal for the History of Science* 6 (1972), pp. 131–51.

Hoskin, Michael, *The Construction of the Heavens: William Herschel's Cosmology* (Cambridge: Cambridge University Press, 2012).

Hoskin, Michael, "The Cosmology of Thomas Wright of Durham," *Journal for the History of Astronomy* 1 (1970), pp. 44–52.

Hoskin, Michael, *Discoverers of the Universe: William and Caroline Herschel* (Princeton: Princeton University Press, 2011).

Hoskin, Michael, "The English Background to the Cosmology of Wright and Herschel," in Wolfgang Yourgrau and Allen D. Breck (eds.), *Cosmology, History, and Theology* (New York: Plenum, 1977), pp. 219–31.

Hoskin, Michael, "Hooke, Bradley and the Aberration of Light," in Hoskin, *Stellar Astronomy: Historical Studies* (New York: Science History Publications, 1982), pp. 29–36.

Hoskin, Michael, "The Nebulae from Herschel to Huggins," in Hoskin, *Stellar Astronomy: Historical Studies* (New York: Science History Publications, 1982), pp. 137–53.

Hoskin, Michael, "Newton, Providence, and the Universe of Stars," *Journal for the History of Astronomy* 8 (1977), pp. 77–101.

Hoskin, Michael, "William Herschel and the Construction of the Heavens," *Proceedings of the American Philosophical Society* 133 (1989), pp. 427–33.

Huggins, William, "On the Spectra of Some of the Nebulae," *Philosophical Transactions of the Royal Society* 154 (1864), pp. 437–44.

Hunt, Bruce J., *Imperial Science: Cable Telegraphy and Electrical Physics in the Victorian British Empire* (Cambridge: Cambridge University Press, 2020).

Hunt, Bruce J., "Maxwell, Measurement, and the Modes of Electromagnetic Theory," *Historical Studies in the Natural Sciences* 45 (2015), pp. 303–39.

Hunt, Bruce J., *The Maxwellians* (Ithaca: Cornell University Press, 1991).

Iliffe, Rob, *Priest of Nature: The Religious Worlds of Isaac Newton* (Oxford: Oxford University Press, 2017).

Jackson, Myles, *Spectrum of Belief: Joseph von Fraunhofer and the Craft of Precision Optics* (Cambridge, MA: MIT Press, 2000).

Jacob, Margaret C., *The Newtonians and the English Revolution, 1689–1720* (Ithaca: Cornell University Press, 1976).

Johnson, George, *Miss Leavitt's Stars: The Untold Story of the Woman Who Discovered How to Measure the Universe* (New York: Norton, 2005).

Jones, Kenneth Glyn, "The Observational Basis for Kant's *Cosmogony*: A Critical Analysis," *Journal for the History of Astronomy* 2 (1971), pp. 29–34.

Joule, J. P., "On the Calorific Effects of Magneto-Electricity, and on the Mechanical Value of Heat," in *The Scientific Papers of James Prescott Joule* (Cambridge: Cambridge University Press, 2011), vol. 1, pp. 123–59.

Joule, J. P., "On the Changes of Temperature Produced by the Rarefaction and Condensation of Air," in *Scientific Papers of James Prescott Joule* (Cambridge: Cambridge University Press, 2011), vol. 1, pp. 172–89.

Kaiser, David (ed.), *Pedagogy and the Practice of Science: Historical and Contemporary Perspectives* (Cambridge, MA: MIT Press, 2005).

Kant, Immanuel, *Allgemeine Naturgeschichte und Theorie des Himmels* (Königsberg, 1755).

Kant, Immanuel, *Universal Natural History and Theory of the Heavens*, trans. W. Hastie (Ann Arbor: University of Michigan Press, 1969).

Keill, John, *An Introduction to the True Astronomy* (London, 1721).

Kevles, Daniel J., *The Baltimore Case: A Trial of Politics, Science, and Character* (New York: Norton, 1998).

Kim, Mi Gyung, *Affinity, that Elusive Dream: A Genealogy of the Chemical Revolution* (Cambridge, MA: MIT Press, 2003).

Klein, Martin J., "The Physics of J. Willard Gibbs in His Time," *Physics Today* 43 (1990), pp. 40–48.

Klein, Ursula, and Wolfgang Lefèvre, *Materials in Eighteenth-Century Science: A Historical Ontology* (Cambridge, MA: MIT Press, 2007).

Koyré, Alexandre, *From the Closed World to the Infinite Universe* (Baltimore: Johns Hopkin's University Press, 1957).

Kragh, Helge, *Cosmology and Controversy: The Historical Development of Two Theories of the Universe* (Princeton: Princeton University Press, 1999).

Kragh, Helge, *Niels Bohr and the Quantum Atom* (Oxford: Oxford University Press, 2012).

Kragh, Helge, *Quantum Generations: A History of Physics in the Twentieth Century* (Princeton: Princeton University Press, 1999).

Krüger, Lorenz, "The Slow Rise of Probabilism: Philosophical Arguments in the Nineteenth Century," in Krüger, Lorraine J. Daston, and Michael Heidelberger (eds.), *The Probabilistic Revolution*, 2 vols. (Cambridge, MA: MIT Press, 1887), vol. 1, pp. 50–90.

Kuhn, Thomas S., *Black-Body Theory and the Quantum Discontinuity, 1894–1912* (Oxford: Clarendon, 1978).

Kuhn, Thomas S., "Robert Boyle and Structural Chemistry in the Seventeenth Century," *Isis* 43 (1952), pp. 12–36.

Kuhn, Thomas S., *The Structure of Scientific Revolutions*, 2nd ed. (Chicago: University of Chicago Press, 1970).

Lakatos, Imre, *The Methodology of Scientific Research Programmes*, ed. John Worrall and Gregory Currie (Cambridge: Cambridge University Press, 1978).

Lankford, John, *American Astronomy: Community, Careers, and Power, 1859–1940* (Chicago: University of Chicago Press, 1997).

Lankford, John, "The Impact of Photography on Astronomy," in Owen Gingerich (ed.), *The General History of Astronomy*, vol. 4, "Astrophysics and Twentieth-Century Astronomy to 1950: Part A" (Cambridge: Cambridge University Press, 1984), pp. 16–39.

Laplace, Pierre Simon, *A Philosophical Essay on Probabilities*, trans. Frederick Wilson Truscott and Frederick Lincoln Emory (New York: Dover, 1952).

Laslett, Peter, *The World We Have Lost* (London: Routledge, 2021 [1965]).

Latour, Bruno, *Science in Action: How to Follow Scientists and Engineers through Society* (Cambridge, MA: Harvard University Press, 1988).

Laudan, Rachel, "Histories of the Sciences and Their Uses: A Review to 1913," *History of Science* 31 (1993), pp. 1–34.

Lavoisier, Antoine, *Elements of Chemistry*, trans. Robert Kerr (Edinburgh, 1790).

Lavoisier, Antoine, Louis-Bernard Guyton de Morveau, and Claude Berthollet, *Méthode de nomenclature chimique, proposée par MM. de Morveau, Lavoisier, Bertholet [sic], et de Fourcroy* (Paris, 1787).

Lavoisier, Antoine, and Pierre Simon de Laplace, *Memoir on Heat*, trans. and ed. Henry Guerlac (New York: Neal Watson Academic Publications, 1982).

Lefèvre, Wolfgang, "The *Méthode de nomenclature chimique* (1787): A Document of Transition," *Ambix* 65 (2018), pp. 9–29.

Lesch, John E., *Science and Medicine in France: The Emergence of Experimental Physiology, 1790–1855* (Cambridge, MA: Harvard University Press, 1984).

Lewis, Sinclair, *Arrowsmith* (New York, 1925).

Lightman, Bernard, "Scientists as Materialists in the Periodical Press: Tyndall's Belfast Address," in Geoffrey Cantor and Sally Shuttleworth (eds.), *Science Serialized: Representations of the Sciences in Nineteenth-Century Periodicals* (Cambridge MA: MIT Press, 2004), pp. 199–238.

Locke, John, *An Essay Concerning Human Understanding*, ed. Peter Nidditch (Oxford: Clarendon Press, 1979).

Lyell, Charles, *Principles of Geology*, vol. 1 (London: John Murray, 1830).

Lyon, John, and Phillip R. Sloan (eds. and trans.), *From Natural History to the History of Nature: Readings from Buffon and His Critics* (Notre Dame: University of Notre Dame Press, 1981), pp. 97–128.

MacKenzie, Donald A., *Statistics in Britain 1865–1930: The Social Construction of Scientific Knowledge* (Edinburgh: Edinburgh University Press, 1981).

Magendie, François, "Quelques idées générales sur les phénomènes particuliers aux corps vivans," *Bulletin des Sciences Medicales* 4 (1809), pp. 145–70.

Maxwell, James Clerk, "Molecules," in Maxwell, *Scientific Papers*, ed. W. D. Niven, 2 vols. (Cambridge: Cambridge University Press, 1890), vol. 2, pp. 46–78.

Maxwell, James Clerk, "On Action at a Distance," *Proceedings of the Royal Institution of Great Britain* 7 (1873–1875), pp. 44–54.

Mazzotti, Massimo, *The World of Maria Gaetana Agnesi: Mathematician of God* (Baltimore: Johns Hopkins University Press, 2007).

McCormmach, Russell, "John Michell and Henry Cavendish: Weighing the Stars," *British Journal for the History of Science* 4 (1968), pp. 126–55.
McCormmach, Russell, *Weighing the World: The Reverend John Michell of Thornhill* (Dordrecht: Springer, 2012).
Meadows, A. J., "The Origins of Astrophysics," in Owen Gingerich (ed.), *The General History of Astronomy*, vol. 4, "Astrophysics and Twentieth-Century Astronomy to 1950: Part A" (Cambridge: Cambridge University Press, 1984), pp. 3–15.
Merton, Robert K., *The Sociology of Science: Theoretical and Empirical Investigations* (Chicago: University of Chicago Press, 1973).
Michell, John, "An Inquiry into the Probable Parallax, and Magnitude of the Fixed Stars, from the Quantity of Light which They Afford us, and the Particular Circumstances of Their Situation," *Philosophical Transactions* 57 (1767), pp. 234–64.
Miller, Arthur I., *Albert Einstein's Special Theory of Relativity: Emergence (1905) and Early Interpretation (1905–1911)* (Reading, MA: Addison-Wesley, 1981).
Miller, David Philip, "A New Perspective on the Natural Philosophy of Steams and Its Relation to the Steam Engine," *Technology and Culture* 61 (2020), pp. 1129–48.
Morell, Jack, "The Chemist Breeders: The Research Schools of Liebig and Thomas Thomson," *Ambix* 19 (1972), pp. 1–46.
Morrell, Jack, and Arnold Thackray, *Gentlemen of Science: Early Years of the British Association for the Advancement of Science* (Oxford: Clarendon, 1981).
Morus, Iwan Rhys, *When Physics Became King* (Chicago: University of Chicago Press, 2005).
Nasim, Omar W., *Observing by Hand: Sketching the Nebulae in the Nineteenth Century* (Chicago: University of Chicago Press, 2013)
Newcomb, Simon, "The Place of Astronomy among the Sciences," *Sidereal Messenger* 7 (1888), pp. 65–73.
Newman, William R., "Robert Boyle, Transmutation and the History of Chemistry before Lavoisier: A Response to Kuhn," *Osiris* 29 (2014), pp. 63–77.
Newton, Isaac, *The Optical Papers of Isaac Newton*, ed. Alan E. Shapiro, vol. 2 (Cambridge: Cambridge University Press, 2021).
Newton, Isaac, *Opticks*, 2nd English ed. (London, W. and J. Innys, 1717–18).
Newton, Isaac, *The Principia: Mathematical Principles of Natural Philosophy*, trans. I. Bernard Cohen and Anne Whitman (Berkeley: University of California Press, 1999).
Noakes, Richard, *Physics and Psychics: The Occult and the Sciences in Modern Britain* (Cambridge: Cambridge University Press, 2019).
North, John D., *The Norton History of Astronomy and Cosmology* (New York: Norton, 1994).
Nye, Mary Jo, *Before Big Science: The Pursuit of Modern Chemistry and Physics, 1800–1940* (New York: Twayne, 1996).
Nye, Mary Jo, *Molecular Reality: A Perspective on the Scientific Work of Jean Perrin* (London: Macdonald, 1972).
Nye, Mary Jo (ed.), *The Question of the Atom: From the Karlsruhe Congress to the First Solvay Conference, 1860–1911* (Los Angeles: Tomash, 1984).
Ogilvie, Brian W., *The Science of Describing: Natural History in Renaissance Europe* (Chicago: University of Chicago Press, 2006).
Olesko, Kathryn M., *Physics as a Calling: Discipline and Practice in the Königsberg Seminar for Physics* (Ithaca: Cornell University Press, 1991).
Oreskes, Naomi, "Objectivity or Heroism? On the Invisibility of Women in Science," *Osiris* 11 (1996), pp. 87–113.
Outram, Dorinda, *Georges Cuvier: Vocation, Science, and Authority in Post-Revolutionary France* (Manchester: Manchester University Press, 1984).

Outram, Dorinda, "Scientific Biography and the Case of Georges Cuvier," *History of Science* 14 (1976), pp. 101–37.
Outram, Dorinda, "Uncertain Legislator: Cuvier's Laws of Nature in Their Intellectual Context," *History of Science* 19 (1986), pp. 323–68.
Overbye, Dennis, *Lonely Hearts of the Cosmos: The Story of the Scientific Quest for the Secret of the Universe* (New York: Little, Brown, 2021).
Pancaldi, Giuliano, *Volta: Science and Culture in the Age of Enlightenment* (Princeton: Princeton University Press, 2003).
Partington, J. R., *A History of Chemistry*, 4 vols. (London: Macmillan, 1962–70).
Partridge, Derek, "Darwin's Two Theories, 1844 and 1859," *Journal of the History of Biology* 51 (2018), pp. 563–92.
Pearson, Karl *The Life, Letters and Labours of Francis Galton*, 3 vols. (Cambridge: Cambridge University Press, 1914–1930).
Pera, Marcello, *The Ambiguous Frog: The Galvani-Volta Controversy on Animal Electricity*, trans. Jonathan Mandelbaum (Princeton: Princeton University Press, 1992).
Porter, Theodore M., *Karl Pearson: The Scientific Life in a Statistical Age* (Princeton: Princeton University Press, 2004).
Porter, Theodore M., "The Mathematics of Society: Variation and Error in Quetelet's Statisics," *British Journal for the History of Science* 18 (1985), pp. 51–69.
Porter, Theodore M., "Precision and Trust: Early Victorian Insurance and the Politics of Calculation," in M. Norton Wise, *The Values of Precision* (Princeton: Princeton University Press, 1995), pp. 173–97.
Porter, Theodore M., *The Rise of Statistical Thinking, 1820–1900* (Princeton: Princeton University Press, 1986).
Powers, John C., *Inventing Chemistry: Herman Boerhaave and the Reform of the Chemical Arts* (Chicago: University of Chicago Press, 2012).
Price, Derek J. de Solla, *Little Science, Big Science* (New York: Columbia University Press, 1963).
Principe, Lawrence, *The Secrets of Alchemy* (Chicago: University of Chicago Press, 2013).
Principe, Lawrence M., *The Transmutations of Chemistry: Wilhem Homberg and the Académie Royale des Sciences* (Chicago: University of Chicago Press, 2020).
Richards, Evelleen, *Darwin and the Making of Sexual Selection* (Chicago: University of Chicago Press, 2017).
Richards, Robert J., "Darwin's Theory of Natural Selection and Its Moral Purpose," in Michael Ruse and Robert J. Richards (eds.), *The Cambridge Companion to the Origin of Species* (Cambridge: Cambridge University Press, 2009), pp. 47–66.
Richards, Robert J., *The Romantic Conception of Life: Science and Philosophy in the Age of Goethe* (Chicago: University of Chicago Press, 2002).
Riskin, Jessica, "Poor Richard's Leyden Jar: Electricity and Economy in Franklinist France," *Historical Studies in the Physical and Biological Sciences* 28 (1998), pp. 301–36.
Rocke, Alan J., "Atoms and Equivalents: The Early Development of the Chemical Atomic Theory," *Historical Studies in the Physical Sciences* 9 (1978), pp. 225–63.
Rocke, Alan J., *Chemical Atomism in the Nineteenth Century: From Dalton to Cannizzaro* (Columbus: Ohio State University Press, 1984).
Rocke, Alan J., *Image and Reality: Kekulé, Kopp, and the Scientific Imagination* (Chicago: University of Chicago Press, 2010).
Rocke, Alan J., "Origins and Spread of the 'Giessen Model' in University Science," *Ambix* 50 (2003), pp. 90–115.
Roger, Jacques, *Buffon: A Life in Natural History* (Ithaca: Cornell University Press, 1997).
Rogers, G.A.J., "Descartes and the Method of English Science," *Annals of Science* 29 (1972), pp. 237–55.

Rossiter, Margaret W., *Women Scientists in America: Struggles and Strategies to 1940* (Baltimore: Johns Hopkins University Press, 1982).

Rotenstreich, Nathan, "Relativity and Relativism," in Gerald Holton and Yehuda Elkana (eds.), *Albert Einstein: Historical and Cultural Perspectives* (Princeton: Princeton University Press, 1982), pp. 175–204.

Rudwick, Martin, "Darwin and Glen Roy: A 'Great Failure' in Scientific Method?," *Studies in History and Philosophy of Science* 5 (1974), pp. 97–185.

Rudwick, Martin J. S., *Bursting the Limits of Time: The Reconstruction of Geohistory in the Age of Revolution* (Chicago: University of Chicago Press, 2005).

Rudwick, Martin J. S., *Earth's Deep History: How It Was Discovered and Why It Matters* (Chicago: University of Chicago Press, 2014).

Rudwick, Martin J. S. (ed. and trans.), *Georges Cuvier, Fossil Bones, and Geological Catastrophes* (Chicago: University of Chicago Press, 1997).

Rudwick, Martin J. S., *The Meaning of Fossils: Episodes in the History of Paleontology* (Chicago: University of Chicago Press, 1985 [1972]).

Rudwick, Martin J. S., *Worlds Before Adam: The Reconstruction of Geohistory in the Age of Reform* (Chicago: University of Chicago Press, 2008).

Rupke, Nicolaas A., *Richard Owen: Biology without Darwin*, rev. ed. (Chicago: University of Chicago Press, 2009).

Rupke, Nicolaas A., "Richard Owen's Vertebrate Archetype," *Isis* 84 (1993), pp. 231–51.

Ruse, Michael, *The Darwinian Revolution: Science Red in Tooth and Claw*, 2nd ed. (Chicago: University of Chicago Press, 1999).

Ruse, Michael, "Darwin's Debt to Philosophy: An Examination of the Influence of the Philosophical Ideas of John F. W. Herschel and William Whewell on the Development of Charles Darwin's Theory of Evolution," *Studies in History and Philosophy of Science* 6 (1975), pp. 159–81.

Schaffer, Simon, "Accurate Measurement is an English Science," in M. Norton Wise (ed.), *The Values of Precision* (Princeton: Princeton University Press, 1995), pp. 135–72.

Schaffer, Simon, "Herschel in Bedlam: Natural History and Stellar Astronomy," *British Journal for the History of Science* 13 (1980), pp. 211–39.

Schaffer, Simon, "John Michell and Black Holes," *Journal for the History of Astronomy* 10 (1979), pp. 42–43.

Schaffer, Simon, "The Leviathan of Parsonstown: Literary Technology and Scientific Representation," in Timothy Lenoir (ed.), *Inscribing Science: Scientific Texts and the Materiality of Communication* (Stanford: Stanford University Press, 1998), pp. 182–222.

Schaffer, Simon, "Metrology, Metrication, and Victorian Values," in Bernard Lightman (ed.), *Victorian Science in Context* (Chicago: University of Chicago Press, 1997), pp. 438–74.

Schaffer, Simon, "Natural Philosophy and Public Spectacle in the Eighteenth Century," *History of Science* 21 (1983), pp. 1–43.

Schaffer, Simon, "Self Evidence," *Critical Inquiry* 18 (1992), pp. 327–62.

Schiebinger, Londa, *The Mind Has No Sex? Women in the Origins of Modern Science* (Cambridge, MA: Harvard University Press, 1989).

Schiebinger, Londa, "Why Mammals Are Called Mammals: Gender Politics in Eighteenth-Century Natural History," *American Historical Review* 98 (1993), pp. 382–411.

Schmaltz, Tad, "Nicolas Malebranche," in *The Stanford Encyclopedia of Philosophy* (Spring 2022), Edward N. Zalta (ed.), https://plato.stanford.edu/archives/spr2022/entries/malebranche/.

Schofield, Robert E., *The Enlightenment of Joseph Priestley: A Study of His Life and Work, 1733–1773* (University Park: Pennsylvania State University Press, 1997).

Schofield, Robert E., "An Evolutionary Taxonomy of Eighteenth-Century Newtonianisms," *Studies in Eighteenth-Century Culture* 7 (1978), pp. 175–92.

Schofield, Robert E., *Mechanism and Materialism: British Natural Philosophy in an Age of Reason* (Princeton: Princeton University Press, 1970).

Schuster, John A., and G. Watchirs, "Natural Philosophy, Experiment and Discourse in the Eighteenth Century: Beyond the Kuhn/Bachelard Problematic," in Homer E. LeGrand (ed.), *Experimental Inquiries: Historical, Philosophical and Social Studies of Experiment* (Dordrecht: Reidel, 1990), pp. 1–48.

Secord, James A., *Victorian Sensation: The Extraordinary Publication, Reception, and Secret Authorship of* Vestiges of the Natural History of Creation (Chicago: University of Chicago Press, 2000).

Seth, Suman, *Crafting the Quantum: Arnold Sommerfeld and the Practice of Theory, 1890–1926* (Cambridge, MA: MIT Press, 2010).

Shapin, Steven, "Rarely Pure and Never Simple: Talking About Truth," *Configurations* 7 (1999), pp. 1–14.

Shapin, Steven, *The Scientific Life: A Moral History of a Late Modern Vocation* (Chicago: University of Chicago Press, 2008).

Sibum, Heinz Otto, "Reworking the Mechanical Value of Heat: Instruments of Precision and Gestures of Accuracy in Early Victorian England," *Studies in History and Philosophy of Science* 26 (1995), pp. 73–106.

Sloan, Phillip R., "John Locke, John Ray, and the Problem of the Natural System," *Journal of the History of Biology* 5 (1972), pp. 1–53.

Smith, Crosbie, *The Science of Energy: A Cultural History of Energy Physics in Victorian Britain* (Chicago: University of Chicago Press, 1998).

Smith, Crosbie, and M. Norton Wise, *Energy and Empire: A Biographical Study of Lord Kelvin* (Cambridge: Cambridge University Press, 1989).

Smith, George, "The *vis viva* Dispute: A Controversy at the Dawn of Dynamics," *Physics Today* 59, no. 10 (2006), p. 31.

Smith, Robert, *A Compleat Systeme of Opticks*, 2 vols. (Cambridge, 1738).

Smith, Robert, *Harmonics, or the Philosophy of Musical Sounds* (Cambridge, 1749).

Snelders, H.A.M., "Oersted's Discovery of Electromagnetism," in Andrew Cunningham and Nicholas Jardine (eds.), *Romanticism and the Sciences* (Cambridge: Cambridge University Press, 1990), pp. 228–40.

Sobel, Dava, *The Glass Universe: How the Ladies of the Harvard Observatory Took the Measure of the Stars* (New York: Viking, 2016).

Sponsel, Alistair, *Darwin's Evolving Identity: Adventure, Ambition, and the Sin of Speculation* (Chicago: University of Chicago Press, 2018).

Stahl, Georg Ernst, *Philosophical Principles of Universal Chemistry*, trans. Peter Shaw (London, 1730). Originally published as *Fundamenta chymiae* (Nuremberg: Wolfgang Mauritius, 1723).

Staley, Richard, *Einstein's Generation: The Origins of the Quantum Revolution* (Chicago: University of Chicago Press, 2008).

Stanley, Matthew, *Huxley's Church and Maxwell's Demon: From Theistic Science to Naturalistic Science* (Chicago: University of Chicago Press, 2015).

Stanley, Matthew, "The Pointsman: Maxwell's Demon, Victorian Free Will, and the Boundaries of Science," *Journal of the History of Ideas* 69 (2008), pp. 467–91.

Stanley, Matthew, *Practical Mystic: Religion, Science, and A. S. Eddington* (Chicago: University of Chicago Press, 2007).

Staley, Richard, *Einstein's Generation: The Origins of the Quantum Revolution* (Chicago: University of Chicago Press, 2008).

Stigler, Stephen M., *The History of Statistics: The Measurement of Uncertainty before 1900* (Cambridge, MA: Harvard University Press, 1986).

Stocking, George W., Jr., *Victorian Anthropology* (New York: Free Press, 1987).
Strick, James E., "Darwinism and the Origin of Life: The Role of H. C. Bastian in the British Spontaneous Generation Debates, 1868–1873," *Journal of the History of Biology* 32 (1999), pp. 51–92.
Strick, James E., *Sparks of Life: Darwinism and the Victorian Debates over Spontaneous Generation* (Cambridge, MA: Harvard University Press, 2000).
Strickland, Stuart, "The Ideology of Self-Knowledge and the Practice of Self-Experimentation," *Eighteenth-Century Studies* 31 (1998), pp. 453–71.
Thackray, Arnold, *Atoms and Powers: An Essay on Newtonian Matter-Theory and the Development of Chemistry* (Cambridge, MA: Harvard University Press, 1970).
Thackray, Arnold, *John Dalton: Critical Assessments of His Life and Science* (Cambridge, MA: Harvard University Press, 1972).
Theunissen, Bert, "The Relevance of Cuvier's *lois zoologiques* for His Paleontological Work," *Annals of Science* 43 (1986), pp. 543–556.
Todhunter, Isaac (ed.), *William Whewell, Master of Trinity College, Cambridge: An Account of His Writings*, 2 vols. (Cambridge: Cambridge University Press, 2011 [1876]).
Travis, Anthony S., *The Rainbow Makers: The Origins of the Synthetic Dyestuffs Industry in Western Europe* (Bethlehem, PA: Lehigh University Press, 1993).
Turner, Frank M., *Between Science and Religion: The Reaction to Scientific Naturalism in Late Victorian England* (New Haven: Yale University Press, 1974).
Turner, R. Steven, "The Growth of Professorial Research in Prussia, 1818–1848: Causes and Context," *Historical Studies in the Physical Sciences* 3 (1971), pp. 137–82.
Tyndall, John, *Address Delivered before the British Association Assembled at Belfast* (London: Longmans, Green, 1874).
Tyndall, John, *Heat Considered as a Mode of Motion* (London: Longmans, Green, 1863).
Underwood, Ted, *The Work of the Sun: Literature, Science, and Economy, 1760–1860* (New York: Palgrave Macmillan, 2005).
Van Bommel, Bas, "Between 'Bildung' and 'Wissenschaft': The 19th-Century German Ideal of Scientific Education," *Brewminate* (blog), August 6, 2019, https://brewminate.com/between-bildung-and-wissenschaft-the-19th-century-german-ideal-of-scientific-education/.
Van Helden, Albert, "Telescope Building, 1850–1900," in Owen Gingerich (ed.), *The General History of Astronomy*, vol. 4, "Astrophysics and Twentieth-Century Astronomy to 1950: Part A" (Cambridge: Cambridge University Press, 1984), pp. 40–58.
Van Wyhe, John (ed.), "Darwin Online," http://darwin-online.org.uk/.
Veysey, Laurence R., *The Emergence of the American University* (Chicago: University of Chicago Press, 1965).
Vorzimmer, Peter, "Darwin, Malthus, and the Theory of Natural Selection," *Journal of the History of Ideas* 30 (1969), pp. 527–42.
Warwick, Andrew, *Masters of Theory: Cambridge and the Rise of Mathematical Physics* (Chicago: University of Chicago Press, 2003).
Watson, William, *A Sequel to the Experiments and Observations Tending to Illustrate the Nature and Properties of Electricity* (London, 1746).
Weeks, Sophie, "The Role of Mechanics in Francis Bacon's Great Instauration," in Claus Zittel, Gisela Engel, Romano Nanni, and Nicole C. Karafyllis (eds), *Philosophies of Technology: Francis Bacon and His Contemporaries* (Leiden, 2008), pp. 133–95.
Weinberg, Steven, *The Discovery of Subatomic Particles*, rev. ed. (Cambridge: Cambridge University Press, 2003).
Weinberg, Steven, *Dreams of a Final Theory* (New York: Pantheon, 1992).
Werrett, Simon, "From the Grand Whim to the Gasworks: 'Philosophical Fireworks' in Georgian England," in Lissa Roberts, Simon Schaffer, and Peter Dear (eds.), *The Mindful Hand:*

Inquiry and Invention from the Late Renaissance to Early Industrialisation (Amsterdam: Edita, 2007), pp. 325–48.

Whiston, William, *Astronomical Principles of Religion, Natural & Reveal'd* (London, 1717).

White, Paul, *Thomas Huxley: Making the "Man of Science"* (Cambridge: Cambridge University Press, 2002).

Whittaker, E. T., *A History of the Theories of Aether and Electricity*, 2 vols. (rev. ed., London: Nelson, 1951).

Williams, L. Pearce, *Michael Faraday: A Biography* (London: Chapman and Hall, 1965).

Williams, L. Pearce (ed.), *Relativity Theory: Its Origins and Impact on Modern Thought* (New York: Wiley, 1968).

Williams, L. Pearce, "Science, Education and Napoleon I," *Isis* 47 (1956), pp. 369–82.

Williams, Mari, "James Bradley and the Eighteenth Century 'Gap' in Attempts to Measure Annual Stellar Parallax," *Notes and Records of the Royal Society of London* 37 (1982), pp. 83–100.

Williams, Raymond, *Keywords: A Vocabulary of Culture and Society* (Oxford: Oxford University Press, 2014).

Winsor, Mary Pickard, "Darwin's Dark Matter: Utter Extinction," *Annals of Science* 80 (2023), https://doi.org/10.1080/00033790.2023.2194889.

Wollaston, William, "A Synoptic Scale of Chemical Equivalents," *Philosophical Transactions of the Royal Society* 104 (1814), pp. 1–22.

Wootton, David, *The Invention of Science: A New History of the Scientific Revolution* (New York: Harper, 2015).

Wright, Thomas [of Durham], *An Original Theory or New Hypothesis of the Universe*, ed. Michael A. Hoskin (London: Macdonald, 1971).

Yeo, Richard, *Defining Science: William Whewell, Natural Knowledge and Public Debate in Early Victorian Britain* (Cambridge: Cambridge University Press, 1993).

Zakariya, Nasser, *A Final Theory: Science, Myth, and Beginnings* (Chicago: University of Chicago Press, 2017).

Zee, A., "An Old Man's Toy," in Donald Goldsmith and Marcia Bartusiak (eds.), *E = Einstein: His Life, His Thought and His Influence on Our Culture* (New York: Sterling, 2006), pp. 223–36.

Zinsser, Judith, *Dame d'Ésprit: A Biography of the Marquise du Châtelet* (New York: Viking, 2006).

INDEX

Page numbers in *italics* refer to images

Adams, John Couch, 257, 261
adaptations, 152–55, 159, 165, 166, 171–72, 175–76
aether: Albert Einstein on, 290–91; Benjamin Franklin on, 71–72; and magnetic field lines, 227–30; and Michelson-Morley experiment, 286–89; Isaac Newton on, 21, 61, 69
Ampère, André-Marie, 217–18, 247
Analytical Theory of Probabilities (Laplace), 197–98
anatomy, comparative: and correlation of parts, 138–41, 143; Georges Cuvier, as originating with, 134–35; as experimentation, 138; to prove extinctions, 143–45, *144*; and subordination of characters, 141–42
Arago, François, 117, 119, 121, 122, 216
Arbuthnot, John, 48–53, 321n9
Arcueil, Society of, 114, 115, 117, 244
Aristotle: and classification of living things, 77, 138, 318n32; and elements, 23, 90, 92, 100, 101; and mathematics, 46; and natural philosophy, 3, 12, 24, 75
Astronomical Principles of Religion (Whiston), 33, 318n26, 319n14
astronomy: and big science, 271, 275, 284; and Cepheid variables, 280–81, 282–83; and Doppler effect, 273–74, 283; and galaxies, study of, 278–79, 281–82, 283; limits of knowledge within, 273; and measurement improvements, 259–60, 271–72; and nebulae, study of, 260–61, 262–65, *264*, *265*, 269–70, 278–79, 339n19; and Neptune, discovery of, 256–57; observatories, 7, 271, 274, 275, 277–78, 282, 283–84, 340n40; and photography, 271–72, 277; and reflecting telescopes, 261–62, *263*, 277–78; and spectroscopy, 264–70, 267, 271–72, 273–74, 276–77; and stars, measuring distance to, 257–60, *258*, 280–81, 283, 338n3, 338n4, 340n36; and telescopes, increasingly large, 274–75, 277, 282, 283–84; as true science ideal, 256; women in, 41, 279–80, 320n31, 340n33. *See also* astrophysics; cosmology
astrophysics: American dominance in, 276; as big science, 284; and evidence of stellar gravitation, 38–40; and large telescopes, 273–75, 277–78; origin of, 275–76; and photography, 271–72; and spectroscopy, 264; and study of nebulae, 269–70. *See also* astronomy; spectroscopy
Astro-Theology (Derham), 23, 32
atoms: and atomic weights, 243–44, 249; and Avogadro's hypothesis, 244, 246–49, 250, 252, 337n10; and chemical formulas, 243–44, 245, 247, 250, 337n8; and John Dalton's atomic theory, 240–41, *241*; Michael Faraday on, 224–25; and Joseph Gay-Lussac's volume ratios, 244–46; and isomers, 249–50, 252; vs. molecules, 246, 248, 331n24, 337n6; and the nature of matter, 242–43, 250, 252–55; and periodic table of the elements, 250; and positivism, 254–55; and radioactive decay, 300; structure of, 298, *299*; and thermodynamics, 254–55. *See also* corpuscles; elements
Avogadro, Amadeo, 244, 246–49, 250, 337n10, 337n14

361

Bacon, Francis, 3, 326n2
Bates, Henry, 173, 177
Bell, John, 303
Bernoulli, Daniel, 54, 55–56
Bernoulli, Jakob, 47, 51–52, 199
Bernoulli, Nicholas, 52–53, 54
Berthollet, Claude, 113, 114, 115, 121, 238
Berzelius, J. J., 246, 248, 250
Bessel, F. W., 259
Bichat, Xavier, 79–80, 135–36, 328n5
big science, 271, 275, 284, 339n21
binomial distributions. *See* error curves
Biot, Jean Baptiste, 117–22, 215, 216
Black, Joseph, 98–99, 240
Boerhaave, Herman, 22, 91, 92, 94
Bohm, David, 303
Bohr, Niels, 292, 298, 299, 300, 302, 303, 304, 308
Boltzmann, Ludwig, 207, 253
Boscovich, Roger, 37–38, 225, 319n21
Bose, G. M., 66, 67
Boyle, Robert: and Boyle Lectureship, establishment of, 15; and electricity, 60, 65; on elements, 89–90, 91; and ideal gases, 183, 331n7; and natural theology, 23, 152; and universe as mechanical, 4
Boyle Lectureship, 15–17, 22, 23–24, 32
Bradley, James, 257–58, 286, 338n4
Buckle, Henry Thomas, 203–4
Buffon, Georges-Louis Leclerc, comte de, 83–87, 85, 142, 324n15
Bunsen, Robert, 268

calculus of probabilities. *See* probability, mathematical
caloric, 105, 116–17, 182–88, 192, 194, 240
Cannizzaro, Stanislao, 249, 252
Carnot, Sadi, 182–87, 184, 192–93, 205
Cartesianism, 13, 14, 18–19, 20, 27, 66
catastrophism theory of geology, 145–46, 155–56
Cavendish, Henry, 105
Cepheid variables, 280–81, 282–83
Charles, Jacques, 183, 331n7
Châtelet, Émilie du, 19
chemical elements. *See* elements
chemistry: chemical concepts, and development of, 91–92, 109; and chemical industries, 251–52; and combustion, 93–94, 101–5; efforts to mathematize, 114, 115; and elective affinities, 94–96, 95; and gases, 97–99; as natural philosophy, 90–91; and nomenclature, 80, 105–8; organic, 250–52, 253; and oxygen, identification of, 103–5; and oxygen, naming of, 105–6; and phlogiston, 93–94, 101–2, 103, 104, 108; as practical, 90, 91, 92; and states of matter, 99–101; structure of matter, for the study of, 88–89, 238, 324n1, 325n16; and water, separation of, 105, 106. *See also* atoms; elements
Clark, Alvan, 274, 275, 277
Clarke, Samuel, 16–17
classification of living things: as artificial, 78–79, 82–83, 84, 86, 87, 134; by breeding groups, 86–87; by Georges-Louis Leclerc, comte de Buffon, 83–87, 85, 324n15; compared to classification of non-living things, 79–80, 88, 95, 96; by Georges Cuvier, 87, 134–35, 141–42, 143–44, 162; early schemes for, 76–77; as ecological, 85–86; as explanatory, 196; and God, 80, 82, 134; by Carl Linnaeus, 77–79, 87, 134, 323n5; as natural, 79, 80–81, 82–83, 84, 87, 134, 141, 142, 159; need for, 76; by John Ray, 81–83. *See also* taxonomy
Clausius, Rudolf, 193–94
combustion, 93–94, 101–5
conservation of energy: in collisions, 182; development of law of, 187–92; and electromagnetism, 234; origin of idea for, 185; and second law of thermodynamics, 193–94, 205
Copernicus, Nicolaus, 4
corpuscles, 21, 88–89, 91, 92, 95
correlation of parts, 138–41, 143, 152
cosmology: and evidence for stellar gravitation, 38–40; and William Herschel's observations, 40–44; and natural philosophy, 26, 310, 343n7; Newtonians, according to, 31–34; Isaac Newton, according to, 26–31; origin of term, 26; and stellar mobility, 34–38. *See also* astronomy; astrophysics; universe
Coulomb, Charles-Augustin, 73, 323n27
creationism: and classification, 80; Charles Darwin on, 174, 177–78, 330n12; William

Derham on, 23–24, 153, 318n28; vs. evolution, 174, 177–79; James Clerk Maxwell on, 178–79, 309; in natural philosophy, 6, 16–17, 25. *See also* God; natural theology; theology
Curie, Marie, 306, 316n15
current, electrical, 213–15, 336n32
Cuvier, Georges: career of, 133–34, 142, 327n1; and classification, 87, 134–35, 141–42, 162; and comparative anatomy, 134–35, 138–42, 143–45; and the correlation of parts, 138–41, 143, 152; on evolution, 134, 140–41, 145, 152; and extinctions, 143–46; natural history, role in development of, 132, 133, 134, 143, 146, 328n15; and natural experiments, 138; and subordination of characters, 141–42

D'Alembert, Jean Lerond, 19–20, 56–58, 111, 333n5
Dalton, John: atomic theory of, 240–41, 241; and chemical formulas and equations, 242; early work of, 239; reactions to atomic theory of, 242–44, 245–47; and study of gases, 239–40, 336n3
Darwin, Charles, 150; on creationism, 174, 177–78; criticism of work of, 169–71, 172, 178–79, 330n10; early life and marriage of, 147–48, 149; and evolution, early ideas on, 148–52, 153–55; on geology, 150–51, 164–65; on H.M.S. Beagle, 148–49, 153–55; importance of Georges Cuvier to the work of, 132–33, 146; and natural selection, 158–61, 165, 169–73, 174, 175; and origins of biology, 2; as private researcher, 130, 149; and scientific naturalism, 176–80; and sexual selection, 175–76; supporters of work of, 167–68, 172, 173, 176–77, 330n18; and taxonomy over time, 162–65, 163. *See also specific works*
Davy, Humphry, 215–16, 222, 243, 327n12
Derham, William: Boyle Lecture of, 23–24; and catalog of nebulae, 35; on cosmology, 32; on creationism, 23–24, 153, 318n28; on mathematical proof of God, 51, 52. *See also specific works*
Desaguliers, Jean, 66
Descartes, René, 13, 14, 26–27, 317n6

Descent of Man (Darwin), 174, 175, 176
determinism: Albert Einstein on, 297–98; Pierre-Simon de Laplace on, 198–99, 203, 297; James Clerk Maxwell on, 178, 203–5, 212; in probability and statistics, 198–200, 202–3, 204–5, 207; in quantum mechanics, 300
Diderot, Denis, 20, 90
divergence, 158–59, 162, 164
Doctrine of Chances, The (de Moivre), 52
Doppler, Christian, 273–74
Doppler effect, 273–74
Dufay, Charles, 64–66
Dumas, J. B., 248

education: and continuity of science, 305; in France, 112–13, 128–29, 133; in Germany, 123–27; and the professionalization of science, 123, 127, 131; the role of research in, 124, 125, 126–27, 128, 131; and training in experimentation, 123–24; in the United Kingdom, 129–31, 327n12; in the United States, 127–28. *See also* universities
Einstein, Albert: on aether, 290–91, 292; and Brownian motion, 255; and determinism, 297–98; on energy, 182; and general relativity, 283, 295–97; and God, 292, 300; on Michelson-Morley experiment, 291; and natural philosophy, 292, 301–4, 308; and origins of modern physics, 2; as positivist, 292; on quantum mechanics, 300–303, 308, 342n33; and responses to relativity theory, 293–95; and special relativity, 289, 290–91, 293, 294, 341n7, 341n10; on structure of matter, 254–55; and theories of principle, 289, 295–96, 297; and thought experiments, 289–90, 297
elective affinities, 94–96, 95
electrical machines, 66–69, 67, 68, 71–73, 72
electricity: and the amber effect, 60; and conductors and insulators, 66, 223; early demonstrations of, 60–61, 62; and earth as electrical sink, 69, 71; and electrical current, 213–15, 336n32; and electrical machines, 66–69, 67, 68, 71–73, 72; as entertainment, 68–69; Benjamin Franklin's experiments with, 69–74, 71, 72; in medicine, 322n13; as Newtonian,

electricity (*continued*)
69, 71–72, 73, 74; as origin of modern physics, 59; and origin of term, 60; as physiological, 213–14; quantitative study of, 73–74, 323n27; transmission of, 63–64, 65, 70–71, 71; as two types, 65–66, 72, 73–74. *See also* electromagnetism

electrochemistry, 188, 215, 216, 222–23

electromagnetic induction, 216–20, 217, 218, 219, 220, 221–22

electromagnetism: as action at a distance, 217, 222–23, 224, 227, 228, 229, 237, 335n25; and aether, 227, 228–37, 286–89, 290–91; and electrical current, 213–15; and electromagnetic induction, 216–20, 217, 218, 219, 220, 221–22; and electromagnetic waves, generation of, 236; and field lines, mathematizing of, 230–31; and field lines as physical things, 225–28, 226, 230, 237; and light, 234–35, 236–37; and light, velocity of, 290; and magnetic fields, 220–24, 221; and measurement issues, 234–35, 236; as mechanical, 229–34, 237, 290–91; and mechanical aether model for field lines, 231–34, 233, 235–36; and the nature of matter, 224–25; and principle of relativity, 290; as quantized, 298; and the Voltaic pile, 214–15, 216. *See also* electricity

elements: Aristotelian, 23, 90, 92, 100, 101; and atomic weights, 244; as chemical principles, 92, 107, 109, 240, 243; as components of chemical compounds, 240–41, 241, 242, 248; early definitions of, 89–90, 324n1; electrical fire as, 69; hydrogen as basic unit of, 244; and isomers, 249–50; radioactive, 300; spectra of, 266, 267, 268–69, 298. *See also* atoms

Elements of Chemistry (Lavoisier), 106, 107, 324n1

Elements of Newton's Philosophy (Voltaire), 19

entropy, 193–94, 205, 205–7

error curves: in astronomy, 259; for data analysis, 201, 202–3; as distribution curve for error analysis, 197–98, 198; for gases, 204, 206; in life sciences, 207, 209; as probability, 199

Essay on a Principle of Population (Malthus), 160

eugenics, 208, 209, 210–11, 212, 334n28

evolution: and adaptations, 152–55, 159, 165, 166, 171–72, 175–76; adaptedness of species as counterargument to, 152–53, 159; and archetypes, 167; vs. creationism, 174, 177–79; criticism of Charles Darwin's work on, 169–71, 172, 178–79, 330n10; Georges Cuvier on, 134, 140–41, 145; Charles Darwin's early ideas on, 148–52; and divergence, 158–59, 162, 164; in geological record, 164–65; of humans, 174–76; and Charles Lyell's uniformity of nature, 156–58; modern rejection of, 306; and morphology, 166–68, 172; and natural selection, 158–61, 165, 168–73, 174, 175, 178, 209, 211–12; and progressionism, 166; and scientific naturalism, 176–80; and sexual selection, 175–76; supporters of Charles Darwin's work on, 167–68, 172, 173, 176–77; and taxonomy, 162–63; and transmutation of species, 148, 149, 152, 153–55, 163, 164–65

experimentation: by amateurs, 61, 131, 251, 269–70; in demonstrations, 60–61; Albert Einstein's views on, 289–90, 295, 297; electrical, and origin of physics discipline, 59, 182 (*See also* electricity); increasing precision in, 118; mathematical physics, in development of, 58, 110, 116–17, 118–19, 121, 122; vs. mathematical theory, 215–16, 227–28; in natural history, 135–36, 138, 180; in *Opticks* (Newton), 21–22; origins of, 5–6; in physiology, 75, 138, 328n5; training for, 123–24. *See also specific individuals*

extinction, 143–46

Faraday, Michael: and chemical compounds, 249, 252; on electric current, 336n32; and electrochemistry, 222–23; and electromagnetic induction, 216–20, 217, 218, 219, 220, 221–22; as experimenter, 215–16; and magnetic field lines, 224–28, 226, 230–31; and magnetic fields, 220–24, 221; at the Royal Institution (London), 215–16, 327n12

fixed air, 99, 100, 103–4, 240

fossil record: and comparative anatomy, 139; and Charles Darwin's taxonomy, 163–64;

and extinctions, 143–45; homologies in, 166–67, 172; incompleteness of, 157, 164–65, 171, 174; and progressionism, 166
Fourier, Joseph, 121–22
Franklin, Benjamin, 69–74
Fraunhofer, Joseph von, 265–66, 269, 270
French Academy of Sciences, 7, 19–20, 64, 83, 100, 113–14, 129, 133
French Revolution, 111–12, 129, 132
Fresnel, Augustin-Jean, 121–22, 228–29, 238
Fundamentals of Chemistry (Stahl), 91

galaxies, 35–37, 278–79, 281–82, 283. *See also* Milky Way
Galileo, 10, 27, 34, 46, 296
Galton, Francis, 204, 207–12, 211, 334n28
Galvani, Luigi, 213–15
Galvanism, 214
Gauss, Carl Friedrich, 197
Gay-Lussac, Joseph, 244–48, 251
General Anatomy Applied to Physiology and Medicine (Bichat), 136
Geoffroy, Étienne-François, 94–96, 95
Geological Society of London, 130, 147–48
geology: catastrophism theory of, 145–46, 155–56; and Charles Darwin, work of, 147–48, 150–51, 153, 164–65; and evolution, 164–65; Geological Society of London, 130, 147; humanity within the timescale of, 156–58, 329n12; uniformitarian theory of, 155–56, 173
Gillispie, Charles, 309, 343n6, 343n7
God: and archetypes, 167; and classification of living things, 80, 81–82, 134; and conservation of energy, 188; as creator of the world, 6, 8, 25, 31, 32–33, 44, 179; and Charles Darwin on creationism, 174, 177–78; vs. deism, 19; Albert Einstein on, 292, 300; evidence for, 14–16, 22–24, 38, 152–53, 318n26; as fundamental to natural philosophy, 5–6, 8, 17–18, 25, 308, 309–10; and gravity, 17–18, 33, 228, 318n6; in mathematics of probability, 48–51, 49, 52–53, 198, 321n9; in a mechanical universe, 5. *See also* creationism; religion; theology
gravity. *See* universal gravitation, law of
Gray, Stephen, 63–65

Hale, George Ellery, 276–78
Hales, Stephen, 97–98
Halley, Edmund, 32, 34
Hare, Robert, 224–25
Hauksbee, Francis, 60–61, 62
Heat Considered as a Mode of Motion (Tyndall), 191, 332n19
Heisenberg, Werner, 127, 292, 298, 299, 300, 301
Helmholtz, Hermann von, 189–91, 237, 285
Henderson, Thomas, 259
heredity, 208–12
Herschel, Caroline, 41, 280, 320n31
Herschel, John, 203, 235, 263, 333n10
Herschel, William: German origins of, 40–41, 259, 319n20; and nebulae, study of, 41–42, 260–61, 339n19; as telescope maker, 40–41, 262; and universe as dynamic, 37; and universe as evolving, 42–44
Hertz, Heinrich, 236, 237, 286
Hertzsprung, Ejnar, 280–81
Higgs boson, 285, 307
History of Electricity (Priestley), 73
Hofmann, August Wilhelm, 251–52
homologies, 166–67, 168, 172
Hooker, Joseph, 149–50, 151, 167–68, 173, 177
Hubble, Edwin, 282–83
Huggins, William, 236, 269–70, 273–74, 278, 339n19
Humboldt, Wilhelm von, 125
Huxley, Thomas Henry: on natural selection, 169, 170–72, 173; and opposition to religion, 177; on spontaneous generation, 179; and support of Charles Darwin's work, 167–68, 172, 176–77
Huygens, Christiaan, 12, 47
hydrogen, 105, *106*, 108, 241, 243–45, 247, 270

isomers, 249–50, 252

Joule, James Prescott: and conservation of energy, 185–88, 332n10, 335n12; experimental work, importance of, 188–89; as private researcher, 130, 327n12

Kant, Immanuel, 34–38, 263, 323n27
Keill, John, 32
Kekulé, August, 250, 252, 253

Kelvin, Lord. *See* Thomson, William (Lord Kelvin)
Kepler, Johannes, 10, 27, 231, 273
kinetic theory of gases: and atoms, 253, 255; and Avogadro's hypothesis, 249; development of by Rudolf Clausius, 194–95; Albert Einstein on, 289; and laws of thermodynamics, 204–7, 205, 206
Kirchhoff, Gustav, 127, 268, 269, 270

Lamarck, Jean-Baptiste de, 140, 145, 159, 163
Langley, Samuel P., 272–73
Laplace, Pierre-Simon de: and determinism, 198–99, 297; legacy of, 182, 308; on nebulae, 278; and patronage of sciences, 114, 137; as pioneer of mathematical physics, 110, 114–22; political power of, 112–13, 121; and probability theory, 197–200; on the solar system, 156
large numbers, law of, 51–52, 199, 201–2
Lavoisier, Antoine: and chemical nomenclature, 80, 105–8; and chemical substances, 238, 240, 242, 324n1; execution of, 113; and experiments with Pierre-Simon de Laplace, 116–17, 326n13; and natural philosophy, 96–97, 99–100; and origins of chemistry, 2; and oxygen theory of combustion, 103–5; and separation of water, 105, 106; and states of matter, 97, 99–101
law of large numbers, 51–52, 199, 201–2
Lawrence, E. O., 303–4
Le Verrier, Urbain Jean Joseph, 257, 261
Leavitt, Henrietta, 279–80, 316n15
Lick, James, 275–76, 276
Liebig, Justus, 123–24, 250–52, 337n23
Linnaeus, Carl, 25, 77–79, 87, 323n5
Locke, John, 20, 86
Lord Rosse. *See* Parsons, William (Lord Rosse)
Lorentz, H. A., 286, 291–92, 341n10
Lyell, Charles, 151, 155–58, 165, 166, 168, 329n12

machines, electrical, 66–69, 67, 68, 71–73, 72
Magendie, François, 136–37
magnetic fields. *See* electromagnetism
Malebranche, Nicolas, 12
Malthus, Reverend Thomas, 160–61, 178
Malus, Etienne, 115–16

Marx, Karl, 133n11, 203, 311
mathematical physics. *See* physics, mathematical
Mathematical Principles of Natural Philosophy, The (Newton). *See Principia* (Newton)
mathematics: applied, 45, 46, 53; as expression of structure of the universe, 5, 47; mixed, 45–47, 48, 53, 57–58; in Isaac Newton's work, 10–13, 15, 18, 19, 22; as pinnacle of knowledge, 50–51; and planetary motion, 10; pure, 45, 46–47, 53. *See also* experimentation; physics, mathematical; probability and statistics; probability, mathematical
matter, nature of: and atoms, 224–25, 242–43, 250, 252–55; and chemistry for the study of, 88–89, 238, 324n1, 325n16; and electromagnetism, 224–25; and gravity, 18
Maupertuis, Pierre-Louis, 19, 35
Maxwell, James Clerk: on aether, 228–30, 235–36, 286, 307; at Cambridge University, 124, 129, 130; on creationism and God, 178–79, 309; on determinism, 178, 203–4, 212; and field lines, mathematizing of, 230–31; and kinetic theory of gases, 194–95, 249, 253, 307; and light and electromagnetism connection, 234–35; and mechanical aether model, 227, 231–34, 233, 237; and origins of modern physics, 2; as private researcher, 130; and statistical physics, 203–7, 205, 206
measurement standards, 235, 236
Memoirs (French Academy of Sciences), 7, 54, 64, 94–95, 95
Mendeleev, Dmitri, 250, 306
Michell, John, 38–40, 48, 304, 338n3
Michelson, Albert, 286–89
Michelson-Morley experiment, 286–89, 288, 291
Milky Way, 34–36, 42, 263, 281–82, 283
modern science. *See* science, modern
Moivre, Abraham de, 52–53, 197
molecules, 246, 248, 249–50, 252, 331n24, 337n6
Morley, Edward, 288–89
morphology, 166–68, 172
Musschenbroek, Pieter van, 67, 68

natural history: and correlation of parts, 138–41, 143; as experimental, 138; and extinctions, 143–46; importance of Georges Cuvier's work to, 132, 133, 135, 143, 145, 328n15; and The Linnean Society, 130; place of humans in, 156–58; and subordination of characters, 141–42; of universe, 40–44. *See also* classification of living things; evolution; geology; paleontology

natural philosophy: and Aristotle, 3, 12, 24, 75; and Cartesianism, 13, 18–19; and chemistry, 90–91; and classification, 75, 79, 81; and cosmology, 26, 43–44, 310, 343n7; defined, 3; and Albert Einstein, 292, 301–4, 308; end of, 303–4; God as fundamental to, 5–6, 8, 17–18, 25, 308, 309–10; and Antoine Lavoisier, 96–97, 99–100; and life sciences, 75; and mathematics, 45, 46, 111, 116, 122; and mechanical model of the universe, 4–6, 13; and medicine, 316n16; and modern science, 3–4, 304, 305, 306, 307; and natural history, 84, 196; and natural philosophers, 6–8; and natural theology, 25; and nature of matter, 238, 242; and Isaac Newton, 8, 9, 12, 14–15, 21, 22; and physics, 3, 59; and publication practices, 99; as public enterprise, 8; purity of, 311; and theology, 14–15, 23, 25, 53, 152; and underlying laws of nature, 110–11

natural selection: and adaptation, 140; vs. creationism, 177–79; Charles Darwin's publication of idea of, 151–52; environmental factors as triggers for, 165; in human evolution, 174; as mechanism of evolution, 158–61; plausibility of, 168–73, 330n10; vs. saltation, 171–73, 209, 211–12; and sexual selection, 175, 176; Alfred Russel Wallace on, 151, 168. *See also* Darwin, Charles; evolution

natural theology, 22–25, 152–53, 159, 318n28. *See also* creationism; God; theology

Natural Theology (Paley), 152

nebulae: Andromeda, 36, 278, 279, 282–83; as birthplace of stars, 278–79; early explanations for, 35; and globular clusters, 42, 43; as groups of stars, 41–42; modern images of, 36, 43, 279; and nebular hypothesis, 37; planetary, 42;

and spectroscopy, 269–70; as stage of stellar formation, 42, 43; and true nebulosity, 260–61, 262–65, 270, 278, 339n19

Neumann, John von, 303

Newcomb, Simon, 273

New System of Chemical Philosophy, A (Dalton), 240, *241*

Newton, Isaac: and acceptance of work in Continental Europe, 18–22; and acceptance of work in England, 13–16; on cosmology, 26–31; criticism of, 5, 10, 12–13; dominance of world view of, 8; on elective affinities, 94, 95; on electricity, 61; on gases, 239; God in work of, 8, 15, 16, 17–18, 19, 22–23, 24, 38, 318n6; and laws of motion, 5, 19, 137, 229, 231, 234, 296; and mass, inertial vs. gravitational, 296; and natural philosophy, 8, 9, 12, 14–15, 21, 22; and origins of physics, 2, 8; political use of the work of, 16–18, 20; as president of Royal Society of London, 8; prestige of, 9–10, 21–22, 61, 135; problems in work of, 28–31; and Scientific Revolution, 4; and stars, measuring distance to, 258; on states of matter, 97, 98; on structure of matter, 88–89, 90, 91, 341n22; tomb of, *11, 12*, 316n3; and use of senses for acquiring knowledge, 86. *See also* universal gravitation, theory of; specific works

observatories, astronomical, 7, 271, 274, 275, 277–78, 282, 283–84, 340n40

Oersted, Hans Christian, 216–17, 219, 305–6

On the Conservation of Force (Helmholtz), 189, 190

On the Origin of Species (Darwin), 132, 152, *163*, 165–66, 174, 330n12

Opticks (Newton), 21, 40, 61, 94

Original Theory or New Hypothesis of the Universe (Wright), 34–35

Ostwald, Friedrich Wilhelm, 253–55

Owen, Richard, 166–68

oxygen: and chemical compounds, early study of, 240–41, 243–44; and chemical nomenclature, 108; as diatomic, 245–47, 248–49; and Antoine Lavoisier's theory of acids, 106–7; and Antoine Lavoisier's theory of combustion, 103–5; naming of, 104, 105; and respiration, 113, 141; separation of water into hydrogen and, 105, *106*

paleontology, 139, 140, 143–45, 172. *See also* fossils
Paley, Reverend William, 152–53
parallax, stellar, 257–60, *258*
Parsons, William (Lord Rosse), 261–63, *263*, 269, 272, 278, 339n19
particles, elementary, 285, 303–4, 307
Pascal, Blaise, 47, 49
pedagogy. *See* education
Perrin, Jean, 255
Philosophical Transactions of the Royal Society (Royal Society of London), 7, 129–30
phlogiston, 93–94, 101–2, 103, 104, 108
Physico-Theology (Derham), 23, 32, 51
physics: energy as unifying concept for, 191–92; as fundamental discipline, 308; and grand theory of superstrings, 307; and light, aberration of, 286, 287; and natural philosophy, 304; origins of, 59, 111, 181–82; and positivism, 254–55, 292; and quantum mechanics, 255, 285, 292, 293, 298–304, 307, 342n31; as scientific truth, 294; statistical, 203–7, *205*, *206*. *See also* electromagnetism; physics, mathematical; relativity theory; thermodynamics
physics, mathematical: Claude Berthollet's role in developing, 114, 115, 121; at the École polytechnique, 112, 114; experimentation in, importance of, 116–17, 121; and heat flow, 121–22; and origin of physics discipline, 122, 182; origins of, 110–11; patronage of by Napoleon, 112–13; as pioneered by Pierre-Simon de Laplace, 110, 114–22; and precision measurements, 116–17, 118–19, 121; and wave theory of light, 122
physiology: and classification of tissues, 79–80; as different from physical sciences, 135–37; and electricity, 214–15, 322n13; experimental, 75, 138, 328n5; Antoine Lavoisier's interest in, 100, 113
Planck, Max, 255, 295, 298, 300
polarization, 115–16
positivism, 254–55, 292
Priestley, Joseph, 73, 103–4, 108, 225, 239, 319n21, 323n27
Principia (Newton): and action at a distance, 12; Anglican Church's use of, 14–15, 16; electricity in, 61; God in, 22–23, 31; and Isaac Newton's prestige, 10; political use of, 16–18; second edition of, 23, 28, 31, 61, 320n3; translation of, 19; and universe as mechanical, 5
Principles of Geology (Lyell), 155, 157
probability and statistics: as applied to data, development of, 197; in cosmology, 39, 42; and determinism, 198–200, 202–3; and determinism vs. free will, 204–5, 207; and error curve, 197–98, *198*, 201, 202–3, 204, *206*, 209–10, 259; and eugenics, 208, 209, 210–11, 212, 334n28; and God, 24; and heredity, 208–12; and laws of nature as statistical, 207; as new model for science when applied to data, 196; and origin of statistics, 200–201; and quantum mechanics, 298, 300, 303; and social statistics, 201–3; and statistical physics, 203–7, *205*, *206*; and thermodynamics, 194, 254, 255. *See also* probability, mathematical
probability, mathematical: calculus of probabilities and statistical probability, 48; and God, 48–51, *49*, 52–53, 198, 321n9; and the law of large numbers, 51–52, 199, 201–2; origin of, 46, 47, 197; purpose of, 47–48; and quantum mechanics, 300, 303; and St. Petersburg paradox, 54–58. *See also* probability and statistics
Prout, William, 244, 247
Prussian Academy of Sciences, 7

quantum mechanics, 255, 285, 292, 293, 298–304, 307, 342n31
"Queries" (Newton), 21–22, 61, 69, 88, 94, 239. See also *Opticks* (Newton)
Quetelet, Adolphe, 201–3, 204

Rankine, W.J. Macquorn, 190, 191–92
rational mechanics: defined, 46–47; and experimentation, 110, 116; and gases, 194–95; limitations of, 75; mathematical physics, in the development of, 110, 111; and Isaac Newton, 19; and vis viva, 182, 187, 189
Ray, John, 24, 81–83
Reflections on the Motive Power of Fire (Carnot), 183, 331n6
refraction of light, 117–22

relativity theory: and general relativity, 295–97, 304; and quantum mechanics, 307; responses to, 293–95; and special relativity, 289, 290–91, 293, 294, 341n7, 341n10

religion: Charles Darwin's ideas, effects of on, 174; decline of, 343n6; and establishment of Boyle Lectureship, 15; and the French Revolution, 111–12; opposition to, 177, 204, 310, 343n9; promotion of using Isaac Newton's work, 14–15, 16, 18. *See also* creationism; God; theology

Rouelle, Guillaume-François, 96

Royal Academy of Sciences (Paris). *See* French Academy of Sciences

Royal Institution (London), 180, 191, 215–16, *218*, *221*, 327n12

Royal Society of London: amateurs in, 21; exclusion of women from, 7–8, 16n18; and experimental demonstrators, 60–61, 66; Benjamin Franklin's correspondence with, 69–70; members of, 8, 15, 23, 52, 60, 89, 97, 306; origins of, 7; publications of, 32, 35, 38, 48–49, 53, 63, 129–30, 224, 243, 269; and science as practical enterprise, 3–4; and separate, specialized societies, 129–30

Russian Academy of Sciences, 7, 54

Rutherford, Ernest, 298

saltation, 171–73, 209, 211–12

Schrödinger, Erwin, 299–300, 302

science, modern: and continuity, 305; as honest, 310–11, 343n12; as knowledge claims (discoveries), 2; and natural philosophy, 3–4, 304–5, 306, 307; as product of history, 1–2, 307–9; as a profession, 2, 131, 308; as progress toward truth, 285; public support for, 285; as pure vs. utilitarian, 311; and religion, 310, 343n9; and the Scientific Revolution, 4–6; as Western tradition, 1, 305–6; and worldview, 306–11

scientific naturalism, 176–80, 285

Scientific Revolution, 4–6

scientific societies: and continuity of science, 305; exclusion of women from, 7–8; as focus of early science, 7; and science as a profession, 2, 131; for science outside of universities, 7; specialized, 129–30; in the United Kingdom, 129–30, 327n12. *See also specific organizations*

second law of thermodynamics, 193–94, 205, 205–7

sexual selection, 175–76

Shapley, Harlow, 281–82

Smith, Robert, 46, 320n3

societies. *See* scientific societies

Society of Arcueil, 114, 115, 117, 244

spectroscopy, *268*; in amateur astrophysics, use of, 269–70; with astronomical photography, use of, 271–73; and Doppler shifts, 273–74, 283; and galaxies, 278–79, 283; and nebulosity, importance to, 264–65; and quantum mechanics, 298–99, 299; and similarities of molecules, 178–79; and spectral lines, discovery and interpretation of, 265–69; and spectroheliograph, 276–77; and stellar composition, 274

Spencer, Herbert, 177, 285

Stahl, Georg Ernst: chemical concepts, and development of, 91–92, 106–7; and chemical principles, 108, 110–11, 242; and phlogiston, 93, 97, 101

stars: distribution of, *29*, 29–30; and gravitation, universal law of, 28–31, 29, 32–34, 37–40; measuring distance to, 30–31, 202, 257–60, 258, 280–81, 283, 338n3, 338n4, 340n36; movement of, 28, 34–38, 318n6; nebulae as groups of, 41–42, 262; nebulae as origins of, 42, 43, 261, 278–79; as other suns, 4, 27; and spectroscopy, 273, 274; and stellar parallax, 257–60, 258; and stellar vortices, 13, 14

statistics. *See* probability and statistics

stellar parallax, 257–60, 258

Struve, Otto, 259, 266

subordination of characters, 141–42 taxonomy: challenges in, 76, 81, 83, 84; Charles Darwin's, 149–50, 154, 162–65, *163*; genomics and, 308–9; human races in, 175; Carl Linnaeus's, 77–79, 80, 84, 87, 134, 323n5. *See also* classification of living things

telescopes: William Herschel, made by, 40–42; as increasingly large, 260, 271–72, 274–75, 277, 282, 283–84; reflecting, 261–62, 263, 277–78; space, 284, 310; and spectral photography, 271–72, 283 theology: and free will, 178, 198, 203–4, 207; mathematics to support, 50–52; and natural philosophy, 18, 44, 53, 311; science to support, 15–16, 35; and thermodynamics, 188. *See also* creationism; determinism; natural theology; religion

thermodynamics: and Sadi Carnot's heat engine, 182–85, *184*, 186, 187–88, 192–93, 205; and conservation of energy, 182, 185, 186, 187–92; and development of statistical physics, 204–7, *205*, *206*; and energy conversions, 190–92; engineering as origin of, 182; and entropy, 193–94, *205*, 205–7; and God, 188; and heat, dissipation of, 192–93; and James Prescott Joule, work of, 185–88, 332n10, 335n12; and kinetic theory of gases, 194–95, 204–7, *205*, *206*; physics, as central to, 181, 182; and positivism, 254–55

Thomson, William (Lord Kelvin): and electromagnetism, 229–30, 231, 232, 336n28; and evolution, opposition to, 172–73, 330n10; on thermodynamics, 191–93, 207

Toland, John, 17–18

Tournefort, Joseph, 81–83

transmutation of species, 148, 149, 152, 153–55, 163, 164–65

Treatise on Electricity and Magnetism, A (Maxwell), 235

Tyndall, John: on energy, 190–91, 332n19; as material atheist, 180, 331n30; on materialism, 179–80; as scientific naturalist, 285; on spiritualism, 331n28; as supporter of Charles Darwin's work, 177

uniformitarian theory of geology, 155–56, 173

universal gravitation, law of: and action at a distance, 10–11, 13, 15–16, 89, 114, 223–24, 335n25; and astronomy, 256; and cosmology, 27, 37–38; criticism of, 5, 10, 12–13; and general relativity, 296–97; Immanuel Kant on, 323n27; and nebulae, 35, 37; and planetary motion, 231; in *Principia* (Newton), 10; role of God in, 17–18, 33, 156, 228, 318n6; Scientific Revolution, as culmination of, 4; and stars, 28–31, 29, 32–34, 37–40; and stellar formation, 42–43; and work in thermodynamics, 189–90

universe: and the Big Bang, 306, 308; as evolving, 42, 43–44; as infinite, 26, 38; island hypothesis of, 263–65, 270, 278–79, 281–82, 283; mathematics as expression of structure of, 47. *See also* cosmology; universe as mechanical

universe as mechanical: and Cartesianism, 13, 27; and criticism of Isaac Newton's work, 10–12; and determinism, 204–5; and energy, 189, 191; origin of, 4–6; and planetary movements, 13, *14*, 23, 27, 198. *See also* electromagnetism; rational mechanics

universities: in France, 129; in Germany, 124–27, 250–51; and natural philosophy, 3, 6; and professionalization of science, 123, 127, 131; research in, 6–7, 124, 125, 126–27, 128, 131, 250–51; role of professors in, 124, 127; in the United Kingdom, 129; in the United States, 127–28. *See also* education

Venel, Gabriel François, 90–91, 96

vis viva, 119–20, 182, 187–90

Volta, Alessandro, 214–15

Voltaic pile, 214–15, 216, 223

Voltaire, 18–20, 111

Wallace, Alfred Russel, 151–52, 168, 169, 170–71, 173

Watson, William, 68, 69

Watt, James, 182, 185

Whewell, William, 131, 256, 329n8

Whiston, William, 18, 33, 318n26, 319n14

Wollaston, William, 243–44, 266

women in science, 7–8, 41, 279–80, 316n15, 320n31, 340n33

Wright, Thomas, 34–35

X Club, 176–77, 178, 179, 190–91, 204, 208, 330n18, 334n24; as agnostic, 334n24; members of, 177, 204, 208, 330n18; as opposed to creationism, 178–79; as opposed to religion, 204; as supporters of Charles Darwin, 176–77; and thermodynamics, 190–91

Yerkes, C. T., 277

A NOTE ON THE TYPE

This book has been composed in Arno, an Old-style serif typeface in the classic Venetian tradition, designed by Robert Slimbach at Adobe.